Buddhist Architecture in East Asia

Buddhist Architecture in East Asia

Editors

Shuishan Yu
Aibin Yan

MDPI • Basel • Beijing • Wuhan • Barcelona • Belgrade • Manchester • Tokyo • Cluj • Tianjin

Editors
Shuishan Yu
Northeastern University
USA

Aibin Yan
East China University of
Science and Technology
China

Editorial Office
MDPI
St. Alban-Anlage 66
4052 Basel, Switzerland

This is a reprint of articles from the Special Issue published online in the open access journal *Religions* (ISSN 2077-1444) (available at: https://www.mdpi.com/journal/religions/special_issues/Buddhist_Architecture).

For citation purposes, cite each article independently as indicated on the article page online and as indicated below:

LastName, A.A.; LastName, B.B.; LastName, C.C. Article Title. *Journal Name* **Year**, *Volume Number*, Page Range.

ISBN 978-3-0365-5311-5 (Hbk)
ISBN 978-3-0365-5312-2 (PDF)

Cover image courtesy of Shuishan Yu

Contents

About the Editors

Shuishan Yu

Shuishan Yu (associate professor of architecture, Northeastern University) focuses his research on Chinese architecture, modern architecture and its theoretical discourse, literati arts, and Buddhist architecture in East Asia. His book Chang'an Avenue and the Modernization of Chinese Architecture was published in English by the University of Washington Press (2012) and in Chinese by the Sanlian Shudian Press (2016). He has also published articles, book chapters, and exhibition catalogs and presented conference papers on the city and architecture of Beijing, guqin music, Tibetan architecture, Buddhist architecture, Chinese literati art, Chinese garden, and modern architectural historiography. Yu's research projects are mostly case studies aiming for the demystification of a specific historical site, issue, or phenomenon, and highlight the significance, nature, and problem of cross-cultural translation of architectural forms, practices, and theories.

At Northeastern University, Professor Yu has been teaching Architecture and Global Cultures, History of Chinese Architecture, and the Modernization of Chinese Architecture. He has taught Western Architectural History in Beijing, Chinese Architecture in the School of Architecture at the University of Washington, and Chinese Architecture, Buddhist Art, Chinese Art, Japanese Art, and Asian Art Survey at the Oakland University. Yu is also a distinguished qin musician and the current chair of the North America Mei'an Guqin Society. He has been invited for performance, teaching, and lecture on qin music both in the US and internationally. His groundbreaking book Yu Shuishan Guqin Etudes was published by the Zhonghua Book Company in 2018.

Yu's current research projects include case studies of historic streets in China and the role they played in the modernization of Chinese cities, architecture and urbanism of Beijing, literati gardens of the Ming-Qing dynasties, and the fingering motif concept of guqin performance and its application in the study, analysis, and composition of guqin music. He is a key member and contributor to the GAHTC (Global Architectural History Teaching Collaboration), an organization of architectural historians aiming for the integration of global history of architecture and the development of new pedagogical strategy in teaching architectural history.

Aibin Yan

Dr. Aibin Yan is an associate professor from the Department of Landscape Planning and Design at East China University of Science and Technology. His research interests cover architectural history, Chinese classical garden, heritage protection and urban spatial culture. He has published more than 50 academic papers in these fields, participated in editing the Chinese architectural history section of the 21st edition of "Sir Banister Fletcher's Global History of Architecture" (Bloomsbury, 2020), translated and published "A Treatise on the Garden of Jiangnan" (Springer, 2022), and also presided over two national research projects. He was a visiting scholar in Center for Geographic Analysis at Harvard University from 2017 to 2019, focusing on the investigation and mapping of ancient stone architecture in the Song and Yuan Dynasties at the southeast coast of China. He is also currently serving as an associate researcher in Harvard CAMLab (Chinese Art Media Lab), an academic member in China Architectural History Society, member of council of Shanghai Urban Study Committee, and the director of China Urban and Rural Heritage Conservation Research Center in China Urban Construction Research Institute, and won the honorary award of "Shanghai Pujiang Scholar".

Preface to "Buddhist Architecture in East Asia"

This Special Issue on Buddhist Architecture in East Asia offers a collection of new scholarship representing the cutting-edge research on the subject. In the selection and organization for publication of articles by different authors, I discovered shared themes that are of great significance in the research on Buddhist architecture and history. They range from the macroscopic cultural transformation of the East Asian society to the small details in the design and decoration of a stone pagoda, from the formation and reshaping of sacred landscapes to the re-organization of pre-existing urban spaces, and cover all three East Asian countries of China, Korea, and Japan. Regional varieties of Buddhist architecture are also explored, showcasing both famous historic sites with concentrated studies in the past and new survey and documentation of rarely studied local temples. Together, the collection of articles in this anthology samples scholarship on the Buddhist built environments of East Asia with both broadness and depth.

The selected articles are grouped into four thematic sections. The first section explores the Buddhist construction of sacred sites, including three case studies of a sacred mountain with famous historic and religious landmarks, a monastic environment for Buddhist ritual ordination, and objects generating sacred spaces for specific religious practices. The second section investigates the ways Buddhism had transformed urban spaces and social relationships in the past, including those of both imperial capitals and the local cities of provincial and prefectural levels. The third section focuses on the Buddhist reshaping of East Asian cultures and its architectural exemplifications, delving into such significant topics as the Buddhist influence on funeral rituals and architectural typologies, Buddhist secularization, and the architecture of Pure Land Buddhism unique to East Asia. The fourth section concentrates on the formal aspects of East Asian Buddhist architecture, both physical and spatial, offering insights into such significant topics as the stone pagodas in ancient Japan, proportional principles of the timber-framed temples in ancient Korea, detailed survey of Buddhist architecture in a specific region, and the comparative study of architecture between Buddhism and other belief systems.

The contributing authors with diverse academic backgrounds are from all over the world, resulting in a colorful anthology of fresh angles, new perspectives, and innovative approaches. I thank all contributors for their excellent works and their willingness to share research through such a venue. Thanks to all external manuscript reviewers and the involved editorial board members for their selfless support and their safeguard of academic quality and vigor. Special thanks to the Managing Editor Kiki Zhang for her excellent work and steadfast support throughout the entire publication process. I also want to take this opportunity to thank my co-editor, Professor Aibin Yan from the East China University of Science and Technology, for his time and expertise, which are indispensable for the successful selection and production of such a volume. I hope readers find in this volume both valuable information and inspiring discoveries on Buddhist architecture in East Asia.

Shuishan Yu and Aibin Yan
Editors

Article

Legends, Inspirations and Space: Landscape Sacralization of the Sacred Site Mount Putuo

Yiwei Pan [1,2] **and Aibin Yan** [1,*]

1   Department of Landscape Planning and Design, East China University of Science and Technology, Shanghai 200237, China; panyiwei@mail.ecust.edu.cn

2   Shanghai World Expo Land Holdings Co., Ltd., Shanghai 200125, China

*   Correspondence: yanaibin@ecust.edu.cn

Abstract: Mount Putuo in Zhejiang Province, China, is the most important holy land of Guanyin in East Asia. Landscape sacralization is a key modality by which sacred meaning is constructed. This paper takes several examples—the Tidal Sound Cave ("chaoyin dong" 潮音洞), the Well of the Immortal Mei ("Meixian jing" 梅仙井), the Well of Ge Hong ("Ge Hong jing" 葛洪井), the Well of the Immortal ("xianren jing" 仙人井), and Duangu Pier ("Duan Gu daotou" 短姑道頭)—to analyze the three types of processes of sacralization. The Tidal Sound Cave is a re-construction of the founding myths; Well of the Immortal Mei, the Well of Ge Hong and the Well of the Immortal reflect harmony between local legends of Daoist immortals and the sacred Buddhist site; and the Duangu Pier accomplished its sanctification process in the course of local pilgrimage activities. By sorting out the mechanism and process of landscape sanctification and exploring the generation and renewal of landscape meaning, we can observe the logic of the construction of this sacred site.

Keywords: sacred site; Mount Putuo; Guanyin; legends; inspirations; space

Citation: Pan, Yiwei, and Aibin Yan. 2021. Legends, Inspirations and Space: Landscape Sacralization of the Sacred Site Mount Putuo. *Religions* 12: 1050. https://doi.org/10.3390/rel12121050

Academic Editors: James G. Lochtefeld and Todd Lewis

Received: 29 August 2021
Accepted: 18 November 2021
Published: 26 November 2021

Publisher's Note: MDPI stays neutral with regard to jurisdictional claims in published maps and institutional affiliations.

1. Introduction

Mount Putuo, an island in the east of Zhejiang Province, China, is an important pilgrimage site of Guanyin in the East Asian cultural circle and is one of the four famous Buddhist mountains ("sida mingshan" 四大名山) in China, along with Mount Wutai in Shanxi, Mount Emei in Sichuan, and Mount Jiuhua in Anhui. Historical literature dates the legends of Mount Putuo to the Tang Dynasty (618–907) and earlier, but it was in 1080 that the monastery on the island was officially recognized and called Baotuo Guanyin Temple 寶陀觀音寺. From 1080 to the Qing Dynasty (1636–1912), due to the island's special geographical location, monasteries on Mount Putuo have risen and fallen unpredictably, but three stages of changes to general spatial patterns can be seen: Baotuo Guanyin Temple as the single center (1080–1606); Putuo Temple 普陀禪寺 and Zhenhai Temple 鎮海禪寺 as the front and rear centers of the island (1606–1793); and Puji Temple 普濟禪寺, Fayu Temple 法雨禪寺 and Huiji Temple 慧濟禪寺 as the three major temples (from 1793) (see Ni 2018, pp. 128–36). Temples provide places for religious practice, and in parallel with the monastic changes, some landscapes participated in the construction of the sacredness of the holy site (Figure 1).

Over time, legendary stories contribute meaning to a site, as does experience of a site. Stories of sightings of Guanyin on Mount Putuo formed an important founding myth. Subsequently, inspirations have been continuously recorded and spread through gazetteers and other media and taken as solid evidence for the presence of Guanyin on the site. "Inspiration" (linggan 靈感) here means happenings attributed the divine power of Guanyin. Some miracle stories that occurred on various sites may replay in later generations with similar stories about different people, resulting in several centers of inspiration; other sites are considered to be associated with legends because of their naturally unique topographies or how space is experienced. The significance of both natural and manmade landscapes develops with use, becoming sacralized. In this paper, this process is called as "landscape sacralization". How that happens is the topic of this paper.

Several scholars have analyzed the island and discussed its meaning. Yü Chün-fang introduces how Mount Putuo became the sacred site of Guanyin in early times, and she also discusses the founding myths and the miracle of sighting Guanyin (Yü 2001, pp. 383–88). Marcus Bingenheimer points out that the meaning of the sacred site exceeds the location; text and site form a feedback loop (Bingenheimer 2016, pp. 12, 13). He also analyzes some examples, such as the Tidal Sound Cave, the Brahma Voice Cave, and the Sudhana Cave. Ni Nongshui, on the other hand, analyzes the cultural meanings behind the inspiration stories of the site, arguing that the miracles reflect relationships between the holy island and the royal families, officials, and ordinary worshippers (Ni 2018, pp. 190, 191). All these discussions provide useful references for this paper.

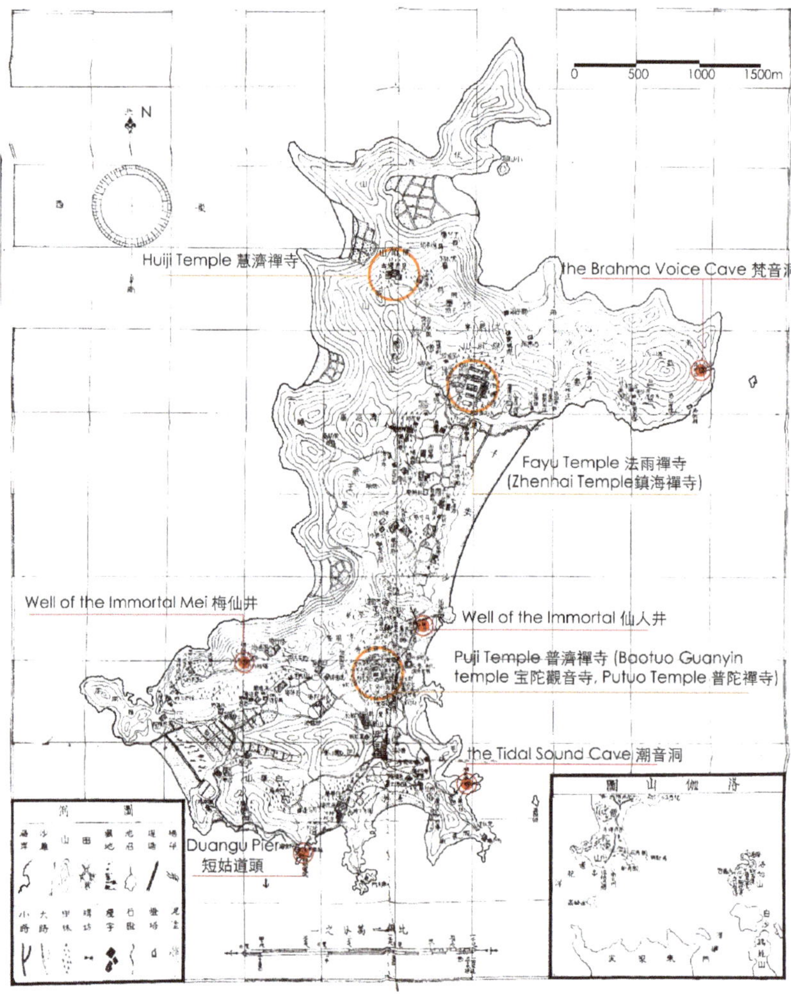

Figure 1. Plan of Mount Putuo, dating to between 1908 and 1949 ("Putuoshan quan tu" 普陀山全圖. "Shi yin ben" 石印本. Harvard Yenching Library Rare Book T 31008728.).

This paper discusses three modes of landscape sacralization. Natural landscape and manmade landscape realize sacralization and three kinds of sacred sites: holy natural site, landscape group or cultural center. Miracle stories are the key link in the process. Three typical cases are used in the following discussion corresponding to the three modes.

This study focuses on stories whose internal sequence of events is often not consistent, however. Miracle stories in texts are often accompanied by specific times or years, but no single story can be said to consistently follow a chronology. For example, iterations of the well-known story "Buddhist Establishment by Monk Egaku 慧鍔" ("Hui E Kai Shan" 慧鍔 開山) are largely the same in detail but those details vary widely in chronology[1]. Therefore, this paper focuses on how legends were carried in specific spaces in gazetteers and other texts, how later inspirations echoed the earlier ones, and how they were developed and fleshed out in the context of the sacred sites, trying to get as close as possible to the content itself without examining chronology.

2. From "Foreign Monk Burning Fingers" and "Buddhist Establishment by Monk Egaku" to the Establishment of the Tidal Sound Cave's Inspiration Status

The Tidal Sound Cave (Figure 2) was the most important inspirational place on Mount Putuo prior to the Kangxi era (1662–1722). It played a key role in the construction of the sacred site of Mount Putuo. A story about the cave records what is regarded as the first miracle of Mount Putuo, about a foreign monk in the Dazhong 大中 period (847–860) of the Tang Dynasty. It appeared in the first gazetteer of Mount Putuo, *Butuoluojia shan zhuan* 補陀洛迦山傳 (1361). The monk in this story burned his ten fingers, one after another, in front of the Tidal Sound Cave, and then he saw Guanyin and was awarded precious stones (*Taishōshinshū daizōkyō* T 51, p. 1136c)[2]. "From then on," we read, "people who pray sincerely may get responses from Guanyin. She sometimes appears in purple and golden appearance, wearing a white dress with silk belt and multi-colored beads; sometimes she appears with a thousand heads and arms, guarded by lokapala [deities who protect Darma]." (Xu 2002, pp. 9b–10a) Although the legend "Foreign Monk Burning Fingers" ("Fan Seng Ran Zhi" 梵僧燃指) cannot be found in firsthand documents of the Tang Dynasty, by at least the Northern Song Dynasty (960–1127), after Buddhism was actually established on the island, the Tidal Sound Cave had become the first natural landscape to be sanctified. In another legend, "Buddhist Establishment by Monk Egaku," according to *Butuoluojia shan zhuan*, the Tidal Sound Cave appears again. The Japanese monk Egaku's boat struck a reef and could not move forward, so he knelt piously in the direction of the Tidal Sound Cave and safely reached the shore (T 51, p. 1136c). This record seems to echo the "Foreign Monk Burning Fingers" legend. The latter story implies that Egaku knew of the direct connection between the Tidal Sound Cave and Guanyin, which is why he chose to kowtow toward that natural site.

In addition to the inheritance relationship between the two stories above, we can also see two different modes of Guanyin's manifestation. In the first story, the foreign monk saw Guanyin and received precious stones; in the second story, the record does not mention whether Egaku saw Guanyin but simply that Guanyin helped him out of danger. Although the precious stones he received did not become part of the legendary system, nor did they develop as material remains or manmade spaces, Egaku enshrined the event by bestowing Guanyin status upon a private house on the side of the Tidal Sound Cave, whose homeowner later donated the house and built a monastery called "Guanyin Who Refuses to Leave" ("Bukenqu Gaunyin Yuan" 不肯去觀音院). This temple has not survived, though it is mentioned in historical texts, and in the 1980s a new temple was built in a similar location with the same name.

The sacralization of the Tidal Sound Cave landscape can be explored through the concept of a palimpsest. In landscape research, discussions about the layering of meaning have accumulated in the West. In the past, when paper was precious, an existing text might be erased to make way for a new one, often leaving traces, and the layering of different traces that could be detected in one manuscript is called a palimpsest. In recent research about landscapes, the idea of treating a landscape as a palimpsest means identifying various physicalities or traces within it (Doherty 2016, p. 29). The influence and interpretation of the above-mentioned two modes of Guanyin's manifestation that are evident in the sacralization of the Tidal Sound Cave can be also seen in other inspiration stories, which together can be understood as one type of palimpsest. In addition, traces of multiple periods can be seen in the area around the Tidal Sound Cave, not in the form of relics but internalized within contemporary spaces. These historical traces correspond to multi-layered miracle tales.

Both characters, the foreign monk and the monk Egaku, are somewhat legendary figures. But in the Song Dynasty, miracles began to occur for a number of officials who were real historical figures. The following will focus on the experiences of Wang Shunfeng 王舜封 and Shi Hao 史浩 (1106–1194), to reveal the characteristics of inspiration stories and their role in spatialization. According to *Butuoluojia shan zhi*, during the Yuanfeng 元豐 period of the Song Dynasty (1078–1085), the emissary Wang Shunfeng prayed when he encountered wind and waves on his journey by water, and he "suddenly saw a golden shimmer, like a full moon beaming with a pearly luster, coming out from the rock cave, so he sailed smoothly (T 51, p. 1137a). This description is quite similar to the legend "Buddhist Establishment by Monk Egaku," but it does not indicate whether the rock cave is the Tidal Sound Cave. Later, in the Wanli 萬曆 period of the Ming Dynasty (1573–1620), the story recorded in *Chongxiu Putuoshan zhi* clearly states that Wang "passed the Tidal Sound Cave when the black wind suddenly rose" and he "kowtowed to the mountain and witnessed Guanyin" (Zhou 1980, p. 136). Not only is the location, the Tidal Sound Cave, specifically mentioned, but the manifestation of Guanyin is concretized as having "witnessed Guanyin," which is obviously mixed with the manifestation mode in the story "Foreign Monk Burning Fingers." Since then, inspiration stories about the Tidal Sound Cave share the similar experience of seeing Guanyin.

The story about Shi Hao says that he came to the Tidal Sound Cave on a morning in March of 1148. He saw auspicious signs in the cup when he served tea for the bodhisattva, but he did not see Guanyin. He came again in the afternoon but still did not see anything. Upon the guidance of a monk, he came to the hollow at the top of the rock cave, and "when he looked down at the cave, he suddenly saw the appearance of Guanyin, with clear features and shining in gold" (T 51, p. 1137a). This story reflects a difference from the above three stories. Shi Hao made a special trip to the cave to witness the manifestation of Guanyin, and his two visits in a single day imply that he wished to see Guanyin in person rather than being satisfied with auspicious signs. This legend initiated a new form of visiting the Tidal Sound Cave—from the top of the rock. The hollow is a called "Sky Window" ("Tianchuang" 天窗) (Figure 2), a new term that became attached to the cave and was specifically marked in gazetteers and also in the painting of Sacred Land of Mount Potalaka (*Butuoluo shan shengjing tu* 補陀落山聖境圖)[3].

The Tidal Sound Cave is located at the southeast corner of the island and is a cave formed by sea erosion. Sea water rushes into the cave during high tide, so it is quite dangerous to go into the cave, and looking down from the "Sky Window" is objectively safer. The area around the cave is full of uneven rocks, a difficult place to walk through. A bridge called "Dashi Bridge"[4] was built in the Southern Song Dynasty. In 1387, however, the government imposed a maritime embargo on Mount Putuo and destroyed almost all the temples on the island. Only one small shrine, with a roof of iron tile ("Tie Wa Dian" 鐵瓦殿), remained to continue the Buddhist tradition. This shrine was located near the Tidal Sound Cave (Tu and Hou 1589), indicating that the cave retained its high inspirational status during the Ming Dynasty.

The "Foreign Monk Burning Fingers" legend seems to indicate that one could see Guanyin by burning one's fingers. Buddhist scriptures, such as the *Wonderful Dharma Lotus Flower Sutra* (*Saddharma Pundarika Sutra*), also point out the merits of burning fingers, arms or bodies as offerings to the Buddha[5] (T 09, pp. 53c–55a). Although there is no record in gazetteers of later generations that worshipers followed the example of the foreign monk and burned their fingers to obtain vision, the idea of sacrificing one's body becomes prohibited in later times, suggesting that this legend led to mimicking. We learn, in an article called "Exhortation of Prohibition against Sacrificing Bodies" ("She Shen Jie" 捨身戒), written by Dong Yongsui 董永燧, a commanding general in the Ming Dynasty, that the general built a pavilion named "Pavilion of Forbidden Living Sacrifice" ("Mo She Shen Ting" 莫捨身亭) near the cave. A memorial stele called "Forbidden Giving Bodies or Burning Fingers" ("Jinzhi Sheshen Ranzhi Bei" 禁止捨身燃指碑) (Figure 2) was erected by the Ming Dynasty officials Li Fen 李分 and Chen Jiusi 陳九思 in front of the temple "Bukenqu Gaunyin Yuan", reflecting the mania of the time of burning fingers at the Tidal Sound Cave. The late Ming literati Zhang Dai 張岱 recorded the scene he saw in the hall of Putuo Temple (the largest temple at that time) on the evening before Guanyin's birthday, February 19 of the Chinese lunar calendar: "Many nuns burned their heads, arms, and fingers on this night, and some young laywomen believers also followed them" (Zhang 1935, p. 47). Although he did not agree, he provided records of burning bodies and fingers on Mount Putuo in the Ming Dynasty. The "Forbidden Giving Bodies or Burning Fingers" stele declared, in the form of a rigid decree, the following: "If there are any foolish people who dare to sacrifice bodies or burn fingers at the Tidal Sound Cave, the abbot monk must forbid them. Repeated offenders will be apprehended and prosecuted." In contrast, Dong Yongsui's "Exhortation of Prohibition against Sacrificing Bodies" exhorted readers in a didactic way: Instead of sacrificing one's body for blessings, one should sacrifice the body for righteousness and practice present filial piety, loyalty and fraternity (Wang 1980, pp. 455–56).

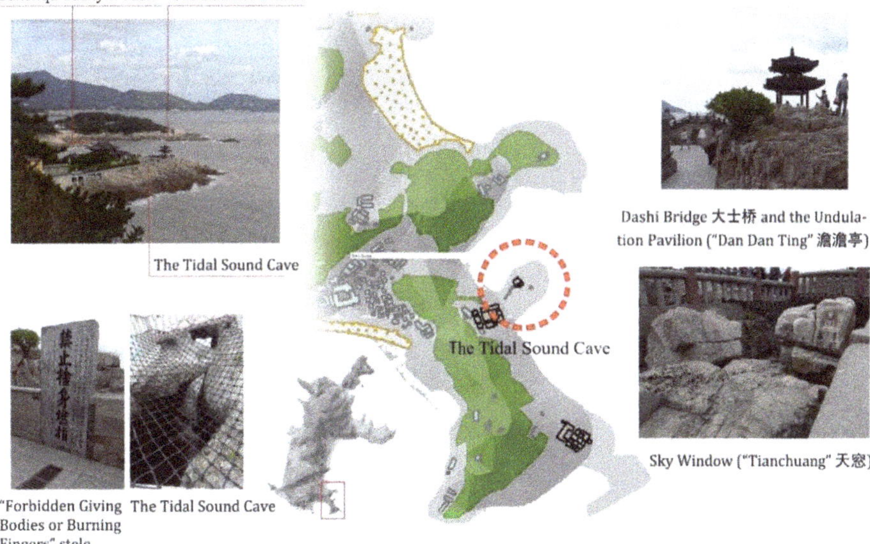

Figure 2. The location of the pier of the Tidal Sound Cave and the surrounding landscape.

The Tidal Sound Cave played a key role in the transformation of Mount Putuo from an ordinary natural island to first a site of inspiration and then to a sacred place of public pilgrimage. As the Tidal Sound Cave became established as a site of inspiration, the textual dissemination of inspiration stories interacted with activities of constructing space. The "Foreign Monk Burning Fingers" legend initiated a direct connection between Mount Putuo and Guanyin; the "Buddhist Establishment by Monk Egaku" legend laid the foundation for the construction of the sacred site. Due to the spread of holy stories, the images of Guanyin seen by previous generations had the potential to affect how later generations experienced them as inspiring. In this process, the specific spaces evolved in three ways. First, the specific spaces from the founding myths of the mountain reappeared, mainly through the construction of "Bukenqu Gaunyin Yuan". Second, spaces developed at different times with specific significance to serve the status and function of Mount Putuo: the Dragon Palace, built at the entrance of the Tidal Sound Cave during the Song and Ming dynasties for rituals to pray for rain; the "Tie Wa Dian", built during the maritime embargo period of the Hongwu era; and the Undulation Pavilion ("Dan Dan Ting" 澹澹亭), built in 1980 for tourists to rest and view the sea. Third, some landscape elements were built to correct the misconceptions spread by inspiration stories, represented by the "Forbidden Giving Bodies or Burning Fingers" stele and the Forbidden Sacrifice Lives Pavilion. These three types of spatial sacralization contributed to the gradual transformation of the Tidal Sound Cave from a single natural cliff cave into a relatively complete landscape group.

3. Wells: "Material Evidence" of Local Legends of Daoist Immortals

The chapter "Buddhist Monks" ("shizi" 释子) in *Chongxiu Putuoshan zhi* (1607) lists the prestigious monks of Mount Putuo before the Wanli period of the Ming Dynasty (1368–1644), followed by four "virtuous Daoist masters": An Qisheng 安期生 in the Qin Dynasty, Mei Fu 梅福[6] in the Han Dynasty, Immortal Ge (Ge Hong) in the Jin Dynasty and Wang Tianzhu in the Yuan Dynasty. Later gazetteers also included the four masters in the legend system, and native miracles about them also participated in the process of spatialization of the sacred site.

The "Mei" of Mount Putuo's old name, "Meicen 梅岑," refers to Mei Fu, the first Daoist master who has been clearly recorded in gazetteers and the only Daoist master who practiced alchemy on Mount Putuo, as recorded in the first gazetteer *Butuoluojia shan zhuan*. The second half of the name, "cen 岑" means the small but high hill. The *Butuoluojia shan zhuan* (1361) says that "according to folklore here is the place where Mei Fu practiced alchemy" (T 51, p. 1136b). There are two scenic names related to Mei Fu in the painting of Sacred Land of Mount Potalaka, the Well of the Immortal Mei ("Meixian jing" 梅仙井) and the Alchemy Platform of the Immortal Mei ("Mei zhen liandan tai" 梅真炼丹台), indicating that, at least in the Yuan Dynasty, the legend of Mei Fu was incorporated into the special creation of the sacred site.

The sixth chapter of *Butuoluojia shan zhuan*, "Poems of Famous Sages," contains a poem by Liu Renben 劉仁本 (1308–1367), which says, "Mei Fu left the magical pellets as red as tangerines/An Qisheng gave jujubes as big as melons" (梅福留丹赤如橘, 安期送枣大于瓜). Here Mei Fu and An Qisheng are paired, but a direct connection between An Qisheng and Mount Putuo is not established. Both the two gazetteers of the Wanli period, *Butuoluojia shan zhi* (1589) and *Chongxiu Putuoshan zhi* (1607), copy in their entirety the first four chapters of *Butuoluojia shan zhuan* by Sheng Ximing, but *Chongxiu Putuoshan zhi* adds an additional passage, written by Liu Renben to the venerable master Seng Rui 僧睿 of Mount Putuo, which records that on October 6 of the Chinese lunar calendar in 1355, Liu visited Mount Putuo and saw White-Robed Guanyin in the Tidal Sound Cave and saw the general and arahants on the wall at the cave entrance. Liu compares Mount Putuo with the mountain on the sea where the immortals live, pursued by emperors of ancient times, saying that "near the place, there are the hometown of An Qisheng and capital of Penglai, and the scenery came clearly into view," thus proving that "what Sheng Ximing said is not lying." Liu points out that at his time, "the hometown of An Qisheng" was near Mount Putuo, an indication that the legend of An Qisheng had been distributed in Zhoushan at that time. According to Ni Nongshui's research, in the Song Dynasty literature, there was "An Qisheng Cave" at Mount Maqin 馬秦山 on Zhujia Jian 朱家尖 and Mount Maji 馬跡山 on Shengsi 嵊泗 Island (Ni 2018, p. 37). Both Wu Lai 吳萊 in the Yuan Dynasty and Tu Long 屠隆 in the Ming Dynasty said that Mei Fu directly practiced alchemy on Mount Putuo and that the legend of An Qisheng occurred in the area somewhere around the island. However, in *Chongxiu putuoshan zhi*, Zhou Yingbin lists An Qisheng as the first virtuous Daoist master of Mount Putuo, saying that he "came to the mountain to avoid turmoil in the Qin Dynasty and sprinkled peach-blossom patterns while being tipsy, so there is a peach blossom mountain at the southwest of the temple" (Zhou 1980, p. 184).

While Mei Fu and An Qisheng are both representatives of the immortals, Ge Hong (283–343) is the representative of the theorists of longevity. Ge Hong is an important historical figure of Daoism, and his book, *Bao pu zi* 抱樸子, records the meditation and alchemy of the time. He proposes that the "outer alchemy" of minerals and the "inner alchemy" of nutritive essence, vitality and the spirit of the body itself may both help people achieve longevity. There are no legends about Ge in gazetteers of the Ming Dynasty, but Ge Hong's Well is spatial evidence of his story. The record about Wang Tianzhu in the Yuan Dynasty is even simpler; he "once practiced austerities in this mountain and later successfully prayed for rain; then he was known to the imperial court and given the assumed name 'Tai xu xuan jing zhi ren' 太虛玄靜志人" (Zhou 1980, p. 184).

The spread and development of immortal legends in Mount Putuo has dual significance, both as place metaphors and for historical construction. On the one hand, the unique natural landscape of Mount Putuo coincides with the place where the immortals of Chinese folklore lived in ancient times—the sacred mountains on the sea were believed to be sites for both immortals and elixirs, so the names of both Mount Meicen and Mount Putuo contain the word "mountain" instead of "island." On the other hand, although no circumstantial evidence connects the legends of Daoist immortals in Mount Putuo to it as a Buddhist sacred site, a connection with the local Daoist tradition had to be established in order to acquire inspiration that would appear inherent and legitimate and become widely recognized as a Buddhist site. Guanyin's residence represents the religious significance of the sacred site, but Buddhist literature is often unclear on how a site is related to the temporal dimension. In the historical imagination of ancient times, the famous mountains are often associated with Daoist legends, and to fully understand the meaning of the site, a local historical lineage perceivable by the Chinese would be useful. For these legends to play a lasting role in Buddhist sacred sites, they need both physical remains and coordination with Buddhism.

Zhou Yingbin used the terms "practices austerities" ("xiulian" 修煉) and "practices alchemy" ("liandan" 煉丹) to describe the behavior of four Daoist masters in Mount Putuo—An Qisheng and Wang Tianzhu "practiced austerities" whereas Mei Fu and Ge Hong "practiced alchemy." No names of scenic places or other spaces in Mount Putuo retain traces of An or Wang, whereas the scenic names associated with Mei and Ge can be seen as the Well of the Immortal Mei (from which later developed the Temple of Immortal Mei) and the Well of Ge Hong. These wells are "material evidence" of legends about Daoist immortals, and they both serve as symbols of the alchemy well ("danjing" 丹井) that echoes the legends and actually contains water that can be drawn. Wells related to Daoist immortals have further contributed their own inspiration stories. For example, the cave in which the Well of the Immortal is located is extremely chilly, like a Daoist retreat room (Figure 3). The water in the Well of the Immortal experiences "no increase or decrease in droughts or floods," and of the Well of the Immortal Mei, it "is said that washing eyes with the well water can make eyes clear" (Wang 1980, p. 124). During the Wanli period, the abbot monk of Puji Temple, Master Ji'an Rujiong 寂庵如迥, set up a hut called Meifu Temple beside the Well of the Immortal Mei, and in the early Qing Dynasty, the literati Lu Bao 陸寶 changed the name to the Temple of the Immortal Mei ("Meixian an" 梅仙庵), to avoid using the sage's full name (Figure 4). The pattern of establishing a related temple near a natural legendary site is quite similar to the formation of a temple around the Tidal Sound Cave.

Figure 3. Well of the Immortal.

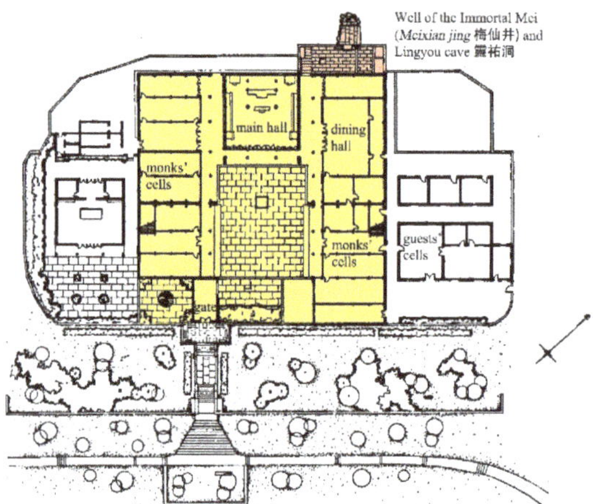

Figure 4. Plan of the Well of the Immortal Mei and the Temple of the Immortal Mei (Zhao and Ding 1997, p. 275).

The stories about Guanyin and the legends of the Daoist immortals represent two contexts. As a sacred Buddhist site, the legend from Daoism must contain some explanation that can be reconciled with Buddhist legends, otherwise it will easily fall into the deviation of attributing the sacred Buddhist site to Daoist origins. Master Yinguang 印光 (1862–1940) wrote a passage called "Stele Record on the Merit of Repairing the Well of the Immortal at Mount Putuo" ("qi Putuoshan Xianrenjing gongde bei ji" 砌普陀山仙人井功德碑記) in 1903 on behalf of Master Jieru 戒如 of Hongfa Temple 洪筏禪院, which began with the statement: "Guanyin bodhisattva always lives on this mountain; the Daoist immortals frequently reside here. Although in the early days the sculptures and Buddhist teaching had not arrived and ordinary people couldn't see the bodhisattva, the response body [Nirmanatkaya] had already resided here and Daoist immortals could always see the Buddha's light." Master Yinguang believes that the ancient monuments about Daoist immortals are spread according to the Buddhist teachings, and the well water is actually the water of great compassion flowing from the bodhisattva's heart, and thus "is known for its efficaciousness" (Yinguang 1938, 4. pp. 14–15).

4. Duangu Pier: From "Divine Trace" to "Renowned Place"

In Wu Hung's work on Chinese ruins, he discusses four kinds of traces in the landscape: divine trace, historical trace, remnant site and renowned place, all of which are marks of the past and can coexist and shift from one to the other (Wu 2012, p. 64). Duangu Pier is a typical landscape that shifts from a divine trace to a historical trace and then to a renowned place, which is both the starting point of the formal journey and a place with multiple layers of imagination.

*Yun lu man chao*雲麓漫鈔[7] by Zhao Yanwei 趙彥衛 in the Southern Song Dynasty and the painting of Sacred Land of Mount Potalaka both referred to the landing place as "Koryo Pier," whereas in gazetteers of the Ming Dynasty the name became "Duangu Pier." According to Wang Liansheng's survey, the locations of the two piers are not the same (Wang 2003). When the name "Duangu Pier" appeared, the name "Koryo Pier" quickly faded from the literature. Later, as the structures and inscriptions around Duangu Pier became more abundant, it formed a renowned landscape space with special significance and served as the beginning of the narrative.

We cannot know from the literature why "Koryo Pier" was replaced with "Duangu Pier", but we can find Koryo saram's activities on the island in early times. *Mo zhuang man lu* 墨莊漫錄 by Zhang Bangji 張邦基 was written at the turn of the Northern and Southern Song dynasties, about half a century after Baotuo Guanyin Temple had been given plaque by the government. An article in the book records that the brass bells in the temple on Mount Putuo was given by Koryo merchants, on which engraved with Koryo's reign title. But records about Koryo saram's activities could not be seen in later literature. With the improvement of island's reputation, the government strengthened its control. Perhaps the change of the pier's name was related to a deep political significance. The name of Duangu Pier is related to an inspiration story about a pilgrimage of a pair of sisters-in-law, which was recorded in one version in *Chongxiu Putuoshan zhi* of the Ming Dynasty and in another in *Chongxiu Nanhai Putuoshan zhi* of the Qing Dynasty. According to *Chongxiu Putuoshan zhi*, there were two sisters-in-law who came to the mountain on pilgrimage. When the boat was about to land, the younger sister-in-law happened to have her menstrual period, so the elder sister-in-law went ashore and visited the mountain alone, leaving the younger one in the boat. After a while, an old woman came to the boat and used her lower garment to cover the stone walkway to guide the younger sister-in-law to the temple hall. When the elder sister-in-law went down the mountain, she found that her sister-in-law was no longer in the boat. After the younger sister-in-law came down the mountain, she told her sister-in-law about the experience. They went back to the temple hall to look for the woman but with no result. Then they realized that the old woman was a manifestation of Guanyin (Zhou 1980, p. 152). The story in *Chongxiu Nanhai Putuoshan zhi*, meanwhile, says that the two sisters-in-law came to Mount Putuo after several years of vegetarian diets, but when the younger sister-in-law had her menstrual period, the elder sister-in-law blamed her, and the younger sister-in-law was too ashamed to go ashore. As the tide rose, the stone path at the pier was flooded, and the younger sister-in-law on the boat was hungry and had no food to eat. At that time, an old woman carrying a bamboo container with food, threw a few stones into the water, walked to the boat, handed food to the younger sister-in-law and left. The younger sister-in-law was surprised and wondered who the old woman was. After a long time, the elder sister-in-law returned and was told the story. She guessed that it was Guanyin who appeared, so she immediately went back to the temple hall to pray and was surprised to see that the bottom of the lower garment on Guanyin's statue was wet. Since then, the pier has been called "Duangu Pier" (Pier of the sister-in-law who was blamed) (Xu 2002, pp. 5. 5a–5b).

The initial formation pattern of the Duangu Pier site is somewhat different from that of the Tidal Sound Cave. First, before it was named, the site was an artificially constructed pier, on the basis of which inspiration stories were created and the pier was given a name. Second, the two main characters of the story, the pair of sisters-in-law, were ordinary female pilgrims who came to the island on their own, without the company of males. This kind of inspiration story of ordinary people is not common in gazetteers, and it is even rarer to develop a fixed name for a scenic spot based on the story of "nameless" female protagonists. Third, new inspiration stories were not repeated or attached to Duangu Pier in later times, and instead, the meaning of the space was enriched by means of inscriptions. The two versions of the story both reflect anxiety about female pilgrimages, and with the help of the female Guanyin, they provide two responses. In the story in *Chongxiu Putuoshan zhi*, the younger sister-in-law still successfully completed the pilgrimage under the guidance of Guanyin, which meant that women were liberated from inherent social prejudices and religious taboos with the help of the female bodhisattva. The story in *Chongxiu nanhai Putuoshan zhi* adopts a more conciliatory stance. The younger sister-in-law failed to complete the pilgrimage, but she nevertheless saw the manifestation of Guanyin and benefited from her presence. Regardless of the versions, the story reflects how Guanyin and Mount Putuo extend kindness to the common people, especially to female worshippers.

According to the story in *Chongxiu Putuoshan zhi*, Duangu Pier is "said to have been built by the bodhisattva herself, and has never been damaged by the pounding of huge waves" (Zhou 1980, p. 110). Zhou regards Duangu Pier as a divine trace left by Guanyin at the sacred site, which has non-historical characteristics. These characteristics can be seen in the woodblock printing "Sacred Relic of Duangu" ("Duangu shengji" 短姑聖蹟) from the series "Twelve Views of Mount Putuo" that appeared in the Qing Dynasty, in which Guanyin appears with a bamboo basket on the shore, rather than on the actual pier protruding from the coastline (Figure 5). The image reflected the imagination of the legend of a Guanyin manifestation rather than conveying the actual scene of the pier.

Figure 5. "Sacred Relic of Duangu" in the "Twelve Views of Mount Putuo" (Qiu 1996, p. 史 239–10).

There is no way to know when the story occurred, but the record in gazetteers and the inscription "Duan Gu gu ji 短姑古跡" (historical trace of Duangu) make this sacred site a historical trace. Inspiration stories recorded in gazetteers of the Qing Dynasty and the Republic of China were often arranged into a special chapter and in chronological order. The corresponding timeframe was pointed out at the beginning of each story. But not only does the story about the sisters-in-law in those gazetteers not include the exact time or the dynasty in which it occurred, it also begins with the phrase, "according to legend." In *Chongxiu nanhai putuoshan zhi* by Xu Yan, the story is placed between the story from 1355 about Liu Renben and the story from 1403 about the manifestation of White-robed Guanyin in the Tidal Sound Cave. Evidently, the time of this story was considered to be the late Yuan to early Ming Dynasty. The stone inscription is another type of textual material. On the west side of the present Duangu Pier, standing in the sea, is a large stone engraved with the words, "Duan Gu gu ji", and the inscription "Shaohai bing Zhang Keda Dinghai dusi Gao Mingqian 紹海兵張可大定海都司高鳴謙" (Zhang Keda, the garrison commander of Shaohai and Gao Mingqian, the army officer of Dinghai) on the right and "Minguo shi'er nian daxun shi chongxiu 民國十二年大汛時重修" (rebuilt after the flood in 1923, the 12th year of the Republic of China) on the left. Both Zhang Keda and Gao Mingqian were from the late Ming Dynasty. The inscription seems to provide evidence that the words "Duan Gu gu ji" were recognized in the late Ming Dynasty. However, this inscription is not mentioned in any of the Ming and Qing Dynasty gazetteers, and the other stone inscriptions around it are all from the Republic of China (Figures 6 and 7). Although the records of the legend and the stone inscriptions are ambiguous in terms of the exact date, the legend of the sisters-in-law is objectively included in the historical narrative.

It is worth noting that the Chinese character "ji" in "Duan Gu gu ji" is "跡" with the "foot" radical (足), whereas the traditional characters for "跡" and "迹" are both unified as "迹" in simplified Chinese. Wu Hung points out the difference between the two characters. "跡" refers to a mark on the land whereas "迹" emphasizes movement, which means to leave one's own footprints when searching for traces of the past (Wu 2012, p. 63). The former character "跡" is used in "Duan Gu gu ji," whereas the latter "迹" is used in the writing of "spiritual traces" ("lingji" 靈迹) of the Tidal Sound Cave in gazetteers.

According to the record of Huang Yingxiong 黃應熊, who came to Mount Putuo in 1730, a wooden ornamental column ("huabiao" 華表) stood on Dangu Pier with the inscription "hai tian er fan 海天二梵" (Buddhist worlds of the sea and heaven). According to Wang Hengyan's record, this ornamental column was later rebuilt and the inscription was changed to "ci hang pu du 慈航普渡" (the barge of mercy ferries all the miserable people to the world of bliss) on the front and "fu hai wu ya 福海無涯" (the boundless sea of fortune) on the back (Wang 1980, p. 471). In addition to the structures and inscriptions, there are nearly ten inscriptions by people of the Republic of China around the stone inscription "Duan Gu gu ji." In addition, many publications about Mount Putuo from the Republic of China to the present have used the photograph of Duangu Pier as their cover. Although Duangu Pier has become a historical trace, it has also gradually become as renowned as a genius loci by regional consensus. As Wu Hung pointed out, the renowned place "cancels the historical specificity of individual traces" and became a place attracting visitors to leave their marks (Wu 2012, p. 86).

Figure 6. Stone inscriptions at Duangu Pier in 1930 (Yinguang and Zhenda 1930, p. 5).

Figure 7. Stone inscriptions at Duangu Pier now.

5. Conclusions

Many legends of Mount Putuo show a reciprocal relationship with spaces of that landscape. The three modes of spatial interpretation discussed in this paper are particularly typical. The first mode is superimposing similar inspiration stories upon natural landscape and giving the natural landscape a special sacred meaning. In the course of the Tidal Sound Cave becoming the sacred center of the entire island, we can see how legends "Foreign Monk Burning Fingers" and "Buddhist Establishment by Monk Egaku" giving sacred meanings to the natural landscape. The second mode is forming a landscape space based on local Daoist legends, which in turn attaches new inspiration stories to the space, to the point where it gradually expands into a monastery. Wells named after Taoist immortals are the most representative. They both enrich the imagination of the history of the site and connect to Buddhist miracle stories. The third mode is inspiration stories based on the existing functional spaces and developing names for attractions and series of views. Duangu Pier is an excellent example. Legendary stories give the name and a unique religious significance to this island's indispensable functional space. In addition to the cases discussed above, other scenic spots on Mount Putuo can be classified according to these three modes.

The process of landscape sacralization requires a long passage of time. The sacred meanings of the landscape are transmitted and renewed through the use of space and the constant reenactment of inspiration. Sacred legends from ancient times to modern times, as listed by Yü in her important work (Yü 2001), are the soul of a sacred landscape. They maintain Mount Putuo as a site with sacred significance beyond its physical space, so that for centuries, throughout the East Asian cultural circle, the island has been recognized as the equivalent to Guanyin's dwelling of Mount Potalaka that is mentioned in the Buddhist sutra.

Author Contributions: Conceptualization, Y.P. and A.Y.; methodology, Y.P. and A.Y.; software, Y.P.; validation, Y.P. and A.Y.; formal analysis, Y.P. and A.Y.; investigation, Y.P.; resources, Y.P. and A.Y.; writing—original draft preparation, Y.P.; writing—review and editing, A.Y. and Y.P.; visualization, Y.P.; supervision, A.Y.; project administration, A.Y.; funding acquisition, A.Y. All authors have read and agreed to the published version of the manuscript.

Funding: This research was funded by [Shanghai Pujiang Program] grant number [2020PJC021].

Conflicts of Interest: The authors declare no conflict of interest.

Notes

1.   *Fo zu tong ji* 佛祖统纪 of the 13th century records the story "Buddhist Establishment by Monk Egaku" took place in 858, while *Butuoluojia shan zhuan* 補陀洛迦山傳 (1361) records the story took place in 916. *Shi shi ji gu lüe* 釋氏稽古略 of the 14th century followed the records of *Butuoluojia shan zhuan*.

2.   T. *Taishōshinshū daizōkyō* 大正新修大藏經. Edited by Takakusu Junjirō 高楠順次郎 and Watanabe Kaigyoku 渡边海旭. 100 vol. Tokyo: Taishō issaikyō kankōkai, 1924–1935. *Butuoluojia Shan Zhuan* 補陀洛迦山傳. T 51, no. 2101. Edited by Sheng Ximing 盛熙明.

3 The painting Sacred Land of Mount Potalaka is preserved at the Jōshō Temple 定勝寺 in Nagano, Japan. Marcus Bingenheimer inferred that it was created between 1334 and 1369, the same period as mentioned by Sheng Ximing in *Butuoluojia shan zhuan* (Bingenheimer 2016, p. 57).

4 Dashi 大士 is a generic term used to call bodhisattva.

5 See "The Former Deeds of Medicine King Bodhisattva" (Chapter 23) of *The Wonderful Dharma Lotus Flower Sutra* (*Miao Fa Lian Hua Jing* 妙法蓮華經).T 09, no. 262. Translated by Kumārajīva.

6 Mei Fu 梅福 (*zi,* Zizhen 子真) was an education officer (*junwenxue* 郡文學) and Nanchang military commander (*wei* 尉). He repeatedly wrote directly to the emperor but was never accepted. In the period of Wang Mang 王莽, he abandoned his wife and children and went to Jiujiang 九江. Since then, he has been regarded as an immortal. Daoism regards Mei Fu as a member of its system of immortals.

7 *Yun lu man chao* 雲麓漫鈔 dates back to 1206, according to the inscription in the preface. This book records the names of landscapes in Mount Putuo from the Song Dynasty in some detail, so it is an important historical source for understanding the landscape during the Song Dynasty.

References

Bingenheimer, Marcus. 2016. *Island of Guanyin: Mount Putuo and Its Gazetteers.* New York: Oxford University Press.

Doherty, Gareth. 2016. Is Landscape Literature? In *Is Landscape . . . ?: Essays on the Identity of Landscape.* Edited by Gareth Doherty and Charles Waldheim. Abingdon and New York: Routledge, pp. 13–43.

Ni, Nongshui 倪浓水. 2018. *Putuoshan Wenhua Shi.* 普陀山文化史. Hangzhou: Zhejiang Daxue Chubanshe.

Qiu, Lian 裘璉. 1996. *Nanhai Putuoshan Zhi Shiwu Juan Shou Yi Juan* 南海普陀山志十五卷首一卷. In *Sikuquanshu Cunmu Congshu (Shibu Di er San Jiu Ce).* 四庫全書存目叢書(史部第二三九冊). Edited by Sikuquanshu Cunmu Congshu Bianzuan Weiyaunhui 四庫全書存目叢書編纂委員會. Jinan: Qilu shushe.

Tu, Long 屠隆, and Jigao Hou 侯繼高. 1589. *Butuoluojia Shan Zhi.* 補陀洛迦山志 (3 vol.). Beijing: Guojia Tushu Guan Cang.

Wang, Hengyan 王亨彥. 1980. *Putuoluojia xin zhi* 普陀洛迦新志. In *Zhongguo Fosi Shi Zhi Huikan (Di Yi Ji Di 10 Ce).* 中國佛寺志彙刊（第一輯第10冊）. Edited by Jiexiang Du 杜潔祥. Taipei: Ming Wen Shu Ju.

Wang, Liansheng 王胜. 2003. *Putuoshan Gaoli Daotou Yizhi Chongxian* 普陀山高丽道头遗址重现. *Zhejiang Haiyang Xueyuan Xuebao (Renwen Kexue Ban)* 浙江海洋学院学报(人文科学版). 20: 26–29, 48.

Wu, Hung. 2012. *A Story of Ruins: Presence and Absence of in Chinese Art and Visual Culture.* London: Reaktion Books Ltd.

Xu, Yan 許琰, ed. 2002. *Chongxiu Nanhai Putuoshan zhi* 重修南海普陀山志. In *Xuxiu Siku Quanshu.* 續修四庫全書, no. 723. *Shanghai Guji Chubanshe Cang Qing Qianlong keben* 上海古籍出版社藏清乾隆刻本. Shanghai: Shanghai Guji Chubanshe, vol. 14, pp. 9b–11b.

Yinguang, fashi 印光法師, and fashi Zhenda 真達法師. 1930. *Putuo Shengji.* 普陀勝績. Shanghai: Shangwu yinshuguan.

Yinguang, fashi 印光法師. 1938. *Yinguang Fashi Wenchao.* 印光法師文鈔. Shanghai: Shanghai Foxue Shuju, vol. 2.

Yü, Chün-fang. 2001. *Kuan-Yin: The Chinese Transformation of Avalokitesvara.* New York: Columbia University Press.

Zhang, Dai 張岱. 1935. *Lang Huan Wen Ji.* 瑯嬛文集. Collated by Dajie Liu 劉大杰. Edited by Zhecun Shi 施蟄存. Shanghai: Shanghai Zazhi Gongsi.

Zhao, Zhenwu 赵振武, and Chengpu Ding 丁承朴. 1997. *Putuoshan Gu Jianzhu.* 普陀山古建筑. Beijing: Zhongguo Jianzhu Gongye Chubanshe.

Zhou, Yingbin 周應賓. 1980. *Chongxiu Putuoshan Zhi* 重修普陀山志. In *Zhongguo Fosi Shi Zhi Huikan (Di Yi Ji Di 9 Ce).* 中國佛寺志彙刊（第一輯第9冊）. Edited by Jiexiang Du 杜潔祥. Taipei: Ming Wen Shuju.

religions

MDPI

Article

Transcending History: (Re)Building Longchang Monastery of Mount Baohua in the Seventeenth Century

Zhenru Zhou

Department of Art History, University of Chicago, Chicago, IL 60637, USA; zhenru@zhenruzhou.com

Abstract: This paper analyzes the roles architectural renovation played in the revival of Longchang Monastery of Mount Baohua (Jiangsu), a major Chinese monastery of the Vinaya School and an ordination center in Late Imperial China. Based on temple gazetteers, monastic memoirs, and modern documentation of monastic architecture and life by Prip-Møller, the author reveals the formation of a spatial system that centered at the threefold ordination rituals. It took the entire seventeenth century for the system to take form under the supervision of a Chan monk-architect Miaofeng and three successive Vinaya abbots, Sanmei, Jianyue, and Ding'an. The spatial practices, comprising a series of reconstructions, reorientations, redesigns, re-demarcations, and refurbishments, have not only reconciled fractures and defects in the monastic architecture but also built a history for the rising institute. This article examines the construction of and the narratives around three centers of the Monastery, namely, the Open-Air Platform Unit where Miaofeng erected a copper hall, the Main Courtyard where Sanmei reoriented the monastic layout to follow the Vinaya tradition, the Ordination Platform Unit where Jianyue rebuilt a stone ordination platform, and again the Open-Air Platform Unit that Ding'an had refurbished and reunited with the later centers. The forces that have driven this seemingly non-progressive history, as the author argues, are not only the consistent efforts to counteract the natural course of material decay, but also the ambition of making a living history without beginning or end.

Keywords: Longchang Monastery; Mount Baohua; the Nanshan Lineage of the Vinaya School; Johannes Prip-Møller; ordination platform; threefold ordination rituals; architectural renovation

Citation: Zhou, Zhenru. 2022. Transcending History: (Re)Building Longchang Monastery of Mount Baohua in the Seventeenth Century. *Religions* 13: 285. https://doi.org/10.3390/rel13040285

Academic Editors: Shuishan Yu and Aibin Yan

Received: 20 February 2022
Accepted: 22 March 2022
Published: 25 March 2022

Publisher's Note: MDPI stays neutral with regard to jurisdictional claims in published maps and institutional affiliations.

These pages are intended to make more vivid the picture of a monastery as an organism living in the present but with its roots deep in the past.

—Johannes Prip-Møller (1889–1943), "Preface", in *Chinese Buddhist Monasteries* (1937).

Every monastery is a living organism which survives only through constant modification.

—Ernst Boerschmann (1873–1949), book review, 1939, p. 292.

1. Introduction

The constant modification in Buddhist monasteries is a major challenge for historians of Chinese architecture. As is often the case, even when a monastery is alleged to have an ancient origin, the buildings in it might be built or rebuilt just decades ago. As one may wonder upon confronting such a monastery, what kind of history do the "repaired" buildings carry?

The practices of "restoration", traditionally referred to as *chongxiu* 重修, range from refurbishment of worn surfaces, reconstruction of lost structures, renovation (which often involves modification), and redesignation of orientation, center, and borders. The restoration of most monasteries in China does not exclude creativity. Such creativity can harm the coherency of the original design, and that is why historians of Chinese architecture are often annoyed by restoration that is destructive of historical monuments. In some lucky instances, however, creativity may well lead to a better realization of an ideal that has been attempted since the remote past.

Longchang Monastery (Longchang si 隆昌寺, The Monastery of Prosperity) of Mount Baohua (Baohua shan 寶華山, The Mountain of Precious Flower) is such a case; restoration has been essential to the survival of the Monastery as a living organism. The Monastery, located about twenty miles east of an old capital city Nanjing (in present-day Jiangsu province), has thrived under the charge of eminent monks and imperial recognition as well as suffered from turmoil resulting from dynastic changes. Mount Baohua, according to the gazetteer of the mountain monastery, *Baohua shanzhi* 寶華山志 (*Gazetteer of Mount Baohua*, hereafter "BHSZ"), compiled by Liu Mingfang 劉名芳 (a. 1751) in 1785–95,[1] was inhabited by an eminent Chanmonk Baozhi 寶誌 (418–514) in as early as 502 CE. Yet the earliest known buildings there were constructed by another group of Chan monks in the late-Ming dynasty (1368–1644). The rise of the monastery as a Buddhist ordination center with a Vinaya (*Lü* 律) specialization during the seventeenth century stimulated building expansions as well as spatial shifts from the pre-existing Chan establishments.[2] Most timber-structured halls had been burnt down more than once during the nineteenth and twentieth centuries, but the following reconstructions stubbornly adhered to an irregularly concentric layout as displayed in a wood-block print illustration from the mid-eighteenth century (Figure 1). Having undergone many hard times, the Monastery resumed its functionality and vigorousness as soon as the same space was reframed by buildings— even just in a modest style. This Vinaya monastery gradually thrived as "the Foremost Ordination Platform under the Heaven" (tianxia diyi jietan 天下第一戒壇) during the Qing dynasty (1644–1911) and the Republican era (1911–1949) (BHSZ [1795] 1975, vol. 14, p. 606; Wen 2004, p. 41). The monastic architecture is testimony to the constant restorative practices that have in effect extended the "life" of the monastery.

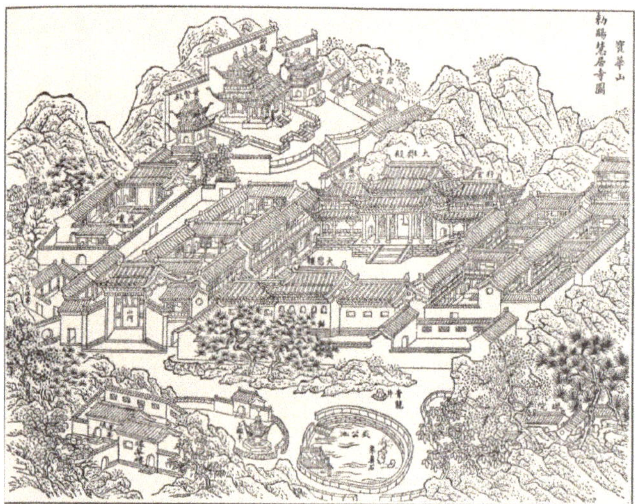

Figure 1. The picture of Longchang Monastery of Mount Baohua. Chinese, wood-block print, book illustration, the eighteenth century (Prip-Møller 1937, p. 197, Figure 237; BHSZ [1795] 1975, vol. 1).

So is Longchang Monastery "a persistent architectural frame around an unchanging religious culture" (Prip-Møller 1937, "Preface")? Perhaps for the defenders of the progressiveness of Chinese architecture, this statement, made by the first modern investigator of the Monastery's building history, may sound derogatory. But a reexamination will reveal that the seemingly non-progressive architecture resulted from a dynamic process of altering the core areas of the Monastery. In the process, a series of renovations were the catalyst of spatial and institutional fractures as well as the reconciliation of them. Commissioned by three successive abbots during the seventeenth century, the almost rebuilt Monastery eventually framed an architectural and spatial system that centered around the threefold ordination, the core activity of the Vinaya center. In addition, the renovation practices helped to build a history for the rising institute. The institutional history "written" in the buildings is a collective work of the Chan monks who laid a keystone, the three Vinaya abbots who decisively reshaped it, and the modern visitors who empathetically translated it. This case study takes architectural renovation as essentially a spatial practice—a non-verbal means of communication, negotiation, and expression.[3] Renovation serves as a collective design in the long durée by means of reconciling fractures and defects in the prototype or the preexisting structures. From this lens, we will see how the mentality underlying restorative practice has defined a transcending history of Chinese architecture.

2. Longchang Monastery: From Miaofeng to the Twentieth Century

A stream that runs a long course must have come from a remote source. Longchang Monastery is no exception to this Chinese belief. As the continuous prosperity of the Monastery in the Late Imperial period proves, the buildings have worked effectively in sustaining a monastic community. What restores the memory of the past is the later reconstructions that seem to comply with the earlier spatial settings. Not only do the reconstructions in the nineteenth and twentieth centuries follow the eighteenth-century layout, but the eighteenth-century layout also, as will be discussed in the next section, resembles an archaic type of monastery layout in China. The literary inspiration of this layout can even be traced back to the Indian prototype of the Buddha's time. Almost all buildings were claimed by those who commissioned them to be reconstruction of some older versions.[4] Indeed, from a historical perspective, a compliance with the tradition legitimates the following generations' actions.

2.1. Miaofeng: Founding the Open-Air Platform Unit

In this context, buildings recognized as "original construction", as opposed to a reconstruction, indicate a watershed in history. The construction of the Open-Air Platform Unit (Figure 2)—the only instance of this sort—planted a seed of spirituality and prestige at Mount Baohua. On a stone platform that would later become the summit of the Monastery, a Chan monk-architect Miaofeng 妙峰 (1540–1612) "founded" (chuangjian 创建) a west-facing architectural triad for the Three Great Bodhisattvas in 1605 (BHSZ [1795] 1975, vol. 3, p. 101). Originally, a double-eave copper hall dedicated to the Guanyin 觀音 (Skt: Avalokiteśvara) Bodhisattva stood on a raised stone platform in the rear center of the courtyard. The two stone vaulted halls, known as "the Beamless Halls" (*wuliang dian* 无梁殿), were respectively dedicated to Wenshu 文殊 (Skt: Mañjuśrī) and Puxian 普賢 (Skt: Samantabhadra) Bodhisattvas. Miaofeng initially planned to dedicate the images and halls to three Famous Buddhist Mountains that were believed to be the abodes of the respective bodhisattvas, namely, Mounts Putuo, Wutai, and Emei. The two projects at Mts. Emei and Wutai were successfully executed. At Xiantong Monastery (Xiantong si 显通寺) of Mt. Wutai, a triad of a central copper hall and two flanking stone halls still stands (Figure 3).[5] However, the Guanyin image could not be distributed due to pirates around the Southern Sea in which this island is located. Thus, after divination, Miaofeng decided to distribute it to Mount Baohua, which was closer to the Southern Capital City Nanjing of the Ming Empire.[6] The copper hall Miaofeng designed for Mount Baohua was almost identical to those of Mounts Wutai and Emei (J. Zhang 2015, p. 297). Sponsored by the Shen Emperor of Ming (r. 1572–1620) and the Cisheng Empress Dowager as well as officials and local communities in Central China, the projects also indicated imperial recognition and support of the statewide Buddhist centers. In other words, the Guanyin image and the Copper Hall won Mount Baohua a special position in the Buddhist landscape of Late-Imperial China.

Figure 2. A detailed image of the copper hall and Beamless Halls of Longchang Monastery: (**a**) Copper Hall; (**b**) Wenshu Hall; (**c**) Puxian; (**d**) Incense Pavilion; (**e**) Ordination Platform Unit. (BHSZ [1795] 1975, vols. 1, 5).

Figure 3. The copper hall and beamless halls of Xiantong Monastery at Mount Wutai, photograph by Jianwei Zhang in 2009 (J. Zhang 2015, p. 305, Figure 8).

2.2. Inevitable Material Decay and Human Counteraction

Marvelous as it was, the Copper Hall was subject to material decay without exception. As soon as it was erected, historical beholders knew the monument would follow the circle of form and deform. And they saw the only way out to be diligently practicing the faith. In a stele text about the construction of the Copper Hall (chijian Baohua shan huguo shenghua si guanyin pusa tongdian bei (勅建寶華山護國聖化隆昌寺觀音菩薩銅殿碑, BHSZ [1795] 1975, vol. 6, p. 208), a literatus-official Jiao Hong 焦竑 (1540–1620) commented after praising the splendid appearance of the hall:

> However, the method of action is essentially relenting. Like carved ice and engraved snow, it will eventually return to non-being. Forming results in being, and being results in deforming. Who knows that something indestructible lies in the process of forming and deforming? [That is,] bowing to the Great Compassionate One, and vowing to safe sentient beings as many as the sand in the Ganges River. May them all attain the diamond body, which is indestructible in a period as long as ten thousand eons.

然此有為法，究竟非堅固。如雕冰鏤雪，終歸于烏有。由成乃得住，由住而為壞。
孰知成壞中，有不壞者存。稽首大悲尊，願度恆沙眾。共證金剛身，萬劫長不毀。

The first half of Jiao's comment was almost a prophecy to the Copper Hall's deformation. The copper and stone monuments were refurbished some eighty years later (BHSZ [1795] 1975, vol. 6, pp. 245–47), as will be discussed in more detail. The Copper Hall eventually perished in the turmoil caused by the Taiping Revolt (1850–1864) and was replaced by a brick-and-wood hall (Figure 4). The second half, meanwhile, although more of an aspiration than a statement of fact, would find echo in the continuing worship of Guanyin—the bodhisattva of Great Compassion—at the site.

Figure 4. Photograph of the Open-Air Platform, with the Surrounding Buildings (Prip-Møller 1937, p. 249, Figure 275).

In the 1930s, the copper hall had been replaced and the stone halls repurposed as the abodes of elderly monks. Nevertheless, residents and pilgrims at the Monastery still "believed that their prayers to Guanyin are more likely to be heard if uttered there (i.e., the Open-Air Platform Unit)" (Prip-Møller 1937, p. 254). The pilgrims believed that the spiritual quality of the Unit was invulnerable to material decay and social reforms. The surviving monuments represent not just the refinement of masonry in seventeenth-century China but also the revival of the Monastery in the subsequence decades and centuries.

2.3. Three Vinaya Abbots: Decisive Roles in Reviving the Monastery

If the architectural monuments erected by Miaofeng mark the eve of the revival campaign, then the substantial works have been conducted by the first three abbots of the Monastery from the Nanshan Tradition (Nanshan zong 南山宗) of the Vinaya School. The three Vinaya abbots, Sanmei Jiguang 三昧寂光 (1580–1645), Jianyue Duti 見月讀體 (1601–1679), and Ding'an 定庵 (aka. Deji 德基, 1634–1700), are disciple and successor of one another.

Institutionalization—in this case, the set-up of a Buddhist monastic order—is not a given thing from the beginning. During Miaofeng's time, a monastic community of considerable scale and communal life seemed not yet to have been established.[7] Miaofeng has left no dharmic heir. The other eminent monk Xuelang 雪浪 (1545–1608) had preached at one of the peaks of Mount Baohua but has no recorded interaction with Miaofeng, despite the fact that they were contemporary Chan monks and known for their expertise in temple construction.[8] Neither Miaofeng, nor Xuelang, nor other contemporary monks whose names we know such as Nanzong 南宗, Tiankong 天空, and Zishan 茲山, were called "abbot" of any monastery. There were teachers and students, but a formed institute that functioned like a community is not known to have existed at Mount Baohua before the seventeenth century.

The Monastery—the framework of an institutionalized community—took three generations to build. Numerous image halls, meditation halls, corridors, and minor units enclose a squarish Main Courtyard in two to three concentric layers. As the sequence has been sorted out (Prip-Møller 1937, pp. 281–96, Figure 5), the first Vinaya abbot Sanmei commissioned the innermost ring around the Main Courtyard, including most important ritual buildings and monks' chambers. Then, Sanmei's successor Jianyue commissioned buildings on the second and third outer rings of the concentric layout. Small cloisters of more independent function, such as the Courtyard of the Abbot, the Courtyard of the Ordination Platform, a courtyard of the Dining Hall, and various kinds of workshops were built. Afterwards, Jianyue's successor Ding'an completed the outermost ring and even renovated the architectural heritage from Miaofeng's days. The difficulty of this transformation is indicated by the irregularity of the layout. For instance, the orthogonal and concentric layout is complicated by a minor unit known as "the Open-Air Platform Unit." Shifting some 45 degrees away from other parts in the Monastery, the unit is connected directly to the main entrance of the Monastery by a relative independent route (Figure 6).

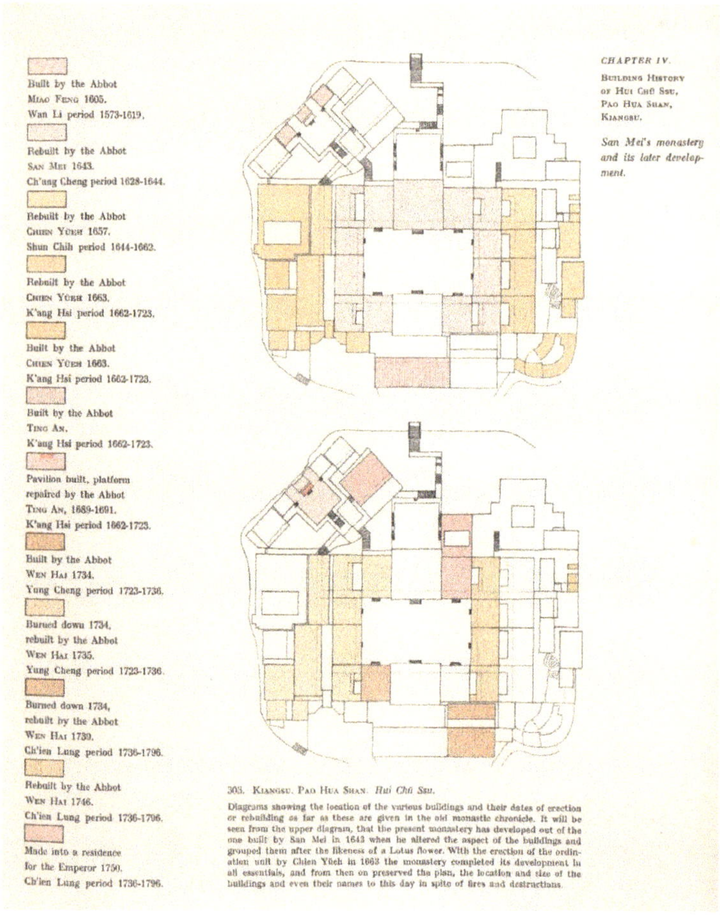

Figure 5. Building history of Longchang Monastery (Prip-Møller 1937, p. 283, Figure 303).

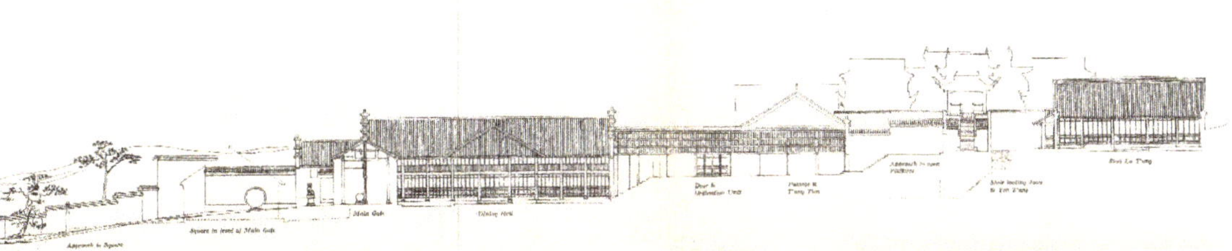

Figure 6. Section of Longchang Monastery, showing the Main Gate, the Ordination Platform Unit and the Open-Air Platform (Prip-Møller 1937, Pl. 4).

After all, the three Vinaya abbots have played decisive roles in the layout of Longchang Monastery as it exists nowadays. They together have determined the orientation of the Monastery, erected the ritual and auxiliary buildings, instituted a paradigm of monastic life, and restored the religious and spiritual pasts for the present.

2.4. Prip-Møller: Recognizing the Extending-Life Phenomenon

The first modern scholar who noticed the "life-extending" phenomenon and the architectural value of Longchang Monastery is Johannes Prip-Møller (Aishuhua 艾術華, 1889–1943).[9] A professional architect, self-trained anthropologist, and Christian missionary, he is among the very few pioneers who conducted extensive architectural and ethnographical surveys on Chinese Buddhist monasteries in the early-twentieth century. Prip-Møller received formal architecture training in the Royal Danish Academy of Fine Arts and Columbia University. Upon arrival in China in 1921, he studied the Chinese language in Beijing and working as an architect in Shenyang in Northeast China for five years.[10] With familial and social ties to the Christian missions in China, Prip-Møller deepened a sympathetic interest in the building practices that have sustained yet another long-lived religious culture–Chinese Buddhism.[11] He decided to study the still functioning Buddhist monasteries, which spread across mountains and towns all over China, yet were concentrated around the Lower Yangzi River Delta.[12] Under the sponsorship of the Carlsburg and the Ny Carlsburg Foundations, Prip-Møller traveled in eleven of the eighteen provinces of China in 1929–1931 and surveyed numerous Buddhist monasteries that were still in practice.[13] He documented the layouts, iconographies, and rituals of the monasteries, among which Longchang Monastery enjoys the most detailed observation.[14] The fieldwork's summative outcome is a monumental monograph titled *Chinese Buddhist Monasteries: Their Plan and Its Function as a Setting for Buddhist Monastic Life*, published in Copenhagen, Denmark in 1937.[15] The account of Longchang Monastery in the book preserves the most comprehensive picture of any Buddhist monastery in China thus far. It comprises systematic documentation of not only the architecture and building history but also the rituals and monastic lives.[16]

It has been well acknowledged that the documentary work in *Chinese Buddhist Monasteries* is invaluable and non-retrievable, especially by historians of Chinese religions (Welch 1967). But Prip-Møller's scholarly approach was largely ignored by the mainstream architectural historians of the time. His interest in the monasteries deviated from most Japanese and European scholars in China as well as the Society for the Study of Chinese Architecture (Yingzao xueshe 營造學社).[17] When others were competing to discover historical remains of as early as possible a date, as well as the lost knowledge of the timber-construction system, Prip-Møller looked at contemporary practices and the built environment around him. Prip-Møller self-consciously chose to study the lately built or rebuilt monasteries that were unstudied by others,[18] but the method and value of his work received serious criticism. Pioneering architectural historian Liang Sicheng (1901–1972), though recognizing a few instances of architectural studies, doubted the relevance of ordination ritual and socio-religious history to the discipline of architecture. Liang regretted that *Chinese Buddhist Monasteries* is a result of "ignoring one's own field but cultivating other's" (Liang 1940, p. 428). Even after three decades, when the book was republished, a pre-eminent art historian Alexander C. Soper (1904–1993) still did not apprehend its relevance to the studies of ancient architecture (Soper 1969, p. 88).[19] German scholar Ernst Boerschmann (1873–1949), whose work had inspired Prip-Møller's, was among the very few scholars who shared the latter's belief in a history-transcending architecture (Boerschmann 1939, p. 294), but he was not well received by the mainstream either.[20] The problem about what significance lies in a non-progressive history lingers on.

3. Sanmei: (Re)Building the Main Courtyard to Establish a Vinaya Monastery

In order to locate the dynamics in this seeming changelessness, it is necessary to revisit the most crucial period in the building history of Longchang Monastery—that is, the seventeenth century under the first three abbots of the Vinaya monastery. The weak state of the institution did not change until Sanmei, the first abbot, took charge. Invited by officials and clergy in Nanjing, Sanmei, as an established monastery founder, accepted the mission of reviving the site. According to a biography written by a literatus-official Chen Danzhong 陈丹衷 (?–17th century) (BHSZ [1795] 1975, vol. 13, pp. 230–31), Sanmei was trained under Xuelang and received teachings from Chan and Pure Land masters in his early career, before he was fully ordained under the thirteenth Vinaya patriarch Guxin 古心 (1540–1615).[21] Sanmei deployed several strategies, including allusion to classical models and ordination rituals, to establish a vigorous institute.

Sanmei's foremost mission was the large-scale construction of an independent system. He set up a main hall on another ground below the Open-Air Platform Unit. The Main Hall, far larger than the copper hall, became the most public ritual building of the Monastery. Furthermore, Sanmei carefully planned the layout of the Main Courtyard surrounding the Main Hall, making the new cloister face the northwest direction instead of the west. The innovations, nevertheless, are not without precedents in the history of Buddhist monasteries.

3.1. Archaic Layout

The core area of Longchang Monastery established by Sanmei comprises a single main courtyard surrounded by several buildings and units. This design does not follow the paradigmatic layout of the Ming and Qing dynasties: the paradigm features a prolonged central axis along which multiple courtyards and major ritual buildings are laid out (Figure 7). Pragmatic buildings, such as the living zone for the monks, would be placed on the lateral sides of the main axis.[22] But Longchang Monastery has neither a prolonged layout nor a main path; instead, most following expansions center around the main courtyard ever since its establishment. What may have helped Sanmei envision this design?

Formally speaking, the layout in which a main courtyard is surrounded by minor courtyards reminds historians of Chinese architecture of the paradigm of monastery layouts in the Tang (618–907). This paradigm follows the ideal monastery prescribed by a scholar-monk Daoxuan 道宣 (596–667), who was the First Patriarch of the Nanshan Lineage of the Vinaya School. According to Daoxuan's *Illustrated Scripture of Jetavana Vihara of Sravasti in Central India* (*Zhong Tianzhu shewei guo qihuan si tujing* 中天竺捨衛國祇洹寺圖經),[23] Jetavana Monastery is formed around the Central Buddha Cloister.. The Buddha cloister, which is a four-sided enclosure preserved for the Buddha and larger than any other units in the building complex, is surrounded by the monks' cloisters. In the outer areas are located a multitude of cloisters for Buddhist education and miscellaneous affairs (Figure 8).[24] As architectural historian Puay-peng Ho points out, this ideal monastery under the name of a holy site in India has effected a spatial symbolism of the Vinaya practices and established a spatial paradigm for Buddhist monasteries in China (Ho 1995). Although no monastery from Daoxuan's time still stands in its primary layout, various historical materials bear witness to the prevalence of this paradigm in Tang China.[25]

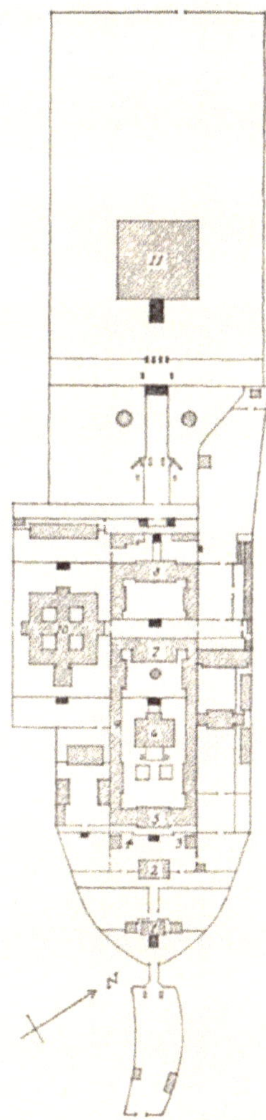

Figure 7. Plan of Biyun Monastery, Mt. Xishan, Beijing, Ming to Qing period, 1368–1911: (1) Mountain Gate; (2) Hall of Heng-Ha (Guardians); (3) bell-tower; (4) drum-tower; (5) Hall of Heavenly Kings; (6) Jade Emperor's Hall; (7) Guanyin Hall; (8) Main Hall, (9) Hell, (10) 500 Arhats' Hall, (11) Pagoda. (Prip-Møller 1937, p. 13, Figure 12).

The Longchang layout (Figure 9) displays a striking similarity to Daoxuan's ideal monastery. The Main Hall, which stands for the presence of the Buddha, is located at the rear center of the Main Courtyard. The Main Courtyard is flanked on the lateral sides by the "Board Halls" (*bantang* 板堂), which are meditation and living quarters for monks in the Vinaya tradition.[26] Further outward one finds the Cloister of the Abbot, dining halls and kitchens, guests' quarters, and workshops.[27] The only significant difference is the peripheral location of the Monastery Gate (Shanmen 山門) and the northwest-facing orientation of the monastery. Yet these decisions were not so much a challenge to the paradigm as a response to the topography.

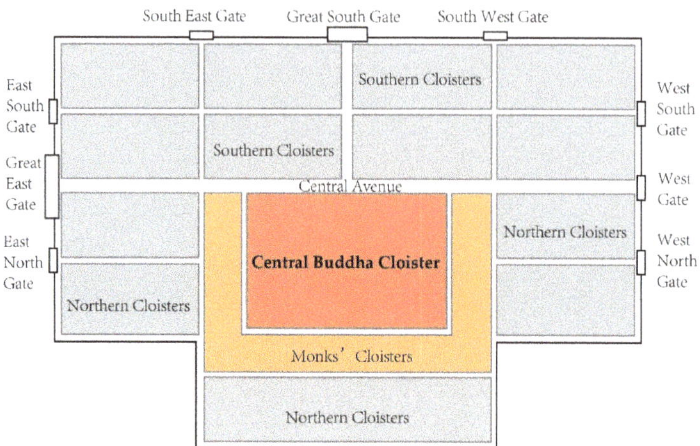

Figure 8. Schematic plan of Jetavana monastery based on Daoxuan's description, reconstruction by Puay-peng Ho (drawing by author, after Ho 1995, p. 6, Figure 2).

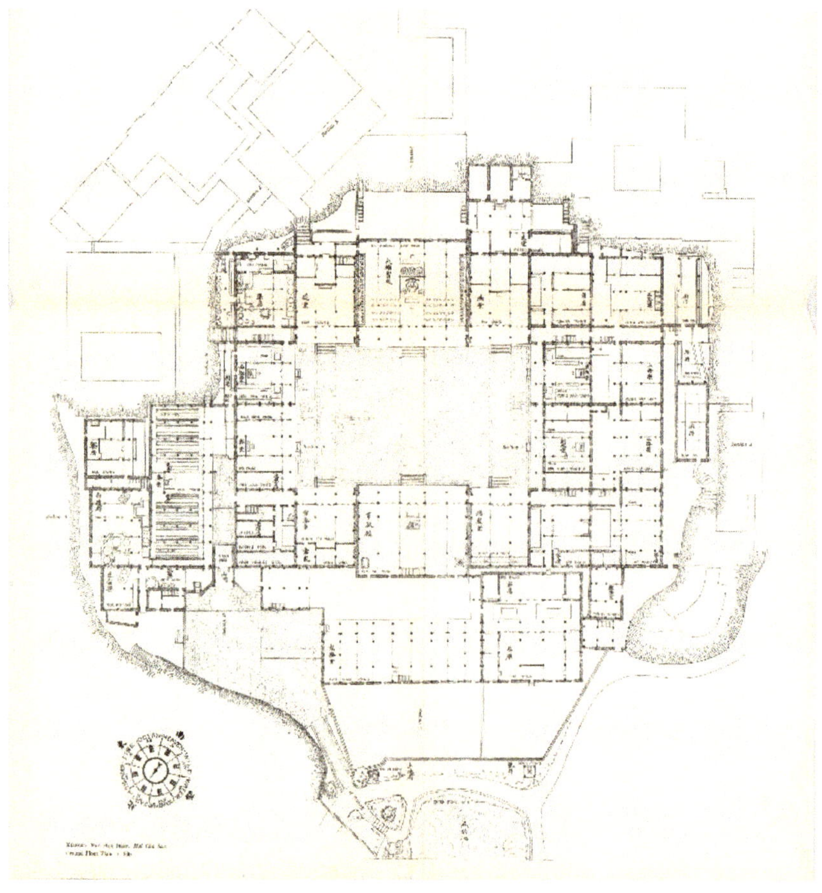

Figure 9. First floor plan of Longchang Monastery in 1930 (Prip-Møller 1937, Pl. 1).

Sanmei has not explicitly addressed the connection between the archaic layout he adopted and the ideal monastery Daoxuan described. It seems natural for the Main Courtyard to echo the topography of the main peak, which is surrounded by many minor peaks. However, it is not unreasonable to consider the archaic layout as an homage Sanmei, a follower of the Nanshan Lineage of the Vinaya School, paid to the first patriarch of his lineage, who represents the remote source of wisdom and authority. Sanmei's followers would soon recognize the historical allusion.

3.2. Experiential Space

Apart from the layout, Sanmei put much effort in creating an adorned and orderly space for believers to experience. The architectural strategies feature four points: the placement of images and instruments, symbolism through the arrangement of functional spaces, an orientation shift, and the erection of an ordination platform.

Firstly, the largest buildings and the major Buddhist images are located at the rear and front centers of the main courtyard. They constitute a central axis that keeps worshippers in a solemn mood (Figure 10). The central icon in the main hall is an image of the Cosmic Buddha Vairocana, who represents the essential body (Chn: fashen 法身; Skt: dharmakāya) of the Historical Buddha Shakyamuni. The buddha icon, as Prip-Møller observed, was a physical and spiritual locus during the morning and evening recitations as well as the annual ordination ritual (Prip-Møller 1937, pp. 309–10). In the opposite side of the courtyard, a Hall of Weituo enshrines the images of Guanyin, Wenshu, and Puxian Bodhisattvas. As mentioned above, the Three Great Bodhisattvas is the main iconography in the Open-Air Platform Unit, yet in the main courtyard the iconography occupies a subordinate position to Vairocana. The hierarchy thus hints at the peerless position of the Main Hall in the monastery. The Main Courtyard is equipped with major sounding instruments used for monastic rituals. The big bell and big drum are placed in the Hall of Weituo, whereas two wooden fish, an inverted bell beaten with a wooden stick (*qing* 磬), and a smaller set of bell and drum are in the main hall.[28] The sounding instruments signify monastic time and schedules. They add an acoustic dimension to the visual order and a temporal dimension to the spatial layout.

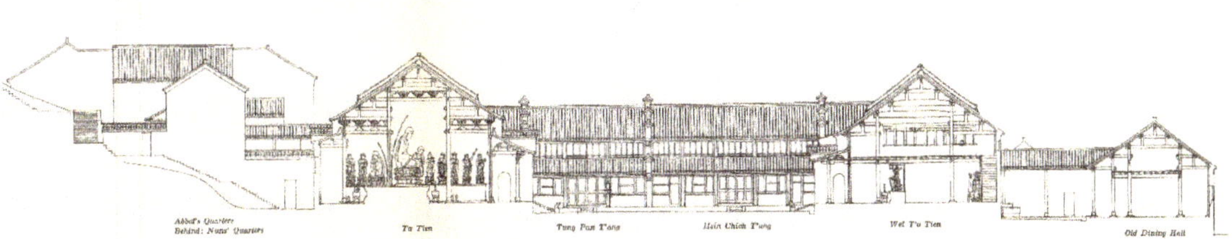

Figure 10. Section of Longchang Monastery, Showing the Main Hall, the Main Courtyard and the Hall of Weituo (Prip-Møller 1937, Pl. 4).

Secondly, the main courtyard accommodates novices and newly ordained monks as a prelude to the ordination rituals.[29] The life and training that happened there helped them prepare for a communal life in a monastery. The buildings surrounding the Main courtyard include the Sutra Pavilion and the Cloister of the Abbot on two sides of the Main Hall, as well as the Board Halls on the lateral sides of the main courtyard connected by long corridors. The former represents the Three Jewels, namely, the Buddha, the Dharma, and the Sangha, whereas the latter connects those in need with the sources of Buddhist teachings and regulations. If the former is symbolic, then the latter is functional. The two aspects constitute a solemn and lively "place for awakening" (Chn: daochang 道場; Skt: bodhimaṇḍa).

Thirdly, the main courtyard shifts orientation about 45 degrees clockwise compared to the Open-Air Platform Unit, and all later constructions followed the inter-cardinal orientation. Yet, because of the orthogonal layout in the new area, beholders would not easily recognize the natural orientation. Locations in the monastery are sensed in a relative reference system. In accordance with the southwest-facing Main Hall, the northwest-southeast running axis was conceptualized as a south-north-running one. By convention, the northwest direction is redefined as "the south" or "the front", whereas the southeast becomes "the north" or "the back" in the independent orientation system. Similarly, the northeast becomes "the west" or "the left", the southwest is regarded as "the east" or "the right" (Prip-Møller 1937, p. 206). A beholder following the relative orientation system would naturally take the Main Hall as the reference point in his or her mental map.

Lastly and most importantly, an ordination platform was erected for performing the ordination ritual, *shoujie* 授戒, or "giving the [monastic] precepts."[30] The ordination platform, which seems to be the only new architectural element that Sanmei introduced, was a careful choice. An ordination platform is the identifying characteristic of any major Vinaya monastery, which Sanmei presumably aspired to found. The ordination platform Sanmei commissioned was a single-level wood platform, and possibly located somewhere in the southeast part of the Main Courtyard. The ordination platform and the main courtyard together compose the stage for the most elaborate set of rituals at Longchang Monastery, namely, the Three-Platform Great Ordination (*santan da jie* 三壇大戒). The threefold ordination includes three precepts-giving rituals that are performed on separate days. The precepts are respectively called "the Śrāmaṇera Precepts" (*shami jie* 沙彌戒), "the Bhikkhu Precepts" (*biqiu jie* 比丘戒), and "the Bodhisattva Precepts/Vows" (*pusa jie* 菩薩戒), following the increasing spirituality of the three kinds of beings, Śrāmaṇera or the novice monk, bhikkhu or the fully ordained monk, and bodhisattva or the enlightened sentient being. The first two precepts are formal ordination rituals and the last one is a repentance and vow-making ritual.[31] At the time when Prip-Møller visited the Monastery, the first and the third precepts were given in the Main Courtyard, whereas the second and most crucial ritual took place in the Ordination Platform Unit, where an ordination platform, albeit not identical with the one built by Sanmei, was located.

3.3. The Main Courtyard as the New Center

In these ways, Sanmei made the main courtyard a new center that met both ritual, symbolic, and pragmatic needs. His fundamental work was well recognized by the following generations. Sanmei was eulogized as one of the three founding patriarchs of the Monastery in a stele erected by Ding'an under the title "Brief Biographies of the Three Great Masters Miaofeng, Sanmei, and Jianyue Who Constructed Longchang Monastery of Mount Baohua" (jianzao Baohua Longchang si Miaofeng Sanmei Jianyue san dashi xinglue 建造寶華隆昌寺妙峰三昧見月三大師形略). The stele text remarks that one of Sanmei's accomplishments was to "select the auspicious [location] and move the orientation" (xuanji qianxiang 選吉遷向) (BHSZ [1795] 1975, vol. 6, pp. 13–18; Y. Liu, forthcoming, p. 102; Wen 2004, pp. 40–41).[32] Concurrently, Sanmei created a problem for his successors; he initiated the tension between the Open-Air Platform Unit—the spiritual core established prior to the institutionalization of Mount Baohua—and the Main Courtyard—the central zone of the newly established institute. He could not control, either, whether the many monks whom he ordained would become supporters or opponents of the order he envisioned and furthered by his favorite disciple and successor Jianyue.

4. Jianyue: Reconstructing the Ordination Platform to Recenter the Vinaya Monastery

Jianyue had been involved in Sanmei's construction projects since the beginning. As a helpful assistant, he was appointed as Monastery Superintendent (*jiansi* 监寺) when Sanmei became the Abbot of Longchang Monastery. Since being appointed the second abbot, Jianyue greatly advanced the mission of reviving Longchang Monastery as a Vinaya monastery that Sanmei founded. Compared with his predecessor, Jianyue was a more pronounced adherent to monastic regulations and order—which are particularly valued in the Vinaya tradition.[33] When further completing the layout Sanmei initiated by adding layers and pinpoints, Jianyue's new vision for the Monastery centered around the ordination platform. Through reconstruction, relocation, and demarcation, he promoted the platform as another spiritual center of the Monastery. In the process, his tough measures intensified the tension between the old and new built environments. The reconstruction of the ordination platform witnessed both the mission and the tension.

4.1. Self-Concious Reference to Daoxuan

Compared with Sanmei's ambiguity, Jianyue took a more self-conscious approach of following the ideal monastery layout of Daoxuan. Jianyue explicated that the placement of the Ordination Platform to the southeast outside of the Main Courtyard (Figure 11) was modeled after the first ordination platform in Buddhist history. Daoxuan describes the prototypical ordination platform in another writing entitled *Illustrated Scripture Concerning the Erection of the Ordination Platform in the Guanzhong Region* (Guanzhong chuangli jietan tujing 關中創立戒壇圖經) (*T* no. 1892, 45). The scripture indicates that Shakyamuni "placed an ordination platform in the south of the east gate of the outer courtyard for the bhikkhu-to-be to receive the precepts" (*T* no. 1892, 45: 0807c08). To proclaim the history and significance of the ordination platform he renovated, Jianyue erected a commemorative stele that bears "the Inscription about the Ordination Platform of Longchang Monastery of Mount Baohua Constructed under Imperial Edict" (chijian Baohua shan Longchang si jietan ming 敕建寶華山隆昌寺戒壇銘) (Appendix A.1, BHSZ [1795] 1975, vol. 6, pp. 12–13; Y. Liu, forthcoming, p. 95; Prip-Møller 1937, pp. 287–88). The inscription begins by accounting the origin of the ordination platform. It traces back to the aforementioned event in which Shakyamuni Buddha "placed an ordination platform in the southeast of the outer courtyard of the [Jetavana] Vinara" of Sravasti in Central India. In effect, Jianyue ambitiously established a conceptual connection between the Ordination Platform of Longchang Monastery and the divine prototype, "the Ordination Platform Made by the Buddha" (fozhi jietan 佛製戒壇) as described by Daoxuan.

4.2. Three Steps to Promote the Ordination Platform

Jianyue upgraded the Ordination Platform in three steps, namely, rebuilding the old wood platform, rebuilding it again with white marble, and demarcating the Monastery's inner, middle, and outer boundaries. Through these steps, the Ordination Platform was transformed into the most spiritual and secluded place in the Monastery.

The first step was to rebuild the old wooden platform made by Sanmei. According to "the Inscription about the Ordination Platform" (Appendix A.1, Lines 1–2),

> Having inherited his master's mission and been told to promote and transmit it, [Jianyue] diligently practiced in person and rectified [the errs]. Starting on the third day of the fourth month of Dinghai, which is the fourth year of Shunzhi era of the current (i.e., Qing) dynasty (1647), ... [Jianyue] constructed a boundary and erected a platform, imitating the past and renewing it in the present ... Therefore his primary goal to revitalizing the Monastery under his arms, at that moment was half-accomplished.[34]

From this text, we learn that Jianyue set out to continue the monastery-reviving work of Sanmei and that the Ordination Platform was the first project he commissioned. In 1647, Jianyue made a new wooden platform based on the pre-existing model, i.e., the wooden platform erected by Sanmei. The platform might have corrected some defects in Sanmei's version, yet there seemed to be an error in the stairways. Indeed, "imitating the past" is only the first half of revitalization and more work is needed to be done.

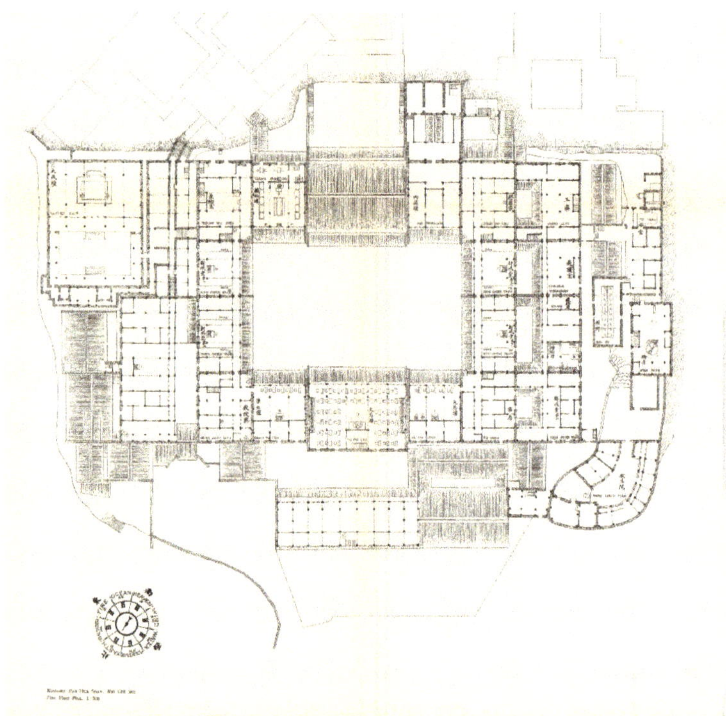

Figure 11. Second floor plan of Longchang Monastery (Prip-Møller 1937, Pl. 2).

Subsequently, Jianyue decided to take a second step to rebuild the platform again. The design updates, as well as the political and religious implications of this event, are thoroughly discussed in the following quote from the aforementioned inscription (Appendix A.1, Lines 3–6),

> On the sixteenth day of the third month of a Guimao year, which was the second year of Kangxi era (1663), on the sound of the board, a communal meeting took place. An unoccupied place in the southeast outside of the Main Courtyard was selected. The old [Ordination Platform] was dismantled and a new one was constructed. The [new designation of] boundary was not constrained by any [conventional] pattern.[35] The stone platform had clearly distinguished upper and lower levels. In the early evening of the twentieth day of that month, clouds and rain dimmed the sky, and mountains and hills were obscured in mist. Suddenly, the Hall of the Ordination Platform radiated lights of five colors which broke through the cloudy sky. The mountain peaks revealed emerald green and ten thousand pine trees embraced the site. The building complex was brilliantly shone as if by the sun in the daytime. Not after a while did the luminosity started to dissolve. The assembly joyfully watched this, and in a single voice praised the wonderful [event]. In the long period since the Vinaya came to exist, an auspicious omen as such is indeed a rare thing in the world of Five Turbidities!

From the quote we know that, in 1663, Jianyue had "the old [Ordination Platform] dismantled and a new one constructed." In the second renovation, significant revisions were made: a white marble material replaced the wood, it was clearly a double-level platform; and meticulous ornamentation was added. The white marble platform (Figure 12) has a dignified appearance clearly distinguishable from the previous version. According to Prip-Møller, this white marble platform likely corrected a design error in the stairs previously made by

Jianyue himself in his first reconstruction from sixteen years ago (Prip-Møller 1937, p. 284). The white marble is a precious material, like copper, only applied to imperial buildings or imperially recognized monasteries. This instance demonstrates that reconstruction could be a good opportunity to optimize the design. The dignified design and the enduring material properly expressed that the ordination platform was both a crucial stage for the threefold ordination rituals and a display of the imperial recognition of a prestigious monastery.[36]

Figure 12. Photograph of the Ordination Platform (Prip-Møller 1937, p. 243, Figure 270).

In addition, the erection of the stele and the recording of the miraculous correspondence (*ganying* 感應) played an important role in the promotion of the status of the stone platform. If the communal meeting, as well as the siting and boundary-making activities, improved the Sangha's knowledge of the ordination platform, which is normally hidden from public view, then the auspicious omen of the five-color light furthered the mythic efficacy of it. By promoting this anecdote, Jianyue tried to convince others of the divine quality of the stone ordination platform, which, upon completion, emitted five-colored brilliant lights into the night sky through layers of clouds. The implied message was that the ordination platform received not only imperial endorsement but also celestial approval. The ordination platform was rendered in all aspects to be the centerpiece of the Monastery.

The third strategy was a threefold demarcation. The Ordination Platform Unit (Figure 13), in which the marble Ordination Platform was permanently retained, was located in the east, outside of the Main Courtyard, and in the north, below the Open-Air Platform Unit. Like the Main Courtyard, the Ordination Platform Unit faces northwest. Because of its content, the Ordination Platform Unit became the third and the newest locus of the Monastery. Compared with the previous two, the Ordination Platform Unit was made even more sacred through the strict control of accessibility, and more central through the redesignation of the Monastery's boundaries.

On the one hand, the Unit could only be accessed during the second and most crucial stage of the threefold ordination ritual. Once a year, every novice monk who wished to enter monkhood passed the narrow corridor at night, ascended the platform in line, and took the Bhikkhu Precepts before the seats of the abbot and other senior monks. The Sangha shared their personal memory of the spot that lay at the very foundation of their career as a fully-ordained monk.

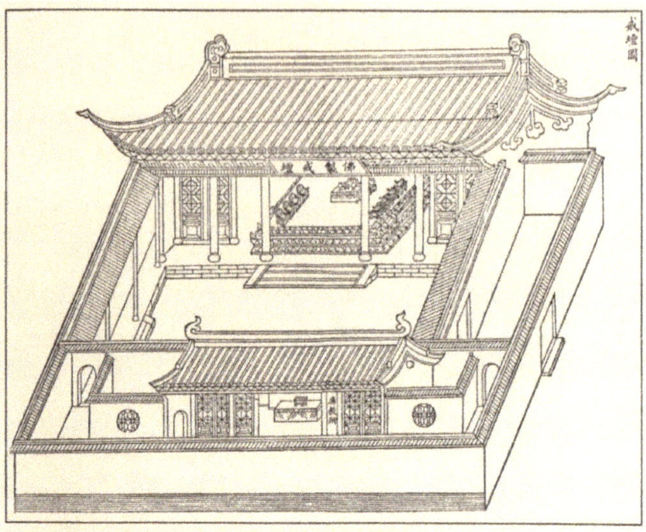

Figure 13. The Picture of the Ordination Platform Unit. Chinese, wood-block print, book illustration, the eighteenth century (Prip-Møller 1937, p. 297; BHSZ [1795] 1975, vols. 1, 6).

On the other hand, Jianyue once gave a sermon to the Sangha, in which he set three concentric layers of boundaries centering at the Ordination Platform. Although it is hard to know how effective the sermon was at that time, its lasting influence is detectable in a stele which was erected seventy-eight years after the sermon had happened. The inscription of this stele, "The Bounds of the Monastery" (Jiexiang zhunze 界相準則) dated 1741, has recorded Jianyue's sermon in 1663 as follows (Appendix A.2, Y. Liu, forthcoming, pp. 106–7; Prip-Møller 1937, p. 296):

> First to be declared is the method of constructing the Field of Ordination Ritual (shoujie chang 受戒場). Its southeastern side is marked by the wall of the Place of Screened-off Teachings (Pingjiao suo 屏教所). Its southwestern side is marked by the wall of the right-side gable wall of the Ordination Platform Hall. Its northwestern side is marked by the wall of the left-side gable wall of the Hall. Its northeastern side is marked by the outer side of the surrounding wall of the Hall. These are the four-sided outer boundaries of the Field of Ordination Ritual.

> Second to be declared is the construction of the Inner Extent of the Larger Boundary. Its southeastern side is marked at two feet (Chinese) from the wall of the Place of Screened-off Teachings. Its southwestern side is marked at two feet (Chinese) from the wall of the right-side gable wall of this Ordination Platform Hall. Its northwestern side is marked at two feet (Chinese) from the wall of the left-side gable wall of this Hall. Its northeastern side is marked at two feet (Chinese) from the [surrounding] wall of this Hall. These are the four-sided inner extent of the Larger Boundary.

> Third to be declared is the construction of the Outer Extent of the Larger Boundary. Its southeastern corner is marked at the crossroad on the further side of the Main Mountain (Zhushan 主山). Its southwestern corner is marked at the crossroad at the foot of the Dragon's Back Mountain (Longbei shan 龍背山). Its northwestern corner is marked at the summit of the West Flower Mountain (Xihua shan 西華山). Its northeast corner is marked at the chestnut tree at the foot of the Bliss Mountain Range (Huanxi ling 歡喜嶺). This is the outer extent of the Larger Boundary.

> Beyond the boundaries [marked by] tied cloth there is no other extent of boundary.
> Precisely adhering to the demonstration, construct up to the outer extend of the
> Larger Boundary.

According to the record, Jianyue defined the boundaries of the Monastery in relation to the Ordination Platform Unit: the innermost boundary encloses the courtyard and halls of the Unit, the middle one is slightly set off from the innermost boundary, and the outermost boundary drastically increases to cover the overall monastic property surrounded by the mountains (Figure 14). As if to emphasize the decisiveness and comprehensiveness of this demarcation, Jianyue uttered that beyond these boundaries there was no other boundary. This historical conception of zoning is surprisingly different from a modern researcher's view that takes the main courtyard to be the core of the Monastery and the smaller units to be the subordinate. Instead, Jianyue constructed a systematic conception of religious space by means of symbolic objects, ritualistic demarcation, and the Monastery's main functionality as a field of ordination.

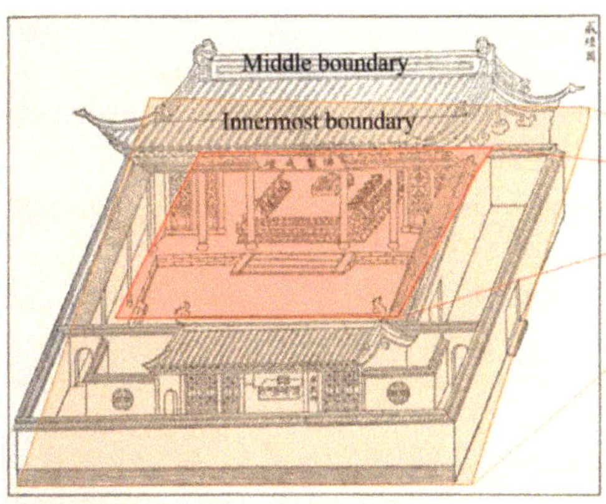

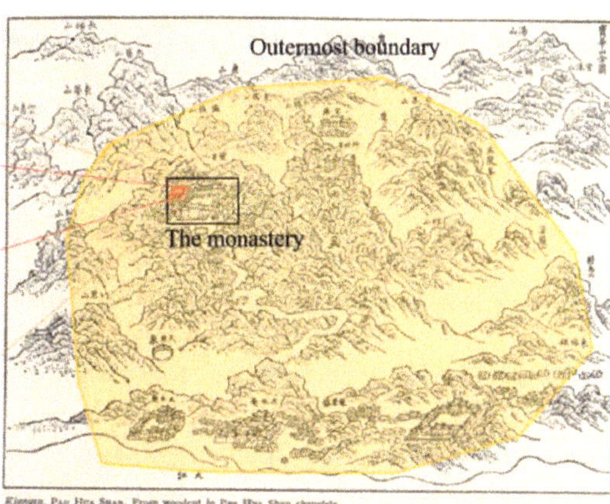

Figure 14. A diagram showing the three boundaries of the Monastery according to Jianyue's sermon (drawing by author, base map: Figure 13 & BHSZ [1795] 1975, vol. 1, pp. 3, 6).

4.3. Aftermath

Jianyue's bold activities and speech left a hazardous impact on building a monastic community. When he was promoting this new centerpiece of the Monastery and reframing the spatial hierarchy, just like the three potential loci confronting one another, differentiation emerged from within the Sangha. Some elder monks were even competing for the position of the Abbot. Although opinions are not settled on whether Jianyue intended to leave or was expelled from the Monastery by some monks unsatisfied with his over-activity,[37] the fact was that, shortly after the second platform reconstruction and the demarcation sermon, he left the Monastery for multiple years because of the conflicts between factions. Jianyue did not enter Mount Baohua once again until the literatus-official Chen Danzhong, acting as mediator, invited him to return as an abbot. Perhaps in order to balance the unsettled forces, Jianyue's last wish was to refurbish the old Open-Air Platform Unit, although he had to leave the unfinished task to his successor Ding'an.

5. Ding'an: Refurbishing the Open-Air Platform Unit to Reunite the Three Centers

Ding'an deeply remembered what Jianyue once signed to him with emotion in old age. Jianyue commented that a human's lifespan is shorter than those of metal and stone, yet even the copper hall and stone platform are also subject to disintegration, and he saw the words "all that has form definitely deforms" in both humans and architecture (BHSZ [1795] 1975, vol. 6, p. 245). To counteract the natural process of material decay and to balance the human forces that may deform the institution, Ding'an took the opportunity to refurbish the Open-Air Platform. Ding'an fashioned himself as a faithful follower of Jianyue without the latter's strong characteristics. He consciously took the role of compiling guidelines and historical accounts, rather than that of a chief architect. Through issuing new placatory rules, Ding'an consolidated the tensions among the monastic sections. During the refurbishment projects, he valued the works of the lay Buddhists and literati in fulfilling the revival mission. He even wrote twelve volumes about the Monastery's settings and institutional history, which became the basis of Liu Mingfang's fifteen-volume gazetteer *BHSZ*.[38] Through resuming solidarity among the monastic and lay communities, the ultimate goal was to reunite the Monastery that had an old center and the two new centers.

5.1. Renewing Placatory Rules

To counteract the separatist thread that had surfaced during Jianyue's time, Ding'an issued "the Rules for the Brotherhood" (gongzhu guiyue 共住規約) and had the rules inscribed on a stele in 1683 (Y. Liu, forthcoming, pp. 97–99; Prip-Møller 1937, pp. 288–91). "The Rules" starts by reasserting the importance of the monastic regulations established by Jianyue. Based on Jianyue's version, the rules Ding'an issued "particularly establish some chapters and check erroneous ideas at the outset." The preventive measures particularly address that nobody should be permitted or forced "to separate himself from the community". The twelve chapters in the Rules cover all aspects about communal life in a monastery, including places for living, clothing, dining, lodging, traveling, and tonsure. The guidelines reflected Ding'an's vision for an inclusive community, which was also applicable in regard to monastery construction.

5.2. Refurbishment of the Open-Air Platform

In 1680–1691, Ding'an commissioned a series of small and delicate stone structures for the refurbishment of the Open-Air Platform. Upon the completion of the refurbishment, Ding'an compiled "A Record of the Renovation of the Copper Pavilion, the Incense Pavilion, and the Stone Platform" (chongxiu tongdian xiangting shitai ji 重修銅殿香亭石台記, BHSZ [1795] 1975, vol. 6, pp. 247–48). The record began by addressing Jianyue's last vow and the merits of several lay artists and sponsors who participated in the projects. Among them, a layman Dajing 大莖 decorated the Copper Hall with the assistance of a Head of the Hall Xingbai 省白. The landscape images on the railings were painted by two artists Julai 巨来 and Zhilai 支来, disciples of a famous painter Gong Xian 龔賢 (1618–1689). The projects were sponsored by "dharma protectors of the Ten Quarters." Ding'an did not assert his leadership and authorship; instead, he humbly wrote that what he was doing was "documenting the events, recording the merits, and passing down for future generations".

Nonetheless, it is hard to believe that the decade-long, comprehensive refurbishments were without a planner, who could have been only Ding'an. Under his supervision, an incense-offering shrine known as "the Stone Incense Pavilion" (shixiang ting 石香亭) was erected in front of the copper hall when the latter was refurbished. A commemorative stele in memory of Miaofeng, Sanmei, and Jianyue was erected on a lower platform outside the Open-Air Platform Unit. In addition, the platform was widened, repaired, decorated with bas-relief friezes, and equipped with a white marble balustrade (Figure 15). In general, platforms of the *sumeru*-throne (*xumi zuo* 須彌座) type were only applied to esteemed buildings such as the base of the main hall in imperial palaces or ancestral temples (Figure 16). In Longchang Monastery, this type of platform had been applied only to the Ordination Platform. The refurbishment of the Open-Air Platform must have been a careful design. While the stone platform may be studied as a work of architectural decoration, this study situates it in the larger context of place-making and history-framing of the Monastery.

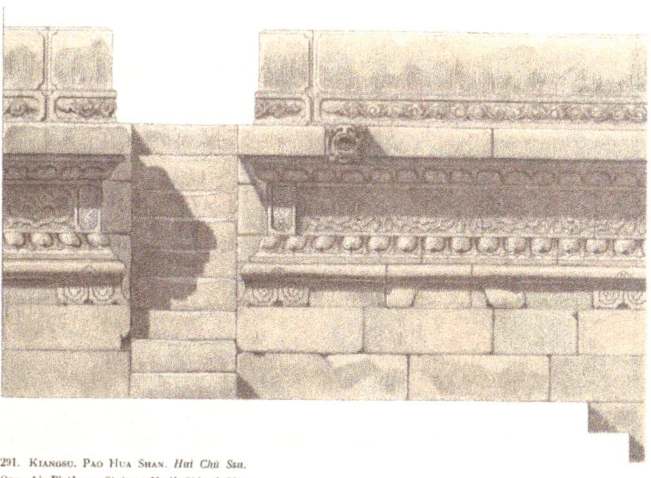

Figure 15. Rendering of the Open-Air Platform (Prip-Møller 1937, p. 236, Figure 291).

Figure 16. A corner of the white marble base of the Hall of Supreme Harmony in the Forbidden Palace, Beijing, China, seventeenth century (Photo by author, 15 October 2021).

5.2.1. Incorporating into the Ordination Rituals

For one thing, Ding'an and others modified the Miaofeng-era monument for the sake of better incorporating it into the spatial system that centered at the threefold ordination rituals. The main stair and two side stairs all point to the newly added Stone Incense Pavilion, which better facilitates the prayer to the Copper Hall (Figures 4 and 17). According to Ding'an's, this project not only fulfilled the last will of Jianyue, but also concerned the ordination ritual. At the end of the Renovation Record, he wrote (BHSZ [1795] 1975, vol. 6, p. 23):

> The precious hall, pavilion, and platform are solemn and splendid. Together with our gem-like Precepts, they shine between heaven and earth and will never exhaust!
> 寶殿亭臺，莊嚴端麗，與吾戒珠並燦爛于天地間而無窮盡也，其庶幾乎！

Only when being inspired by collective efforts of reviving the faith, one dared to vow that the architecture and the "gem-like Precepts" (jiezhu 戒珠) eternally coexist. Although the platform no longer formally carried any function in the threefold ordination rituals when Prip-Møller visited, he assumed that the platform had once been associated with ordination (Prip-Møller 1937, p. 294). The Copper Hall, which enshrined a Guanyin image, was believed to connect with the bodhisattva. Hence, it would well suit the third stage of the threefold ordination, during which the newly ordained monks need to utter the Bodhisattva Vow. Till the modern time, the Copper Hall and the Stone Incense Pavilion were still the "first place after the Great Hall to which the newly ordained monks go to prostrate themselves in thankfulness before the Guanyin image" (Prip-Møller 1937, p. 254).

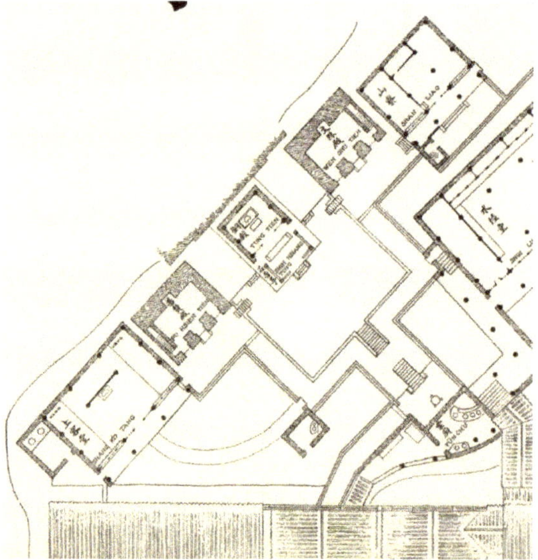

Figure 17. Detail of the third floor plan of Longchang Monastery, showing the Open-Air Platform Unit in the upper half, the Roofs of the Ordination Platform Unit in the bottom-left, and the roofs of a corner of the Main Courtyard in the bottom right (Prip-Møller 1937, Pl. 3).

5.2.2. Alluding to the Ordination Platform

For another, the decoration of the Open-Air Platform was designed to allude to the Ordination Platform. Through some shared visual language, the spiritual core of the past and that of the current constitute a coherent architectural landscape. The two platforms resemble each other's proportions, circulation, and dimensions. The Open-Air Platform sizes are 14.95 m (L.) by 11.27 m (W.) by 2.10 m (H.), whereas the Ordination Platform sizes are ca. 12.8 m (L.) by 11.52 m (W.) by 1.50 m (H.).[39] Besides, the ornate frieze of the open-air platform covers only about half of its total height (ca. 1 m vs. 2.10 m). The design approximates to the height of either tier of the Ordination Platform (66.5 cm and 83.5 cm).[40] By means of this design, the single-level Open-Air Platform is subdivided into two levels—a decorated upper level and an undecorated lower level. Thus, the façade proportion of the two platforms becomes closer to each other. Furthermore, the frieze design of the Open-Air Platform subtly echoes the carvings on the lower tier of the ordination platform. The narrowed waist is flanked above and below by stylized lotus-petal decorations, and both the relief "feet" of the frieze and those of the ordination platform are decorated with pairs of spiral clouds. The cloud-and-thunder pattern on the railings of the ordination platform reoccurs on the narrowed waist of the frieze, showing similar double-lined edges, layering treatments, and configuration of spirals (Figure 18). Upon seeing the Open-Air

Platform, a monastic beholder would probably recall the Ordination Platform, on which he received the Bhikkhu Precepts during the second ordination ritual.

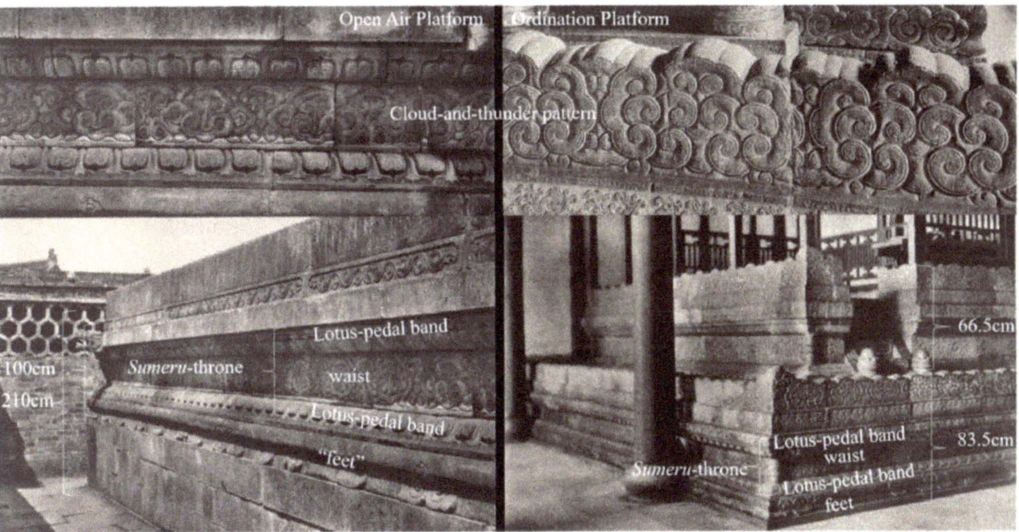

Figure 18. A comparison between the Open-Air Platform (**left**) and the Ordination Platform (**right**), with author's annotations (drawing by author, based on Prip-Møller 1937).

5.3. The Union of the Two Platforms and a Collapse of History

In short, as the final stage of the collective design, the two platforms were united as a coherent setting for the threefold ordination; the main courtyard and the two units, respectively, accommodated the three stages (Figure 19). The tensions among them were raised, intensified, and finally resolved through the accumulation of drastic and subtle adjustments. The Open-Air Platform Unit resumed its spirituality as if it had never changed.

The union of space fed a belief in the correlation of various periods in the history of Mount Baohua. The emerging belief is evident in "the Prose of the Ordination Platform" (jietan fu 戒壇賦, BHSZ [1795] 1975, vol. 10, pp. 409–12) compiled by an artist-scholar Wang Gai 王槩 (aka. Wang Anjie 王安節, 1645–1707), who was another student of Gong Xian and in the social circle of Ding'an. The prose traced the tradition of the Longchang Ordination Platform to a time prior to Daoxuan. Presumably informed by the monks or their abbot directly, Wang associated the Ordination Platform's prototype with an ancient diagram that was allegedly made by the fifth-century Chan monk Baozhi. The fabricated narrative also perfectly connected Ding'an with all the masters prior to him: it suggested that the diagram was discovered by Xuelang, copied and preserved by Guxin, primarily constructed by Sanmei, perfected by Jianyue, and its story transmitted by Ding'an. Whether the ordination platform began with Baozhi or Daoxuan did not really matter in the believers' eyes. What they looked for was a story without a beginning and, hence, without an end.

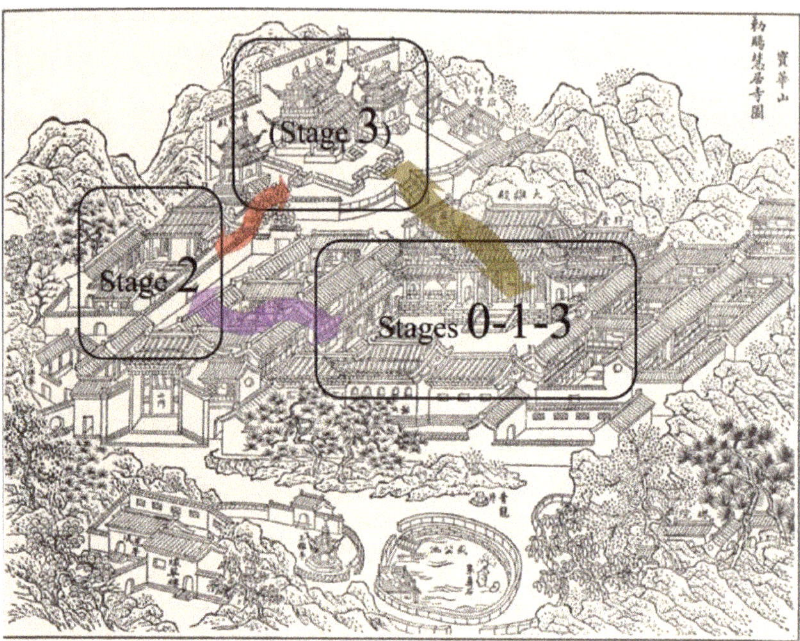

Figure 19. A diagram showing a coherent spatial system in which the main courtyard, the Open-Air Platform Unit, and the Ordination Platform Unit are associated with the prelude and the three stages of the threefold ordination rituals (Drawing by author, base map: Figure 1).

6. Coda: A Methodological Reflection

Through the eyes of Prip-Møller, we see how exactly the repaired monuments have connected the past and the present. That Prip-Møller chose to render the side view of the Open-Air Platform in a meticulous way (Figure 15) is not without any reason: The represented platform was designed to allude to the stone Ordination Platform, which improved twice what it replaced. Moreover, the side view is identical with what one sees in front of one of the beamless halls designed by Miaofeng (Figure 20), and from there an elderly monk living in the beamless hall often came out to guide the modern visitors to Longchang Monastery.

The built environment of Longchang Monastery successfully maintained a dynamism and eventually a balance in the process of constant and sometimes contradictory modifications. During a crucial period of establishing an institution, the three abbots of the Vinaya lineage consciously relocated the spiritual core of the monastery from the Open-Air Platform to the Main Courtyard, then to the Ordination Platform, eventually creating a coherent spatial system. They achieved this by means of modifying the masonry structures and the multi-layered boundaries. Sanmei moved the main monastic activities around the Main Courtyard, which was set up as a new, functional center as opposed to the old, symbolic unit established by Miaofeng. Jianyue redesigned and reconstructed an ordination ritual altar and promoted it as the new spiritual center. In this way, the supreme status of the old one designed by Miaofeng was further eliminated. Eventually, Ding'an reunited the three centers by refurbishing the old Open-Air Platform, alluding to the Ordination Platform, and related both to the Main Courtyard through staging the threefold ordination. Therefore, the monk-architect and the three abbots together designed a spatial system of unprecedented spiritual symbolism.

Figure 20. Viewing the Open-Air Platform from the doorway of one of the Beamless Halls, showing the West Side Stair and part of the Stone Incense Pavilion (Prip-Møller 1937, p. 249, Figure 276).

Through cultural and constructional practices, the "life" of the monastery has been extended in two directions along the timeline. The repeated accounts of its remote origin drive the memory of it into the past, and the practices of persistent renewal promise its existence in the future. Through human efforts, the monastic architecture as a whole is emancipated from a tangible beginning and a predictable ending. The apparatus of immortality is precisely located in the contemporary reenactment of historical prototypes: it shapes the social memory, uncovers forgotten pasts, and sustains communal life.

In the current century, the field has adopted a more tolerant attitude toward Prip-Møller's interdisciplinary approach to architecture. Inspired by scholars of social and religious histories, architectural historians joined the blooming studies of the "living" religious landscapes.[41] Scholars observe the socio-political forces that drove the developments of sacred mountains and urban monasteries. In ways echoing Prip-Møller's, they fruitfully recognize the built environment as a carrier of the local history.[42] In retrospect, some even raise Prip-Møller to the historical position of "representing a modern, international perspective of 'earth-timber/construction'" (J. Wang 2018). Recent scholars have revived the interests Prip-Møller showed in the stone monuments and environmental design of Longchang Monastery (J. Zhang 2015; Wang and Tur 2016). Notably, architectural historian Yan Liu has recognized the methodological significance of Prip-Møller's work for architectural anthropology, conducted a large amount of archival research and field work, translated *Chinese Buddhist Monasteries* into Chinese (Y. Liu, forthcoming), and reintroduced this work to the field.[43]

This case study furthers our understanding of Prip-Møller by adopting his unique view of history. A historian normally seeks to sort things in temporal sequence, whereas Prip-Møller seeks the instances when the sorting loses efficacy.[44] This is because, in Prip-Møller's view, persistent religious and building practices can cross the unbridgeable gap between modern and pre-modern times. Every contemporary monastery that deeply intrigues him is "an organism living in the present but with its roots deep in the past" (Prip-Møller 1937, "preface"). The architecture of Chinese Buddhist monasteries hence is "the frames around a religion which has endured without any fundamental change as a strong spiritual force in the country ever since the days of its introduction there" (ibid.). This view, as I believe, has the potential to resolve an enduring problem—i.e., the so-called "non-historical" character of Chinese architecture.[45] This problem of style has provoked studies about the evolution of layout and building technology. The abovementioned discourse about the dynamics of cultural production is yet another way of arguing for a history more progressive than stationary. But Prip-Møller's solution is different in nature; he finds historical value in the common practices that made Chinese architecture seemingly non-progressive. Through this lens, we see the repaired monasteries not as an architecture that loses traces of the past, but as one that attempts to transcend history.

Funding: A presentation of the research paper at the AAS (Association of Asian Studies) 2019 Annual Conference was supported by a conference travel grant from the Center for the Art of East Asia, the University of Chicago.

Institutional Review Board Statement: Not applicable.

Informed Consent Statement: Not applicable.

Data Availability Statement: Not applicable.

Acknowledgments: The author initiated the case study during a graduate seminar taught by Wei-Cheng Lin at the University of Chicago, from whom she received theoretical fuel for this project. She thanks Delin Lai of Louisville University for his incisive comments on an earlier draft of this paper, which was presented at the AAS 2019 Annual Conference in Denver, CO, March 23, 2019. She is in debt to Yan Liu of Kunming University of Science and Technology for generously sharing a manuscript of his Chinese translation of *Chinese Buddhist Monasteries* with the author. She also thanks the two anonymous reviewers for their constructive feedback that led the paper to the current form.

Conflicts of Interest: The author declares no conflict of interest.

Appendix A. Excerpts of Chinese Texts

Appendix A.1. 敕建寶華山隆昌寺戒壇銘

　　既承師命，委囑以弘傳，密不躬行而匡正。始於國朝順治四年歲次丁亥四月三日，…… 結界立壇，仿古更今。…… 由是復興於中懷之初志，今遂乎半。…… 又於康熙二年歲在癸卯三月十六日，鳴椎共議，擇院外東南閑處，解舊結新，界相無文系。石壇層級上下分明，於是月廿日初夜分，陰雨暗暝，山嵐迷障。驟然壇殿交光五色，直沖雲霄，峰崚顯翠，萬松環抱，群樓朗如白晝，經時始散。一衆瞻欣，同聲贊善。毗尼久住，瑞兆若斯，誠五濁之希有也。

Appendix A.2. 界相準則

　　先明結受戒場法：東南角至屏教所墻標，西南角在本戒場右山墻標，西北角至本戒場左墻標，東北角至本戒場圍墻外標。此是受戒場四方外相一周訖。
　　次明結大界內相：東南角離屏教所墻二尺標，西南角離戒場右山墻二尺標，西北角離戒場左山墻二尺標，東北角離戒場[圍]墻二尺標。此是大界四方內相一周訖。
　　後明結大界外相：東南角至主山後分路標，西南角至龍背山下分路標，西北角至西華山頂標，東北角至歡喜嶺下栗樹標。此是大界外相一周訖。
　　其攝衣界更無別相，表顯即準大界外相結之。

Notes

1. This version is a substantial revision of an earlier version compiled by Ding'an in 1690. For the circulation history of the BHSZ, see Zhu (2018).

2. The Chan Meditation or Intuitional School received its name from the meditative practice that was understood to be its basis. Developed gradually in China since the fifth century, the school thrived in the eighth century and in the subsequent centuries

made its own quasi-historical accounts about a lineage of patriarchs. The school disregarded ritual and sūtras and depended upon the inner light and personal influence. The Vinaya School is a precept-centered branch of Buddhist studies that specializes in the study and practice of the rules of discipline for the clergy, or Vinaya in Sanskrit. It developed during the early periods of transmission of Buddhism in East Asia, and the most influential tradition of the school is the Nanshan zong (Southern Mountain School).

[3] This study considers the relationship between space and social practices and the various kinds of space which have been discussed in (Lefebvre [1974] 1991).

[4] The temple gazetteer records that, for instance, the Guests' Hall and the Busa Hall were "rebuilt" by Sanmei, the main hall and the Hall of Great Compassion were "rebuilt" by Jianyue, and the Hall of Abbot and the Sutra Storing Pavilion were "rebuilt" "after the Old Mode" (zhao jiushi 照式). (BHSZ [1795] 1975, vol. 3, pp. 101–14).

[5] Because of its completeness and exquisite design, the Xiantong triad has been well studied from the perspectives of building prototype, construction technologies, imperial, and communal patronage (Boerschmann 1925, pp. 38–39; Bodolec 2005, pp. 170–75; J. Zhang 2015, pp. 289–322).

[6] For the biography and architectural works of Miaofeng, see (Prip-Møller 1937, pp. 275–82; J. Zhang 2015, pp. 292–96; Bodolec 2005, pp. 145–58).

[7] Before Miaofeng's arrival, two generations of local monks had been involved in construction work, but neither their names nor their works have been preserved.

[8] For two biographies of Xuelang, see (BHSZ [1795] 1975, vol. 7, pp. 255–73).

[9] For the biography and architectural work of Prip-Møller in China, see (Faber 1989; L. Zhang 2012, pp. 40–54). For the architectural surveys, see Madsen (2003).

[10] Both Beijing and Shenyang had been capital cities of the former Qing dynasty (1644–1911) and the latter was a thriving center of modern architecture and urbanism in the early years of the Republican era before it fell into Japanese colonization in 1932.

[11] Prip-Møller was born in a Christian family of and befriended Christian missionaries in Hong Kong and mainland China. For his manifesto about how architecture should serve religion, see Prip-Møller (1939). For his historical interests in the exchanges between Christian and Buddhist architecture in China, see Prip-Møller (1935).

[12] For his keen interest in the Chinese temples, see Prip-Møller (1931). For a case study of this kind of functioning monastery, see Prip-Moller (1936).

[13] The provinces and cities cover from Beijing and Shanxi province in Central North China to Guangdong and Yunnan provinces in the south, and from Jiangsu province in the East to the old Tibetan borders in the West. Prip-Møller (1937), "preface".

[14] Prip-Møller refers to the monastery as "Hui Chu Ssu 慧居寺" (Huiju Monastery), by which name it was known at that time.

[15] Before this book was published, Prip-Moller also published several case studies of Buddhist monasteries near Hangzhou and Nanjing (Prip-Møller 1935, 1936).

[16] This book consists of six chapters that respectively cover spatial, substantial, and intangible aspects of the monasteries. They respectively discuss the ordination ritual and the monastic life, picturing the apparatus of Buddhist institutions in the early years of the Republican era (1911–1949) that were somewhat continuous since earlier times. The first one third of the book discusses the typical layout of the massive monasteries he surveyed. Chapter One concerns the ritual architecture along the central axis and Chapter Two presents the auxiliary architecture in the lateral sides. The following two thirds of the book center around Longchang Monastery. Chapters Three and Four respectively deal with the built environment and building history of the Monastery, constituting an in-depth case study. Chapters Five and Six are comprehensive studies about the main dwellers in monasteries, i.e., the individual monks and the Sangha. Sangha (Sengqie 僧伽) refers to the Buddhist monastic community and is one of the Three Jewels in Buddhism.

[17] Some representative scholarship of the mainstream of Chinese architectural studies in the first half of the twentieth century include (Tokiwa and Sekino 1924; Sirén 1929; Itō 1931; Zhongguo Yingzao Xueshe 1930–1945).

[18] When Prip-Møller started his survey, he was highly aware that this kind of work had been done by Japanese scholars such as Ito Chuta (1867–1954), Tokiwa Daijō (1870–1945), and Sekino Tadashi (1868–1935), and the Society for the Study of Chinese Architecture, as well as that of Ernst Boerschmann (1873–1949). Thus with self-consciousness Prip-Møller chose to study Chinese monasteries from an alternative stand and on less studied materials. Prip-Møller (1937), "preface".

[19] Soper comments: "the author ... had only the most meagre opportunity to study early architectural history".

[20] In his book review, Boerschmann supports Prip-Møller's conception of the Monastery by suggesting the continuity between the traditional and the current religion: "In the study of religions, Chinese religion occupies a peculiar place between historical religions, no longer wholly to be reconstructed, and current religions, too near for understanding. In China, ancient origins and rich development are in combination with a living present. There, both the former and the present cult are documented so that the religious attitude of every epoch can be understood in relation to the history of Chinese culture. Buddhist monasteries are perfect cases in point".

[21] Therefore, like most of his contemporaries and followers, Sanmei's Buddhist practices were syncretic, infusing Chan meditation, Pure Land rituals, and Vinaya regulations.

22   The typical layout of the Buddhist monasteries of the Ming and Qing periods have been extensively discussed by scholars including Prip-Moller. For recent scholarship, see (Qi 2011, p. 64; G. Wang 2016, vol. 3, pp. 1777, 2047, 2082; Steinhardt 2019, p. 249).

23   See (T 1924–1933) (hereafter "T") no. 1899, 45. For the biography of Daoxuan, see (Zanning[988] 1987, pp. 323–30).

24   The 29 subsidiary cloisters in the south half of the monastery are associated with Buddhist education, whereas the other 19 cloisters surrounding the Central Buddha Cloister are more intimatedly related to the Buddha's daily life. (Ho 1995, p. 9).

25   Examples of similar layouts range from the archaeological remains of a monastery site in the Tang Chang'an city to the pictorial representation of a palatial complex in Dunhuang mural paintings and textual records. For further discussions of these historical materials, see (Xi'an Tangcheng Gongzuo Dui (Working Team of the Tang City of Xi'an) 1990, pp. 46–51; Xiao 1989, pp. 61–94; Gong 2006; Qi 2011, pp. 50–51).

26   For detailed discussion of the Board Halls in Longchang Monastery, see (Prip-Møller 1937, pp. 221–23).

27   For discussions of these buildings, see (Prip-Møller 1937, pp. 224–49).

28   For discussion of these sounding instruments, see (Prip-Møller 1937, pp. 204–8).

29   Before ordination, a novice should live in the monastery where they will receive the precepts for about two months. For the activities during this period, see (Prip-Møller 1937, p. 324).

30   Prip-Møller assumes that Sanmei erected a wooden ordination platform. There is no textual record of Sanmei's ordination platform, but, since the ordination rituals had already taken place in the Longshang Monastery, a proper stage for ordination, namely, an ordination platform, persumably existed. Yan Liu also notices supporting evidence in the BHSZ. According to the record, it was because the pre-existing wooden ordination platform was used for a long time that Jianyue decided to rebuild it. It would be reasonable to suggest that the old platform was a heritage from Sanmei's era. (Prip-Møller 1937, p. 282; Y. Liu, forthcoming, p. 90, note 7).

31   The process of the threefold ordination is summarized as follows: before the ceremony, the novice has to live in the ordination-giving monastery for two weeks, for the sake of learning how to live a regulated monastic life and to rehearse the ceremonies that will last from thirty to fifty days. The long process includes three stages. Each are separated from the next by intervals of 8–10 days. For a detailed explanation of the three stages of the threefold ordination, based on Prip-Møller's observation, see (Prip-Møller 1937, pp. 312–17, 324–26). For a slightly different version, based on Welch's interviews with refugee monks once ordained at Longchang Monastery, see (Welch 1967, pp. 287–96).

32   The stele was still located on the lower platform below the Open-Air Platform when Yan Liu visited it in 2012.

33   Even though he was also known for diligently practicing *bozhou sanmei* 般舟三昧 (Skt: *pratyutpannasamādhi*)—a prolonged ritual during which the practitioner ceaselessly chants Amitabha Buddha's name while walking—in his later years, he always insisted that monastic regulations were the best weapon for maintaining a monastic community and fought against his strong rivals.

34   Unless otherwise noticed, translations are the author's from the Chinese text.

35   This sentence is difficult to understand. Prip-Møller's translation is "Its limits were clearly defined" (Prip-Møller 1937, p. 287). But in Hanyu dacidian, "Wenxi 文系" refers to the circumstance in which one's verbal expression is constrained by the conventional rules of grammar. My tentative take is that the boundaries set by Jianyue are not constrained by conventions, meaning, they are new and unconventional. This take is supported by the following discussion of the shape of the boundaries.

36   Ordination platforms had been used in Chinese Buddhist practices for about a thousand years before this case, and the material remains of stone platforms in the Yangzi area can date back to the twelfth century. But not every monastery could have an ordination platform unless they were allowed by the government offices in charge of religious affairs. This rule was particularly strict in the Ming dynasty. For example, Gulin-an 古林庵, a Vinaya monastery in Nanjing, had to receive an imperial order before they could construct an ordination platform in 1613. For Gulin-an and the examples of the extant earlier platforms, see (Prip-Møller 1937, pp. 345–51).

37   Prip-Møller believes that Jianyue was expelled whereas, Yan Liu gives more agency to Jianyue. In the historical accounts, Jianyue's disciples used ambivalent words such as "rejected" (xie 謝) and "departed" (qu 去) instead of "forced to flee", as used by Prip-Møller. For Jianyue's own accounts, see (BHSZ [1795] 1975, vol. 6, pp. 233–34; Jianyue Duti[1675] 1978, pp. 77–78; Prip-Møller 1937, p. 285; Y. Liu, forthcoming, p. 100, note 6).

38   For how Liu's work evolved from Ding'an's work, see Zhu (2018).

39   The data is translated from BHSZ. Besides, the stone ordination platform made by Jianyue was destroyed in the latter half of the twentieth century and the current one in the Monastery is a replacement.

40   The data is from (Prip-Møller 1937, p. 349).

41   The works tending toward the cultural history side of the spectrum include Susan Naquin's study of the temples and city life in Beijing and James Robson's study of the Southern Peachmount, while those more toward the art and architecture side may include Wei-Cheng Lin's study of Mount Wutai. See (Naquin 2000; Robson 2010; Lin 2014).

42   These works, nevertheless, focus mainly on the major urban and mountain settings, which are more like cult centers and pilgrimage destinations, rather than more-or-less self-sustained and self-regulated entities like in the case of Longchang Monastery.

Besides, it is increasingly difficult for the current scholars to collect oral history or to observe traditional practices, since the Buddhist institutes in contemporary China have significantly changed, if they were not already during the 20th century.

[43] Dr. Yan Liu's Chinese translation is about to be published by Cultural Relics Press (Beijing) in 2023. All primary texts are quoted from his unpublished manuscript as of Aug 2018.

[44] Historians of Chinese architecture have proposed various ways of periodization. For example, one of the classical periodizations is based on the styles of the timber-structured buildings and consists of "The Period of Vigor"(ca. 850–1050), "The Period of Elegance" (ca. 1000–1400), and "The Period of Rigidity" (Liang 1984). In addition, a more recent periodization is based on social developments from the Marxist perspective and consists of architecture of "the primitive society", "the slave society", and "the feudal society" in dynastic successions (D. Liu 1984). For a brief overview of the historiography and the contemporary challenges for the narratives, see (Steinhardt 2014).

[45] For an early account of Chinese architecture as a "non-historical style" in a global history of architecture, see (Fletcher[1898] 1901, pp. 461–62). For how this kind of account stimulated the establishment of the subfield of architectural study in China, see (Steinhardt 2014; Lai 2016).

References

Baohua shan zhi 寶華山志 *(Gazetteer of Mount Baohua)*. 1975. *Compiled by Mingfang Liu*. Taipei: Wenhai Press. First published around 1795.

Bodolec, Caroline. 2005. *L'architecture en voûte chinoise: Un partimoine méconnu*. Paris: Maisonneuve & Larose.

Boerschmann, Ernst. 1925. *Chinesische Architektur (Chinese Architecture)*. Berlin: Ernst Wasmuth.

Boerschmann, Ernst. 1939. Review of *Chinese Buddhist Monasteries* by Johannes Prip-Møller. *The Art Bulletin* 21: 292–94.

Faber, Tobias. 1989. En dansk arkitekt i Kina: Johannes Prip-Møller, 1889–1943 (A Danish architect in China: Johannes Prip-Møller, 1889–1943). *Architectura (Architecture)* 11: 7–66.

Fletcher, Banister. 1901. *A History of Architecture on the Comparative Method for Students, Craftsmen & Amateur*. London: B.T. Batsford, Ltd. First published 1898.

Gong, Guoqiang. 2006. *Suitang chang'an cheng fosi yanjiu* 隋唐长安城佛寺研究 *(Research on the Buddhist Monasteries in the Chang'an City during the Sui and the Tang)*. Beijing: Cultural Relics Press.

Ho, Puay-peng. 1995. The Ideal Monastery: Daoxuan's Description of the Central Indian Jetavana Vihāra. *East Asian History* 10: 1–18.

Itō, Chūta. 1931. *Shina kenchikushi* 支那建築史 *(History of Chinese Architecture)*. Tokyo: Yūzankan.

Jianyue Duti. 1978. *Yimeng manyan* 一夢漫言 *(Divagating Words in a Dream)*. Taipei: Buddhism Press. First published around 1675.

Lai, Delin. 2016. *Zhongguo Jindai Sixiang Shi Yu Jianzhu Shixue Shi* 中国近代思想史与建筑史学史 *(Changing Ideals in Modern China and Its Historiography of Architecture)*. Beijing: China Architecture Publishing & Media Co., Ltd.

Lefebvre, Henri. 1991. *The Production of Space*. Malden: Blackwell. First published 1974.

Liang, Sicheng. 1940. Review of Chinese Buddhist Monasteries: Their Plan and Its Function as a Setting for Buddhist Monastic Life. *Tushu Jikan (Seasonal Journal of Books)* N2-2: 426–28.

Liang, Sicheng (Liang Ssu-Ch'eng). 1984. *A Pictorial History of Chinese Architecture: A Study of the Development of Its Structural System and the Evolution of Its Types*. Cambridge: MIT Press.

Lin, Wei-Cheng. 2014. Introduction: Early Buddhist Monastic Architecture in Context. In *Building a Sacred Mountain: The Buddhist Architecture of China's Mount Wutai*. Seattle: University of Washington Press, pp. 1–18.

Liu, Dunzhen. 1984. *Zhongguo gudai jianzhushi* 中國古代建築史 *(History of Premodern Chinese Architecture)*, 2nd ed. Beijing: China Building Industry Press.

Liu, Yan. forthcoming. Chinese Translation of *Chinese Buddhist Monasteries: Their Plan and Its Function as a Setting for Buddhist Monastic Life*. Beijing: Cultural Relics Press, The book is to be published in 2023.

Madsen, Hans Helge. 2003. *Prip-Møllers Kina: Arkitekt, Missionaer og Fotograf i 1920rne og 30rne (Prip-Møller's China: Architect, Missionary, and Photographer in the 1920s and 30s)*. Copenhagen: Arkitektens Forlag.

Naquin, Susan. 2000. *Peking: Temples and City Life, 1400–1900*. Berkeley: University of California Press.

Prip-Møller, Johannes. 1931. *About Buddhist Temples*. Beijing: North China Union Language School Cooperating with California College in China.

Prip-Møller, Johannes. 1935. *A Fourteenth Century Church in China*. Artes, 3 vols. Copenhagen: P. Haase & Fits.

Prip-Moller, Johannes. 1936. On the Building History of Pao Shu T'a, Hangchow. *Journal of the North China Branch of the Royal Asiatic Society* 67: 50–57.

Prip-Møller, Johannes. 1937. *Chinese Buddhist Monasteries: Their Plan and Its Function as a Setting for Buddhist Monastic Life*. Copenhagen: G.E.C. Gad, London: Oxford University Press, Hong Kong: Hong Kong University Press, 1967 (second publication). Hong Kong: Hong Kong University Press, 1982 (third publication).

Prip-Møller, Johannes. 1939. Architecture: A servant of foreign missions. *International Review of Mission* 28: 105–15. [CrossRef]

Qi, Shan. 2011. Xuexiu tixi sixiang xia de woguo xiandai fosi kongjian geju yanjiu 学修体系思想下的我国现代佛寺空间格局研究 (Research on the Space Paradigm of Modern Monastery under the System of Learning and Practice). Ph.D. dissertation, Tsinghua University, Beijing, China.

Robson, James. 2010. 'Neither too far, nor too near': The historical and cultural contexts of Buddhist monasteries in medieval China and Japan. In *Buddhist Monasticism in East Asia: Places of Practice*. Edited by James A. Benn, Lori Rachelle Meeks and James Robson. London and New York: Routledge, pp. 1–17.

Sirén, Osvald. 1929. *A History of Early Chinese Art*. vol 4: "Architecture". London: E. Benn, Limited.

Soper, Alexander C. 1969. Book Review of *Chinese Buddhist Monasteries* (2nd edition) by Johannes Prip-Møller. *Artibus Asiae* 31: 87–88. [CrossRef]

Steinhardt, Nancy Shatzman. 2014. Chinese Architectural History in the Twenty-First Century. *Journal of the Society of Architectural Historians* 73: 38–60. [CrossRef]

Steinhardt, Nancy Shatzman. 2019. *Chinese Architecture: A History*. Princeton: Princeton University Press.

Taishō Shinshū Daizōkyō 大正新脩大藏經 (*The Tripiṭaka Newly Edited in the Taishō era*). 1924–1933. 100 vols. Takakusu, Junjirō, and Watanabe Kaigyoku, eds. Tokyo: Taishō/Issai-Kyō Kankō kwai.

Tokiwa, Daijō, and Tadashi Sekino. 1924. *Shina Bukkyō Shiseki Hyōkai* 支那佛教史蹟評解 (*Buddhist Monuments in China*). Tōkyō: Bukkyō Shiseki Kenkyūkai.

Wang, Guixiang. 2016. *Zhongguo hanchuan fojiao jianzhu shi: fosi de jianzao, fenbu yu siyuan gejv, jianzhu leixing ji qi bianqian* 中国汉传佛教建筑史: 佛寺的建造, 分布与寺院格局, 建筑类型及其变迁 (*The History of Chinese Buddhist Architecture: The Construction, Distribution, Spatial Layout, Building Prototypes and the Developments of Buddhist Monastery*). Beijing: Tsinghua University Press.

Wang, Junyang. 2018. Lishi de yu fei lishi de: 80 nian hou zaikan foguang si "历史的" 与 "非历史的"—80年后再看佛光寺 (Historical and Ahistorical: Revisiting the Foguang Monastery in 80 years). *Jianzhu xuebao* 9: 1–10.

Wang, Weiqiao, and Sofía Melero Tur. 2016. Patrimonio de templos budistas en china wei y sofía (Buddhist monasteries in China). Available online: https://issuu.com/proposal-upm/docs/patrimonio_de_templos_budistas_en_c (accessed on 5 May 2020).

Welch, Holmes. 1967. *The Practice of Chinese Buddhism, 1900–1950*. Cambridge: Harvard University Press.

Wen, Jinyu. 2004. Lüzong diyi mingshan: Baohua shan 律宗第一名山律宗第一名山—宝华山 (The First Famous Mountain of the Vinaya School: Mount Baohua). *Zhongguo zongjiao* 11: 40–41.

Xi'an Tangcheng Gongzuo Dui (Working Team of the Tang City of Xi'an). 1990. Tang Chang'an Ximing si yizhi fajue gongzuo jianbao 唐长安西明寺遗址发掘工作简报 (Brief Report on the Excavation of the Remains of Ximing Monastery in Tang Chang'an). *Kaogu (Archaeology)* 1: 45–55.

Xiao, Mo. 1989. *Dunhuang jianzhu yanjiu* 敦煌建筑研究 (*Research on the Architecture of Dunhuang*). Beijing: Cultural Relics Press.

Zanning. 1987. *Song Gaoseng zhuan* 宋高僧传 (*Biographies of Eminent Monks Compiled in the Song*). Beijing: Zhonghua shuju. First published 988.

Zhang, Jianwei. 2015. A Study of the Three Buddhist Copper Hall Projects, 1602–1607. *Frontiers of History in China* 10: 289–322.

Zhang, Lingling. 2012. Linglei quanshi: Jindai yuwai jianzhushi zaihua zhi zhongguo minjian fengge jianzhu chuangzuo yanjiu 另类诠释:近代域外建筑者在华之中国民间风格建筑创作研究 (Alternative Interpretation: Research on the Folk Character in Foreign Architects' Works in Modern China). Master's thesis, Nanjing University, Nanjing, China.

Zhongguo Yingzao Xueshe (Society for the Study of Chinese Architecture). 1930–1945. *Zhongguo Yingzao Xueshe Huikan* 中国营造学社汇刊 (*Journal of the Society for the Study of Chinese Architecture*). Beijing: Zhongguo Yingzao Xueshe.

Zhu, Fuyi. 2018. Baohua shanzhi kaolue 山志考略 (A Brief Examination of Records of Mountain Baohua). *Foxue yanjiu (Buddhist Studies)* 2: 264–84.

 religions

 MDPI

Article

Generating Sacred Space beyond Architecture: Stacked Stone Pagodas in Sixth-Century Northern China

Jinchao Zhao

Center for Global Asia, New York University Shanghai, Shanghai 200122, China; jz1689@nyu.edu

Abstract: A large number of stone blocks, stacked up in diminishing size to form pagodas, was discovered in northern China, primarily eastern Gansu and southeastern Shanxi. Their stylistic traits and inscriptions indicate the popularity of the practice of making stacked pagodas in the Northern dynasties (circa the fifth and sixth centuries CE). They display a variety of Buddhist imagery on surface, which is in contrast with the simplification of the structural elements. This contrast raises questions about how stone pagodas of the time were understood and how they related to contemporaneous pagoda buildings. This essay examines these stacked pagodas against the broader historical and artistic milieu, especially the practice of dedicating Buddhist stone implements, explores the way the stacked pagodas were made, displayed, and venerated, and discusses their religious significance generated beyond their structural resemblance to real buildings.

Keywords: pagoda; stūpa; miniature; China; Northern dynasties; Buddhism

Citation: Zhao, Jinchao. 2021. Generating Sacred Space beyond Architecture: Stacked Stone Pagodas in Sixth-Century Northern China. *Religions* 12: 730. https://doi.org/10.3390/rel12090730

Academic Editors: Shuishan Yu and Aibin Yan

Received: 14 July 2021
Accepted: 24 August 2021
Published: 6 September 2021

Publisher's Note: MDPI stays neutral with regard to jurisdictional claims in published maps and institutional affiliations.

1. Introduction

During the twentieth century, a number of miniature stone pagodas and hundreds of fragmented pieces were discovered at monastic sites and hoarding pits located in northern China (Figure 1). Among them, a particular group comprises pagodas formed by a series of cubical or trapezoidal stone blocks that are once stacked up in diminishing sizes (hereafter referred to as "stacked pagodas," or "stone blocks" if only a single block is under discussion). This group of pagodas is excavated primarily in the Nannieshui 南涅水 County of Shanxi 山西 province (Figure 2), and several sites in Gansu 甘肅 province (Figure 3). According to dedicatory inscriptions, as well as styles of relief carvings on these stone blocks, they were commissioned over the course of the sixth century, when northern China was successively reigned by the Northern Wei 北魏 (386–534), Eastern Wei (534–50), and Western Wei (535–57), Northern Qi (550–77) and Northern Zhou (557–81). Historically, eastern Gansu was named "Longdong" 隴東, meaning "to the east of the Mount Long". Once a center of Buddhism, the region is home to several cave-temple sites and numerous Buddhist statues and steles that date to the Northern dynasties (Gansu Sheng Wenwu Gongzuodui, and Qingyang Bei Shiku Wenwu Baoguansuo 1987; Cheng 1998; Cheng and Yang 2003; Dong 2008; Song 2009; Wei and Wu 2009; Gansu Beishikusi Wenwu Baohu Yanjiusuo 2013; Zheng et al. 2014).[1] Nannieshui in Shanxi, although never a significant local center in history, is located on the path connecting major political centers in the sixth century (Guo 1959, 1979; Shanxi Sheng Kaogu Yanjiusuo 1994, pp. 313–18; Zhang 2005, pp. 51–68).

Besides excavation reports, and several catalog entries, there is almost no extensive scholarly discussion of the phenomenon of making pagodas by stacking stone blocks. These pagodas' display of rich Buddhist imagery on the surface, in contrast to the simplification of structural elements, raises pointed questions about religion, imagery, and architecture. Why were stacked pagodas made? How to understand their regional flourish in Shanxi and Gansu? How were stacked pagodas understood in their production, consumption, and veneration? How did they relate to the construction of pagoda buildings and other pagoda-centered Buddhist activities of the time period?

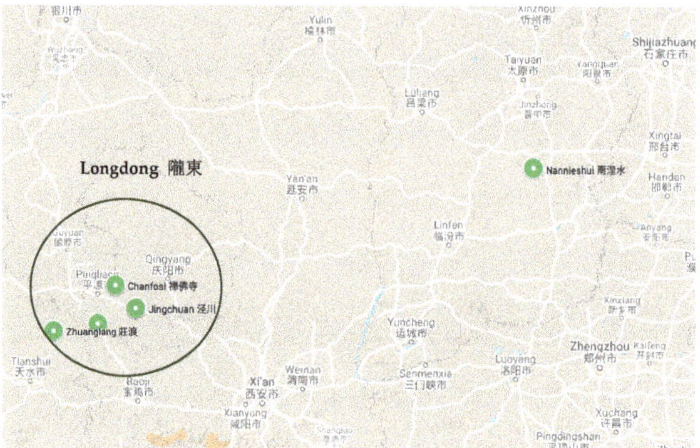

Figure 1. Map of Eastern Gansu, Shaanxi, and Shanxi province.

Figure 2. A Nannieshui pagoda in display. Early sixth century, Northern Wei. Stone. Nannieshui Museum. Photograph taken by the author.

The pagoda is usually considered the reinterpretation of the hemispherical stūpa, but features a tall, multistory tower-like body instead. In the third century BCE, Buddhism adopted the mound structure to house the relics of the Buddha or to mark the places consecrated as sites of his acts, constituting the earliest presence of stūpa.[2] Despite the scarcity of material remains dated prior to the fifth century, literary sources record that the earliest stūpas were produced in China no later than the second century CE, along with the eastwards spread of Buddhism.[3] From the third century, Buddhism rapidly filtered into Chinese society through increasing contacts with Central Asian Buddhist kingdoms, translations of Buddhist texts into Chinese, and constructions of Buddhist monasteries (Zürcher 1972; Seishi 2017). The tower-shaped and multistory type of stūpa appeared and later became the predominant form in East Asia. Modern scholars usually referred to this type as "pagoda" in distinction from the hemispherical stūpa. Both types are generally referred to as *ta* 塔 in Chinese scholarship.

Figure 3. Zhuanglang Pagoda. Gansu Province. Early sixth century, Northern Wei. Gansu Provincial Museum. Photograph courtesy of Wang Xiaoshu.

The period in the following fifth and sixth centuries was marked by the conquest of northern China by non-Chinese regimes. Buddhism continued to flourish regardless of the political chaos and social upheaval. Both imperial and ordinary patrons sponsored the construction of pagoda buildings. Pagoda and stūpa imagery in reliefs and murals burgeoned in Buddhist cave-temples and on Buddhist statues and steles. Surviving historical texts from the time, especially *Luoyang qielan ji* 洛陽伽藍記 (A Record of Buddhist Monasteries in Luoyang) by Yang Xuanzhi 楊衒之, and *Shuijing zhu* 水經注 (Commentary on the Water Classic), compiled by Li Daoyuan 酈道元, record a myriad of Buddhist pagodas being erected within the precinct of Buddhist monasteries in Luoyang (*Luoyang qielan ji jiaoshi* 1963; *Shui jing zhu* 2007). An upsurge of archaeological excavations of pagoda foundations of the Northern dynasties since the second half of the twentieth century CE has brought to light information on pagoda constructions dated to this period, such as the *Siyuan fotu* 思遠佛圖 at Pingcheng 平城 (present-day Datong 大同, Shanxi Province), *Siyan fotu* 思燕佛圖 at Chaoyang 朝陽, Pagoda of the Yongning Monastery 永寧寺 at Luoyang 洛陽, the Taihe Pagoda at Dingxian 定縣 County, Hebei, and pagodas in temples located at the City of Ye 鄴, Hebei (Hebei Sheng Wenhuaju Wenwu Gongzuodui 1966; Xia 1966; Dingxian County Museum 1972; Du 1981; Xu 1994; Zhongguo Shehui Kexueyuan Kaogu Yanjiusuo 1996, 2010, 2013; Su 2011).

In comparison to their Indian predecessors, surviving Chinese counterparts exhibit much greater diversity in forms, functions, and contexts. The earliest surviving group of stone pagodas/stūpas includes the thirteen miniature pieces that were discovered in the last century in major Buddhist sites of northwestern China.[4] They share an almost identical design of domed top and three registers on body. There are six bear inscriptions dated to between 426 and 436, when the region was ruled by the Northern Liang 北涼 (397–439).[5] However, the form of the Northern Liang group did not achieve currency in the following period. Instead, around twenty multilevel stone pagodas became predominant, which have a square, tower-like body of three, five, seven, or nine stories, adorned with roofs, brackets, and columns imitating wooden structures, and feature a central niche on each side of each story (Figure 4). For instance, the Cao Tiandu 曹天度 pagoda consists of a square pedestal, a nine-level tower-like body, and a top, with parallel roof rafters rendered on each roof tier beneath an imitation ceramic-tile roof.[6] Other works similar to the Cao Tiandu Pagoda from the fifth and the sixth century were usually discovered near Pingcheng, Shanxi, and in western Gansu.

Figure 4. Cao Tiandu 曹天度 Pagoda. Shuozhou, Shanxi Province. 466, Northern Wei. Stone. H. 211.7 cm (with chattra top). National Palace Museum, Taipei (*Chattra* preserved in the Chongfu Monastery, Shanxi). Photograph courtesy of Tu Shih-yi.

In contrast to these typical multistory stone pagodas discussed above, stacked pagodas of the sixth century, which are formed by a series of cubical or trapezoidal stone blocks that are stacked up in diminishing sizes, become particularly intriguing in terms of the absence of any architectural elements (see Figures 2 and 3). On surviving stacked pagodas, there are barely any pieces that display components imitating wooden structures.[7] However, we know that these sixth-century stone pieces were made as pagoda, according to inscriptions and the vertical rise of their multilevel structure. Several bear inscriptions that refer to themselves as *shi fotu sanjie* 石佛圖三劫 (three-story stone pagoda), or *shi futu sanji* 石浮圖三级 (three-story stone pagoda).[8] *Futu* 浮屠/浮圖, and in some cases *fotu* 佛圖, are synonyms of *ta* in history.[9] The emphasis of the three levels in these inscriptions, indicating the multilevel structure of the stacked pagodas, is universally used in epigraphs describing the dedication of pagoda buildings.

Furthermore, each surviving stone block of stacked pagodas displays rich Buddhist imagery on each of the four sides, revealing that imagery plays a quintessential role on these pagodas. However, the very act of adorning stacked pagodas with imagery and the arrangement of these images have not yet been discussed. Previous scholarship focused primarily on examining individual depictive scenes (Guo 1959; Guo 1979; Cao 2011). A comprehensive evaluation of the iconography and style of images depicted on stacked pagodas is much needed to better understand the mechanism of producing stacked pagodas.

The pondering of the mechanism of dedicating stacked pagodas is particularly important in view of the broader historical context of constructing pagoda buildings and commissioning stone pagodas in fifth- and sixth-century China. Miniature stone pagodas from the time, often studied only as imitations of actual wooden buildings, and as evidence to reconstruct the look of real buildings that barely survived, were rarely examined for their own instances. Major scholarship in art and architectural history has focused mainly on reconstructing a linear development based on formal analysis of the stūpa/pagoda's structure.[10] This interest in an evolutionary narrative of pagoda buildings understands surviving visual materials of stūpas and pagodas, including miniature pieces and the pictorial representation in reliefs and murals, primarily as evidence for reconstructing monastic architecture, resulting in limitations in further inquiry of other valuable aspects. Despite the contribution to the advancement of the field, this evolutionary approach becomes problematic in two aspects in the study of non-built forms: the "origin" narrative, and a fabricated evolutionary scheme. Meanwhile, research in the field of Buddhist studies lends itself to abstract discourses in history and doctrine, rarely contributing to analyses of specific images or objects.

While I am indebted to the previous scholarship conducted on the architectural aspects of pagodas, this essay aims to further explore how stone pagodas, particularly stacked pagodas from sixth-century Gansu and Shanxi, display their function and significance beyond their architectural form. Why were they made in the first place? What do we make of the images lavishly adorning their surfaces? How were they situated in the artistic, historical, and religious milieu of the time? How were they related yet differentiated from the contemporaneous practice of building pagodas and dedicating Buddhist sculptures?

The first section of this study offers a survey on the style, iconography, and patronage of surviving stacked stone pagodas. The second section examines the regional feature of making stacked pagodas in eastern Gansu and Shanxi, respectively, and proposes an eastern Gansu origin given the historical context. It also highlights the local reception and promotion of hybridity, which is exhibited by the integration of styles, motifs, and pictorial programs of diverse origins in both regions.[11]

The third section pinpoints the absence of architectural elements imitating wooden structures on stacked pagodas, which is in contrast with the rich Buddhist images covering every corner of their surface. In the fourth section, I further examine the broader historical milieus of producing stone pagodas in preceding periods, showing a successive tradition of communicating Buddhist teachings through imagery depicted on the surface of stone pagodas. On stacked pagodas, however, the well-configured pictorial programs featured on fifth-century stone pagodas are dissembled, with each individual image exhibiting independent significance. Such an abstraction and distortion of their shapes exhibits a tension between form and meaning.

The fifth section explores factors that shaped this new emphasis on individual images on stacked pagodas, arguing for the ignored correlation between the configuration of images and the organization of patronage, which deeply shaped the way how stacked pagodas were commissioned and venerated. According to surviving inscriptions, an individual image located on one side of a stone block from Shanxi, instead of the stone block or the stacked pagoda, becomes the targeted unit to be donated by a single patron. Following this emphasis on the perception of stacked pagodas among the local community, the sixth section studies the reference to pagodas in surviving inscriptions, and discovers a transition from *futu* 浮圖 (pagoda) to *xiang* 像 (image). I contend that it is the growing emphasis on individual image, which was formed by the very making-process of these stacked pagodas in module, that gradually altered the perception of them from pagodas to images.

2. Stacked Pagodas from Gansu and Shanxi

2.1. Gansu

There are two primary geographical loci for the discovery of stacked pagodas: Pingliang 平涼 Prefecture in eastern Gansu province and Nannieshui at Qin 沁 County in southeastern Shanxi province.[12] Eastern Gansu in its modern administrative division includes two sub-areas: Qingyang 慶陽 and Pingliang. Once a center of Buddhism, the region is home to several cave-temples, such as the Northern and the Southern Cave-temples 南北石窟寺 at Qingyang, Wangmu Palace 王母宮, and Luohan Cave 羅漢堂, and numerous Buddhist statues and steles that date to the late fifth century.[13] The development of Buddhist art in Pingliang relates closely to two centers at the time. The first is Chang'an 長安, Shaanxi 陝西 province, which is adjacent to eastern Gansu, and has been a political and cultural center since the fourth century. Another center is the Maijishan 麥積山 Cave-temples at Tianshui 天水, located to the southwest of Pingliang.[14] Overall, Pingliang Buddhist art exhibits distinctive regional features but also possesses a blend of stylistic traits that had already circulated in the adjacent area of Chang'an and Tianshui, as well as Pingcheng, the Northern Wei capital city and art center of the fifth century.

Among the dated stacked pagodas in eastern Gansu, the earliest one was commissioned in 503 (Figure 5).[15] A square stone block from this pagoda was discovered in 1982 among over forty similar ones, as well as Buddhist steles and statues, at the site of Chanfosi 禪佛寺 in the Caowan 曹灣 Village, Kongtong 崆峒 District, Pingliang. Three other stone blocks respectively bear dates of 514,[16] 518,[17] and 519. They are identifiable as parts of stacked pagodas, given their trapezoidal shapes, the varied sizes, and the traces of mortise and tenon on the top and the bottom. Each face of the stone blocks was carved with images of Buddhas, bodhisattvas, narrative scenes, or decorative motifs in recessed niches. On each side of the 503 block, there is a seated Buddha and two flanking bodhisattvas depicted within niche (See Figure 5). The Buddhas are dressed in monastic robes with symmetrical folds and sleeves with sharply defined edges splayed below the hands. The 518 Pagoda and other undated pieces from Chanfosi show a more diverse iconography. The Twin Buddhas motif (a pair of Śākyamuni and the Past Buddha Prabhutaratna), which was created in China based on the *Lotus Sūtra*, was becoming popular. Representing the Buddhas of the Present and the Past together, the motif underscores the notion that more than one Buddha can exist at the same time in the cosmos.[18]

Another group of Buddhist stacked pagodas was discovered in 1990 along with over 20 other pieces of statues in a hoarding pit located at Xiejiemiao 谢家庙, Huating 华亭 County.[19] Three sets of stacked pagodas, each formed vertically by three stone blocks, are identified. All the three sets bear inscriptions: two sets were commissioned in the years 516 and 534 of the Northern Wei; and one set in 558, Northern Zhou (Figure 6).[20] The Northern Wei pieces exhibit hybridity of styles from different origins. On the 534 piece, the main Buddha image on one side dresses in a typical Han-style robe with sashes. The Buddha's right foot protrudes with the sole facing outside, representing a particular feature that is usually found in cave-temples of Longmen bsmi龙门 and Gongxian 巩县 of the 520s near Luoyang.[21] Another Buddha triad depicted on the same piece renders petal-like edges of draperies in symmetry, exhibiting the local style of Gansu. Pieces of the second half of the sixth century instead show a more simplified iconography echoing other contemporaneous sites across the north, while their style is distinguished from the squarish style of other Northern Zhou sites.

Figure 5. Stone Block from Chanfo Monastery, Kongtong District, Pingliang City, Gansu Province. 503, Northern Wei. H. 44 cm, W. 33 cm. Pingliang Museum. Reprinted with permission from ref. [Gansu fojiao shike zaoxiang]. 2000, Zhang, p. 99.

A number of undated stone blocks discovered at Zhuanglang 莊浪 embrace dimensions and styles similar to the above-discussed cases.[22] A specific group was excavated in the courtyard of a Northern Wei Monastery, Baoquansi 寶泉寺, including one individual stone block and five that form a stacked pagoda (see Figure 3, hereafter the Zhuanglang Pagoda).[23] There are also several pieces found scattered in cave-temple sites near Zhuanglang. At the Wangmugong 王母宮 Cave-temple, a stone block in trapezoidal shape was once found on display. This piece, together with six others in situ, was brought to the cave-temple in the 20th century without much information of its original recorded provenance.[24]

2.2. Shanxi

At Nannieshui, Shanxi, cubical stone blocks used to form stacked pagodas also survived in large number.[25] They were discovered in 1957 from several hoarding pits (see Figure 2). In 1990, another excavation led by the Archaeological Institute of Shanxi and the County Museum at Nannieshui discovered the foundation of a temple dated to the Tang Dynasty, as well as over 100 pieces of stone Buddhist statues of the Song Dynasty (Shanxi Sheng Kaogu Yanjiusuo 1994, p. 313). The relatively well-preserved condition of all surviving pieces without any apparent pattern suggesting deliberate breakage, in contrast to the broken nature of statues discovered in some other hoarding pits discovered in Hebei and Shandong, suggests that the Nannieshui sculptures were not damaged before their interment in these hoarding pits. Nevertheless, the lack of records of the first excavation in 1957 makes it difficult to explore the nature of the burial and the original deposit of these sculptures.[26]

Figure 6. Pagoda from Xiejiamiao, Huating, Gansu Province. 558, Northern Zhou. Stone. H. 78.6 cm, W. (bottom) 26 cm. Huating Museum. Reprinted with permission from ref. [Gansu guta yanjiu]. 2014, Gansu Sheng Wenwu Ju, pp. 55–56.

Totaling 1373 blocks, each side of these cubic stones depicts Buddhist images, including Buddhas, bodhisattvas, guardian kings, narrative scenes from the Buddha's life story, as well as images of lay donors. Almost all the sculptures were carved from the grey, fine-grained sandstone that could be quarried locally. Based on surviving inscriptions, their execution dates range from 510, Northern Wei, to 1031, Northern Song 北宋. Most of the sculptures date to the sixth century, from the late Northern Wei through the Eastern Wei and the Northern Qi dynasties.

Previous scholarship on Nannieshui tends to introduce the materials among other Buddhist sculptures from southern Shanxi, highlighting the provincial style of the region but leaving space for more in-depth research. Yagi Haruo examines the depiction of intertwined dragons, the sun and the moon, and acrobatics among Nannieshui discoveries. Hinako Ishimatsu argues for a provincial style that has been shared by southern Shanxi and eastern Gansu and developed independently from the capital styles at the Yungang 云冈 Cave-temples near Pingcheng and Longmen 龍門 near Luoyang. Gao Meng's dissertation on Nannieshui pagodas provides a survey without much analytical discussion in connection to contemporary Buddhist art (Yagi 2004, p. 84; Ishimatsu 2005, pp. 185–86; Gao 2012). In *Shanxi kaogu sishinian*, a four-period chronology is provided as an overview of the Nannieshui materials (Shanxi Sheng Kaogu Yanjiusuo 1994). However, this chronology has failed to thoroughly recognize the complexity of Nannieshui materials that is shown by the hybridity of styles and images with various origins and arranged in an array of ways.

The current study suggests a relative chronology for the Northern Wei materials based primarily on style, in combination with an extensive discussion of the change in iconography over the sixth century at Nannieshui. The proposed chronology divides Nannieshui materials into three groups: the Northern Wei group, the Northern Qi group, and the Tang and Song group.[27] The complexity of the Nannieshui materials lies primarily in the Northern Wei group, with multiple styles and rich iconography carved. The later

Northern Qi group shows stylistic traits that echo the idiom of contemporaneous Hebei and Shandong, which features a fuller body of the Buddha and a dressing mode that features thin draperies clinging to the body tightly. Those of the Tang and Song dynasties only constitute a small portion; the very fact that they were discovered along with those dated to the Northern dynasties in the same pits suggests that these works could not have been buried earlier than the tenth century, and those of the Northern dynasties might have been venerated throughout the five centuries in between.

Among the Northern Wei group, Buddha figures are modeled with slender bodies, squarish heads, and thick robes with draperies cascading on the throne, appearing to be ramifications of the late Northern Wei style popular in other areas. Nevertheless, one finds apparent regional variations between Nannieshui and eastern Gansu in their craftsmanship, the richness of decorative details in the background, and specific details with various origins. For instance, image no. 742 employs a typical Han style in rendering the sash of the Buddha's inner robe, with a rarely seen flower-like edge, as well as the thickly folded draperies that were popular at Luoyang and Majishan (Figure 7). Meanwhile, in the same scene, the Buddha sitting on the right protrudes the left hand to the other Buddha, exhibiting an idiosyncratic way of depicting the twin Buddhas motif of which one finds no examples beyond the area. In addition, the niche is surrounded by floral petals supporting Mani jewels half of which is cut off, two beings symmetrically arranged in between the petals, and one climbing onto the petal on the right, forming a design that is not found in other areas. The broad variety of details exhibited by Nannieshui pieces dated to the first half of the sixth century shows the strong individuality of each piece of work. Their variations suggest the coexistence of various factors that are involved in modulating each individual work, and challenges the traditional approach that examines them as one group.

Figure 7. Twin Buddhas. On a Nannieshui Pagoda, scene no. 742. Early sixth century, Northern Wei. Shanxi province. Nannieshui Museum. Photograph taken by the author.

2.3. Between Gansu and Shanxi

It remains under-exploration about the relation between eastern Gansu and Shanxi in making stacked pagodas. While it is difficult to pinpoint the exact origin of the practice of stacking stone blocks to form pagodas, current evidence alludes to a greater likelihood of eastern Gansu origin. The earliest dated stone block is found in eastern Gansu, in the year 503, predating the Nannieshui pieces. Meanwhile, Lushui hu 卢水胡 (barbarians from Lushui), the ethnic group based at Jingchuan 泾川 (Anding 安定 County in Northern Wei), eastern Gansu, has gradually migrated eastward to Shaanxi since the third century, constituting a major power in the area.[28] By the early fifth century, a series of rebellions were led by the Lushui hu residing in Shaanxi and eastern Gansu. Their fervent support of Buddhism is considered by scholars as a factor accelerating Emperor Taiwu's (r. 423–452 CE) decision of prosecuting Buddhism. Despite the failure of these rebellious actions and the elimination of the Juqu 沮渠 clan who was behind most of the revolts, ordinary Lushui hu retained their stay in eastern Gansu and Shaanxi. Meanwhile, the Xue 薛 clan of the Hedong 河東 Prefecture (present-day southeastern Shanxi), which lies in between the region where Nannieshui is located, Shaanxi, and eastern Gansu, maintained to be an ally of the Lushui hu through the fifth century, allowing for frequent traffic across Shanxi, Shaanxi, and Gansu (Liu 2005, 2008). The Xue Clan, originated from Shu 蜀 in present-day Sichuan, migrated to Hedong in the third century, and gradually became a predominating clan active in the area. The Xue clan retained its prosperity in the Hedong prefecture until the fall of the dynasty in 530s, and maintained contact with Lushui hu from the west along the way (Shanxi Sheng Kaogu Yanjiusuo 1994). Despite the absence of any stone blocks from the Hedong area, its geographical location at the crossroads connecting other parts of Shanxi with Shaanxi and Gansu, and the stability provided by the Xue clan during the Northern Wei, provided a path through which the practice of making stacked pagodas was transmitted in the north. Furthermore, numerous cubical stone blocks discovered in Shaanxi resembled the Gansu and Shanxi cases in size, style, and iconography (Abe 2001). The Shaanxi group's provenance in the Chang'an area—which is located right between eastern Gansu and southern Shanxi—connects Nannieshui and eastern Gansu, the two areas seemingly isolated from each other (Figure 8).

Figure 8. Stone block dated to 501, Northern Wei. Beilin Museum, Xi'an, Shaanxi province. Photographs taken by the author.

3. Architectural Elements in Absence, Buddhist Imagery in Presence

Among surviving stacked pagoda pieces from eastern Gansu and Shanxi, an intriguing phenomenon appears to be the lack of any structural components that imitate wooden building. In addition, the emphasis on adorning every side of pagodas with images is consistent of the fifth-century tradition. In the following discussion, I argue that the particular emphasis on pictorial programs adorning stone pagodas has already achieved prominence since the fifth century, showing the underlying mechanism of stone pagodas to generate significance and communicating with viewers beyond its architectural form. On stacked pagodas of the sixth century, structural components were further eliminated, leaving the veneration foci more focused on the images adorning pagoda surfaces, and showing a growing emphasis on images. Moreover, the relatively unified pictorial program of the fifth-century pagodas was also abandoned. Instead, images on stacked pagodas feature a more individualized and independent arrangement, constituting another piece of evidence to assert the importance of images.

To delve into the details depicted on stacked pagodas, a fallacy should be cleared at the outset. In several studies, a common approach is to study the vertical alignment of the images located on each of a pagoda's four sides, according to their current installation in museums (see Figures 2 and 3) (E 2011). It is natural to study a pagoda by looking into each of its facets, however, this approach is obviously not suitable for stacked pagodas, since the original sequence for arranging the stone blocks, if there is any, is lost. It is beyond any possibility to reach an affirmable reconstruction of the vertical connection among those cubic stone blocks as they were scattered even before being buried underground. The current logic of restacking these unearthed stone blocks relies primarily on their stylistic similarities. In other words, the way how stone blocks are displayed in museums should not be considered as their original arrangement, but an attempt to reconstruct. What we are certain about is the configuration of the four images located on the same cubic block. In the following discussion of images, any pictorial program is presented based on their content and their connection with the three other sides of the same block.

Taking the Zhuanglang pagoda as an example of the eastern Gansu group, images featured on each side of each story show a great variety of iconography, some of which enjoyed particular popularity (see Figure 4). In addition to Buddha triads, the most frequently depicted subject, scenes from the Buddha's Life Story (Ch. *fozhuan gushi* 佛傳故事), appears to be a preferred topic. The Buddha's Birth and Bath are combined on the same pictorial space on one side of the fourth stone (Figure 9a).[29] On the lower half of the side, the space on the right is occupied by the scene that represents the Buddha's Birth as it follows the contemporary convention that depicts the baby Buddha emerging from the right flank of Queen Māyā. Depicted in the middle of the space is the infant Buddha taking a bath; he stands naked on a three-legged stool, under a canopy formed by the nine-headed nāga (serpent divinity translated into Chinese as a dragon), with two attendants pouring water from a vase. On the left we see the infant Buddha held by an ascetic astrologer invited to the palace to tell the future of the prince. The similar grouping of episodes from the Buddha's Life Story has been popular since the mid-fifth century in Chang'an and Pingcheng area.

Another episode from the Life Story of the Buddha is the Great Departure, which represents Prince Siddhārtha departing the palace, acquiring importance for its highlighting of the historical Buddha leaving his princely life for a journey to seek enlightenment. In the scene on the first story of the Zhuanglang Pagoda, two standing figures and a horse in the foreground represent the moment when the prince is leaving with his steed (Figure 9c). Another figure raising hands in a building in the background shows the upset emotion of the prince's family in the palace.[30]

Figure 9. (a–d) Five-story Pagoda from Baoquansi, Zhuanglang, Gansu province. Early sixth century, Northern Wei. Stone. H. ~45 cm, W. 28–34 cm. Gansu Provincial Museum. Reprinted with permission from ref. [Gansu fojiao shike zaoxiang]. 2000, Zhang Baoxi, p. 25.

The Parinirvāṇa scene is carved on the fourth story of the pagoda (Figure 9c). Referring to the Buddha's "death", the episode is represented with reclining Buddha with the head pointing to the left and feet to the right, and an accompanying group of mourners crying in the background.

An interesting question arises: why do the episodes of Parinirvāṇa and the Great Departure appear independently while the other scenes are clustered in the same scene? A contextualized survey of precursors of these scenes shows that the two have gradually achieved independence since the late-fifth century for their respective teachings that constitute self-sufficient importance. The earliest Parinirvāṇa image with a reclining figure in the center is found in Yungang Cave 11, reliefs within which dated to the 480s and 490s, as well as Phase III Yungang cave-temples.[31] Although the scene at Yungang is not prominent given its modest size and marginal location below a small niche, it stands out because of its separation from the sequence of the life story cycle. Similarly, the Parinirvāṇa image on the Zhuanglang Pagoda is not aligned with other narrative episodes.[32]

The particular independence achieved by the image of Parinirvāṇa and the Great Departure suggests a phenomenon of iconization, or the making of icons, which is also attested by the depiction of the Aśoka story. Located on the third story of the Zhanglang Pagoda, the Aśoka story represents the teaching that good karmic practices will lead to favorable reincarnation (Figure 9a). In the narrative, the Buddha and his disciple Ananda encounter several children playing outside during their trip. One of the children, in the hope of making offerings to the Buddha, takes a handful of soil and climbs upon another person's shoulder in order to reach the Buddha's alms-bowl. The Buddha accepts the earth and predicts that the boy would be reborn as King Aśoka. In China, this story is commonly represented by three children making offerings to a standing Buddha. The boy who stands closest to the Buddha will become King Aśoka. The story originally belonged to the category of avadāna tales that correlates the virtuous deeds of the Buddha's past lives to subsequent lives' events. The story was accounted in *Aśokāvadāna*, which was translated into Chinese by An Faqin 安法欽 (active at Luoyang from 281 to 306), at the turn of the fourth century as *Ayu wang zhuan* 阿育王傳, and later by Sanghapāla 僧伽婆羅 (460–524 CE) in 512 CE as *Ayu wang jing* 阿育王經.[33] The story was also found in *Xianyu jing* 賢愚經 (the Sutra of the Wise and the Fool), a collection of tales translated into Chinese in the fourth century.[34] Already shown in relief around the second century CE in Gandhāra, the Aśoka story began to be depicted in the Phase II Yungang Cave 12 of the 480s, and further gained wide popularity in Phase III Yungang cave-temples and steles of the early sixth century (Strong 1983; Behrendt 2003, 2007; Brancaccio and Behrendt 2006).[35] The story is also found depicted on some of the Chanfosi stone blocks and cave-temple sites in eastern Gansu (Zhang 2000, p. 104, Figure 109; Gansu Sheng Wenwu Gongzuodui 1987, pp. 11–15). The Aśoka story depicted on stone blocks follows the convention of depicting three children approaching a standing Buddha, while featuring additional elements including flying apsaras and two attendant bodhisattvas.[36]

All these narrative scenes, the Aśoka story and those from the Buddha's Life Story, are also carved frequently among the Nannieshui group. The Aśoka story at Nannieshui shows the Buddha standing, with his right hand touched by a child who is supported by another one (Figure 10). The Great Departure includes all the major elements we see in contemporaneous examples: a horse at his feet, and a tilting tree behind the prince (Figure 10). The composition within the square space reveals a possible influence from Yungang, as indicated by the modeling of the horse in high relief. The pensive Buddha image is shown sitting below a tree in the background and additional decorative elements carved in low relief surrounding the niche, including thatched huts that are usually associated with the venue of meditation, and Brahminic figures shown with knotted hair, naked upper bodies, and sometimes holding staffs. The highlight of the Buddha sitting in pensive posture among the Nannieshui cases distinguishes it from the depiction of the Buddha riding a horse in eastern Gansu. This regional difference might result from a growing importance of the pensive figure in the sixth century.[37]

The Parinirvāṇa image, meanwhile, is represented in a completely new idiom at Nannieshui, which combines the typical scene showing a reclining Buddha side by side, with another scene that depicts a coffin (Figure 11). Both scenes feature a pictorial space formed by two intertwined tree crowns in the background. Such a combination reminds us of the renowned stele preserved at the Art Institute of Chicago (Lee 2010, Figure 1.1). The top register of the stele's reverse side renders two Parinirvāṇa scenes almost identical to the Nannieshui case. This depiction of Parinirvāṇa in two images is only found in sculptural remains dated to periods from the late sixth century. In her case study of the stele, Sonya Lee argues that the introduction of the coffin helps expand the temporal scope of the actions of mourning in depiction and establish a relationship of symbolic equivalence with the reclining Buddha (Lee 2010, chp. 1). Due to the survival of sculptures by chance, we do not know what the source is for this provoking invention of a two-scene representation of the Parinirvāṇa. However, the provenance of the Chicago stele in southern Shanxi along the Fen River clearly indicates the circulation of this double depiction of a reclining Buddha and a coffin in the area.

Figure 10. Scenes no. 993–96, stone block from Nannieshui. Early sixth century, Northern Wei. Shanxi province. Photographs taken by the author.

Figure 11. Nirvāṇa. Scenes no. 430 and no. 431. Stone block, Nannieshui, Shanxi province. Mid-sixth century. Nannieshui Museum. Source: (Cao 2011, Figures 6 and 7).

Meanwhile, the Northern Wei group from Nannieshui shows some idiosyncratic traits. Although the most commonly depicted images on a stone block include the seated Buddha, Maitreya, and the Twin Buddhas, we find numerous new elements that are distinctive from the traditions flourished in other places in late Northern Wei, as well as unidentifiable subjects. In addition, the arrangement of images on these Nannieshui pagodas is arbitrary to a degree. To better examine the vast number of stone pagoda blocks surviving from the area, Gephi, the network analysis tool that uses the direction and frequency of links between nodes, is employed.[38] The nodes of the tool in this survey refer to imagery depicted on each side of Nannieshui stone blocks. It reveals that the Aśoka story and the Great Departure usually are aligned horizontally with a seated Buddha, the Twin Buddhas, and Maitreya.

Overall, a survey of the narrative scenes in individual pictorial space on stacked pagodas from eastern Gansu and Shanxi reveals their prototypes developed in the late fifth century Yungang. The quick absorption of these individual narrative scenes on stacked pagodas shows recognition of their significance and popularity among the worshippers, and further alludes to a parallel correlation with steles and statues made around the same time. Each of these prototypes found at Yungang is also depicted independently from any other narrative scenes. This preference of Yungang tradition in eastern Gansu might relate to the historical context of the proliferation of Buddhist cave-temple sites at Longdong. An important site on the crossroads of trades and military campaigns connecting the Central Plains to the Hexi Corridor, the Longdong area has been kept close with the political center of Pingcheng. Dowager Empress Hu, who held the political whip in the early sixth century, has family origins in Jingzhou, Longdong (Gansu Sheng Wenwu Gongzuodui 1987, pp. 19–20). Many of the cave-temple sites of Longdong also show a preference for depicting the Great Departure episode and the Aśoka story (Gansu Sheng Wenwu Gongzuodui 1987, p. 95).

As briefly mentioned in the above discussion of Zhuanglang Pagoda, new questions also arise. Why were these scenes selected and why are they shown independently without connections to their surrounding images? A possible answer lies in the iconization of each of the three scenes. One notices that narrative scenes rely on the law of the series, which subordinates the individual image to the whole, or partial course of the narrative. This law is followed in images depicted on one side of the Zhuanglang Pagoda's second story, where the Buddha's birth, bath, and the prophecy with astrologers are grouped together in the same pictorial space. However, the Aśoka story, the Great Departure, and Parinirvāṇa episodes are shown individually, independently, and aligned arbitrarily with images of the Buddha triad on the other sides of the same stone block. Thus, the law of series is suspended as an individual scene becomes an independent motif in isolation from the narrative cycle.

In her study on Parinirvāṇa images, Sonya Lee points out the Parinirvāṇa image being adapted as a Buddhist icon at Yungang, best exemplified by the separation of the Parinirvāṇa scene from the life cycle of the Buddha in late Northern Wei. The final moment of the Buddha's demise shown in independent placement is also found in Cave 132 at Binglingsi of the early sixth century and in Cave 5 at South Xiangtangshan of the mid-sixth century (Lee 2010, pp. 45–48). Lee attributes the independent depiction of the Parinirvāṇa scene to its inclusion into the new thematic setting of the Buddhas of Three Ages (Past, Present, and Future), which stressed the infinite, continuous presence of Buddhas. Her emphasis on the thematic setting of an image can be employed to interpret the iconization of the Aśoka story and the Great Departure. The Great Departure indicates the start of the journey that leads to the prince's enlightenment. The Aśoka story carries an unmistakable symbolic meaning of the future Buddhahood of the universal ruler. Despite their different narrative contexts, both scenes fit into the thematic setting about the infinity of the Buddhahood. This growing emphasis on each individual narrative further addresses an intentional selection of images to be depicted on stacked pagodas, as well as a recognition of the religious significance that can be communicated through the selected images.

4. Pictorial Programs on Stone Pagodas of the Northern Dynasties

Nevertheless, to better understand the particular emphasis on images over architectural elements on stacked pagodas from Gansu and Shanxi, it is also necessary to examine a broader context of the production of stone pagodas at the time. This section addresses that the interest in depicting images that carry specific Buddhist teachings on stacked pagodas is a continuation of the fifth-century tradition in making single-piece pagodas (see Figure 4). The single-piece pagodas also comprise evidence for the crucial role played by pictorial programs of images in generating symbolic significance beyond architectural elements that imitate wooden buildings.

The single-piece pagodas of the fifth century are carved with Buddhist images popular at the time, including the Historical Buddha Śākyamuni, the Future Buddha Maitreya, the Fasting Buddha, narrative scenes from the Buddha's Life Story, and the Twin Buddhas motif. A comparative study of both the single pieces and the stacked pagodas showcases the exploration of how to embed meanings through the execution of images on pagodas.

One of the most representative examples of the single-piece pagodas, the Cao Tiandu Pagoda of 466, features a Buddha niche in the center on each side of the lowest story of the pagoda's main body. Clockwise, the four niches house images of Śākyamuni in dhyāna mudrā (meditation gesture) on two sides, Bodhisattva Maitreya with legs crossed at his ankles, and the Twin Buddhas motif (see Figure 4) respectively.[39] On the remaining eight stories, each of the upper six depicts two rows of seated Buddha images on each side, while the lower two depict three rows of seated Buddha images, representing the Thousand-buddhas motif (Caswell 1975). Another three pieces showcase a similar design. One is a stone pagoda preserved in the Daiwangcheng 代王城 Museum, in Yu County 蔚縣, present-day northwestern Hebei 河北 Province (Figure 12).[40] The body of the pagoda has eight stories; there should have been nine stories, as pagoda usually have an odd number of stories, and the top of this pagoda is severely damaged. The bottom story is also cut in half, with the lower half and pedestal missing. Each story of the pagoda features a roof with protruding rafters and brackets, suggesting its imitation of a wooden prototype. In the center of each side of each story, there opens a niche with a seated Buddha image, flanked by smaller Buddha images in two registers on either side. The space between brackets under the roof is further carved with a small, seated Buddha image. Despite the lack of epigraphic traces that bear an exact date, the Buddha image on the Daiwangcheng Pagoda exhibits a strong stylistic resemblance to those from Phase II Yungang cave-temples of the 480s (Chen 1996).[41] Two other pagodas from Shanxi can be dated to the 460s and 470s. One is a three-story stone pagoda found in Datong (hereafter named the Datong Pagoda).[42] The other five-story pagoda was found at the Nanchan Monastery 南禪寺 on Mount Wutai 五臺 in northern Shanxi province. On its second story, three sides depict seated Buddha images, while the fourth displays two Buddhas sitting side by side, representing the Twin Buddhas motif.[43] An approximate time range for these two works extends from the 450s to the 470s, slightly earlier or later than the Cao Tiandu Pagoda.[44]

Examining the alignment of the various images pictured on these stone pagodas, one finds an arrangement that lines up the historical Buddha Śākyamuni, Maitreya Bodhisattva as the Future Buddha, and the Twin Buddhas motif horizontally or vertically, representing the concept of the Buddhas of the Three Ages theme. The Buddhas of the Three Ages theme was already developed in Gandhāran tradition, with the past Buddha commonly represented by Dīpaṃkara or the Seven Buddhas (including six Buddhas of the Past and Śākyamuni of the present era). Therefore, the fifth-century miniature pagodas represented the theme in a new form, which replaced the Gandhāran tradition, with the Twin Buddhas motif employed to indicate the Past. The particular significance of the Twin Buddhas motif in underscoring the timelessness of Buddhahood (coexistence of the present and the past) indicates a further emphasis on the Mahāyāna teachings in Chinese Buddhist art.[45] In addition to the theme of the Buddhas of Three Ages, the lowest story of some of these miniature pagodas further depict scenes from the Buddha's Life Story (Ch. *fozhuan gushi* 佛傳故事) in a sequence different from the sporadic depiction of similar scenes on stacked pagodas (see Figure 9).[46]

Figure 12. (**a–d**) Daiwangcheng 代王城 Pagoda, late fifth century, Northern Wei. (Yuxian 蔚县, Hebei province. Stone. H. 150 cm, W. (base) 37.5 cm. Daiwangcheng Museum. Photograph courtesy of Zhang Jianyu. Diagram by the author).

In addition to single-piece stone pagodas discovered in the Pingcheng area, those found in another cultural hub, the Hexi Corridor in northwest China, exhibit variations that correlate to Buddhist teachings circulated regionally. While retaining the horizontal scheme in depicting the Buddha's Life Story on the bottommost story, surviving miniature pagodas from the Hexi region display Maitreya Bodhisattva in combination with a Fasting Buddha image on the highest story (Figure 13). The Cao Tianhu 曹天護 Pagoda of 499 features narrative scenes from the Buddha's Life Story on its lowest story, resembling the Pingcheng tradition.[47] However, on its topmost story, two sides feature a seated Buddha image within a horseshoe-shaped niche (Figure 13a,d); the third side depicts Maitreya (Figure 13b), which is identifiable by his crossed ankles and the trapezoidal arch above; and the fourth side renders a skeleton-like Buddha sitting beneath twin trees, which is unmistakably the image of a Fasting Buddha (Figure 13c).[48] The combination of images of Maitreya and the Fasting Buddha is quite unique, without any similar specimen found in Indian and Central Asian traditions. However, a further inquiry of both textual and visual sources regarding the fasting image reveals that the significance of the combination of fasting Buddha image with Maitreya lies in its representation of the Buddha's enlightenment since its inception in Gandhāra.[49] Given the Fasting Buddha's Gandhāra lineage, and Maitreya's reference to future Buddhahood, the new configuration of the two on the topmost story of miniature pagoda attests a new emphasis on the attainment of enlightenment in fifth-century Buddhist practice in the Hexi Corridor.

Figure 13. (**a–d**) Cao Tianhu 曹天護 Pagoda. Jiuquan, Gansu Province. 499, Northern Wei. (Stone. H. 38 cm, W. (base) 16 cm. Jiuquan Municipal Museum. Diagram by the author based on photographs in Chen 1986, Figures 1–4).

To conclude, the examination of stone pagodas made earlier than stacked pagodas in northern China shows a continuous tradition of depicting images on their surfaces. The images are imbued with Buddhist teachings popular at the time, echoing the development of the image worship found on other media, such as steles and statues. That being said, Buddhist images adorning the surface of stone pagodas have been central to the display of meaning since early on.

5. Individual Image, Collective Patronage

Despite the continuation of generating significance with images on pagodas through the fifth and the sixth century, on stacked pagodas of the sixth century we see a dissolution of pictorial programs that once organize every image adorning the surface of fifth-century single-piece pagodas, as well as a growing complexity and independence in individual images. As examined in the third section, each image depicted on stacked pagodas is not particularly related in content with any adjacent ones. Factors shaping this growing independence of image include the trend of iconization of narrative scenes from the Buddha's Life Story and the growing importance of particular significance carried by individual narrative episodes. Nevertheless, based on surviving inscriptions at Nannieshui, we find that a stacked pagoda was often commissioned by a group of patrons. Usually, a single patron usually donates an individual image rather than a stone block or a stacked pagoda. In other words, patrons donated in unit of the individual image. To understand this specific way of dedicating stacked pagodas, this section examines the correlation between the production of stacked pagodas and the organization of patronage based on surviving epigraphical evidence. I contend that the very fact that these stacked pagodas being assembled by individual stone blocks further allows or responds to the way how they were commissioned by patrons.

The eastern Gansu and the Nannieshui material exhibit differences in terms of patronage. The eastern Gansu stacked pagodas were usually commissioned in whole by local donors. In several examples one also finds depictions of the deceased family members of the donor. For instance, the surviving character of 亡, which refers to "the deceased", on the 518 pagoda, suggests that the work was dedicated to gaining merits for the deceased family members.[50] Another stacked pagoda dated to 536 bears the images and inscriptions of a number of donors on the pedestal.[51]

On the Zhuanglang Pagoda, a procession scene of donors occupies the lower two of the three horizontal registers on one side of the bottom block (see Figure 9b). The upper register depicts a preaching scene that features a central seated Buddha. On the lower two registers, three male figures are pictured on each level riding on horses toward one direction, with cartouches indicating their identities as the donor's deceased family members.[52] The procession scene depicting horses and carts is commonly found on sixth-century steles from Shanxi, Shaanxi, and eastern Gansu.[53] Donor figures began to be carved in an arrayed manner along the bottom of the statue and stele pedestals in the late fifth century, usually standing or kneeling towards the center. In the sixth century, donor figures usually appear in a standing position, sometimes within a frame.[54] A similar representation of horses and carts is found on several other pieces from the same region.[55] The tradition continued to be found on steles from eastern Gansu throughout the sixth century.[56]

The rich corpus of donor images on the Zhuanglang Pagoda may relate to the fact that its donors are from the same family clan. Such a feature of organizing patronage exclusively by family ties is prevailing in works of eastern Gansu of the sixth century. It distinguishes from the contemporaneous practice in other regions, where the *yiyi* 邑邑 society became the dominant form of patronage. By the early sixth century in northern China, the collective patronage of Buddhist sculptures began to flourish, and it was usually organized through the *yiyi* society (Michihata 1967; Hou 1998, 2005, 2007; Lingley 2006, 2010). Referring to a form of Buddhist socio-religious organizations, *yiyi* is usually organized by lay Buddhists living in the same village in rural areas of northern China under the leadership of at least one monk or nun (called *yishi* 邑師), to fulfill the activities of building Buddhist steles or temples, copying or chanting Buddhist sūtras for mass circulation, or holding religious rituals together.[57] On statues from eastern Gansu, the *yiyi* society is not seen in inscriptions until the Western Wei, suggesting a relatively later absorption of the new collective patronage form in the region.

In comparison to the eastern Gansu tradition, stacked pagodas from Nannieshui intriguingly represent a new form of collective patronage, which allows each individual donor to claim ownership over one image or a stone block. First, we do not find any Nannieshui stone blocks that are inscribed with lengthy dedicatory texts that are usually common for single-piece pagodas as well as stacked pagodas from eastern Gansu. Instead, inscriptions on Nannieshui pagodas are located around the edge of each niche, comprising only several characters in a very short length showing the donor's name and title. The surnames of the patrons suggest that they belong to different family clans. However, little historical documentation exists to identify more information on donors who have their names inscribed on the Nannieshui group.

Second, a stone block from Nannieshui usually features on each the four sides a separate inscription dedicated to different donors. An individual donor usually claims just one side of a stone block rather than an entire block or a pagoda. For instance, scenes no. 233, no. 234, and no. 236 on the same stone block are respectively accompanied by three patrons, Gao Wen 高文, Wang Daoqu 王道渠, and Li Xiao 李小. Each side of a stone block can be commissioned by more than one donor. Scene no. 673 is accompanied by the names of Li Hanren 李韓人 and Li Andu 李安都. Meanwhile, one person can be the donor of multiple scenes. Both scenes no. 174 and no. 175 are commissioned by Wang Niusheng 王牛生. The carving of the same donor's name twice on two scenes under his patronage still pinpoints the mechanism of patronage at Nannieshui; no matter how many donors are involved, the unit of donation is an individual scene out of the four sides on a stone block. This arrangement of inscriptions further speaks to the argument that the commissioning and making process of each stone block that constitutes stacked pagodas shows an image-centered system.

Although there is no sufficient written material regarding the local community at Nannieshui, the practice of inscribing a donor name by a specific image is not rare in contemporaneous Buddhist statues and steles that feature a complex pictorial program. Sixth-century patrons of Buddhist steles would inscribe the work with their own names or those of deceased family members in cartouches located right by a Buddhis niche or image. For instance, a mid-fifth-century stele discovered not far from Nannieshui features nearly two dozens of short donor cartouches directly below pictorial niches in addition to names of hundreds of donors that are listed on the bottom register on the stele (Lee 2010, chp. 1). The stele is also renowned for two niches picturing the Parinirvāṇa scene in the identical way to the Nannieshui tradition (see Figure 11).

In general, such a practice is understood on two levels. First, the inscription of donors' names and social affiliations by pictorial niches provides a chance for them to join the same league as those Buddhist deities. As addressed by Sonya Lee, with their names inscribed side by side to pictorial space, the patrons gained access to a sacred realm where they could communicate with Buddhist deities as well as their local society. Second, a correlation is constructed between these donors and the local community, in view of the way in which steles and statues were used after production—to be displayed in public space, most likely in monastic complexes or at the intersection of routes. The stacked pagodas thus become meaningful by providing the devotees with a platform, a space, to demonstrate their religious piety to a larger group of audience.

6. Dissolving the Structure: From Multilevel Pagoda to Stone Image

The image-centered making mechanism of stacked pagodas, as address above, further sheds light on underexplored issues regarding their perception and veneration among the local community. In previous studies, there is another strand of scholarship to define the stacked pagodas under discussion, which considers each individual stone block as stele, *simianxiang* 四面像 (four-sided image), or *zaoxiangshi* 造像石 (image stone), instead of pagodas.[58] The reference to these pagodas as *simianxiang* or *zaoxiangshi* is primarily a created appellation based on scholarly interpretation of the depiction of images on four sides of a stone block. Nevertheless, the discussion on the definition of stacked pagodas also echoes some inscriptions found on stone blocks, which intriguingly refer to themselves as *shixiang* 石像 or *xiang* 像.

Why are some pieces of stacked pagodas denoted as *xiang* 像 instead of *futu* or *ta*? What is the difference between *xiang* and *futu* in the context of making stacked pagodas? How do these different denotations relate to the perception and veneration of stacked pagodas? This section contends that the self-reference of *xiang* in inscriptions found on stacked pagodas is a later phenomenon that has not taken place until the later sixth century, replacing the earlier self-reference, *futu*. The primary factor that contributes to this shift of denotation, lies in the specific emphasis on individual images or stone blocks, or *xiang*, which was fundamentally shaped by very making and commissioning process of stacked pagodas. Meanwhile, this shift also echoes with the broader historical context of the second half of the sixth century, when the growing popularity of the hemispherical stūpa form posed challenges to the dominance of the multilevel pagoda form. In addition, the growing understanding of the miraculous deeds of *xiang* might also contribute to the transformation from *futu* to *xiang* among stacked pagodas.

As discussed in the introduction, several stacked pagodas bear inscriptions that refer to themselves as *shi futu san jie* 石佛圖三劫 (three-story stone pagoda), or *shi futu san ji* 石浮圖三级 (three-story stone pagoda).[59] Both *sanjie* 三劫 and *sanji* 三级 mean "three-story." The emphasis of the three levels in the inscriptions is almost universally employed in contemporaneous epigraphs that describe the construction of pagodas, therefore revealing these stacked pagodas' formal resemblance to pagoda buildings. For instance, the inscription on the Quan Pagoda of 536 also records *sanjie shi yiqu* 三劫石一区 (a three-story stone).[60] The inscription of the Huisuisi 暉福寺 stele of Duke Dangchang 宕昌公 reads, "I commissioned two three-storied *futu* for the two emperors at my old houses, one in the south and another in the north in my hometown."[61] The inscription on the construction of *futu* by Chang Huan 常煥 and the others reads, "There is a five-story *futu* inside the temple yard."[62] The multilevel verticality of these stacked pagodas, as implied by their inscriptions, further reinforces the perception of them as pagodas. Wei-Cheng Lin, in his recent article on the vertical rise of Chinese pagodas in the Middle Period (10th–14th century) contends that the importance of the verticality of pagodas derives from their performative aim in drawing the attention of the faithful performative function (Lin 2016).

However, a few other inscriptions found on stacked pagodas feature the term *shixiang* 石像 (stone image) to refer to stone blocks that comprise these stacked pagodas. The Xiejiamiao Pagoda of 558 records *shixiang yiqu* 石像一区 (a stone image). Inscriptions located on the side marked as no. 392 of the Nannieshui pagoda of 553 mention the construction of *wu shixiang* 五石像 (five stone images), which might refer to five stone blocks including the one bearing the inscription. In addition to *shixiang*, we also find *zaoxiang* 造像 in some inscriptions. Yet a distinction between the two terms is rarely spotted. *Zaoxiang* is found frequently in inscriptions found on contemporaneous steles, highlighting the very act of commissioning an image. The term *shixiang*, distinctively, denotes the entirety of the three-dimensional sculpture.[63] In the case of the Xiejiamiao pagoda of 558, we know that *shixiang* (stone image) in its inscription refers to the sculptural entity, also according to the term *yiqu* 一区 (a piece). Inscriptions on the Nannieshui no. 392 emphasize its entirety by adding *wu* 五 (five) before *shixiang*, showing the number in total.

Comparing all surviving inscriptions, one finds that *futu* is found among the earliest pieces dated to the early sixth century, while *shixiang* is only spotted among those dated to the third quarter. This periodical gap reveals that the *futu* and *shixiang* might not be interchangeable, but successive. That being said, inscriptions reveal a transition of self-referral terms to these stacked pagodas from "three-story stone pagodas" to "five stone images" in the third quarter of the sixth century, although their stacked forms retain the same. The new term, "stone image", pinpoints a shift in perception of these stone sculptures from pagodas to sculptural images.

How did this transition happen? An important factor lies in the way these pagodas were made with a series of stone blocks that are fully adorned with images, or the image-oriented system of commissioning stacked pagodas. In other words, the very making of stacked pagodas with richly adorned stone blocks gradually shaped, altered, and transformed the way they were perceived among the worshippers. From *futu* to *xiang*, the veneration foci shifted from the entire pagoda to individual images. From "three-story" to "five," the emphasis on the vertical rise is replaced by the number of individual stone blocks. The purpose of stacking these stone blocks to erect pagodas was gradually fading away through the decades of the sixth century. The mechanism of making pagodas from individual stone blocks, instead, seized the decisive power of defining its ontological status.

The dissolution of the vertical rise of stacked pagodas might also echo a development which took place from the mid-sixth century—one-story stūpa images with the archaic hemispherical dome began to flourish, primarily in the territory of Eastern Wei and Northern Qi and the south.[64] Although the stūpa image never achieved currency in eastern Gansu or Shanxi, the flourish of the domed stūpa image from the mid-fifth century nevertheless indicates a shifted perception of the form and meaning of pagodas.

Meanwhile, another trend involved is the growing prevalence of image-making on free-standing stone surface through the fifth and the sixth century, and the awareness of image-making. *Xiang* 像, or image, became more recognized as the primary means of inscribing Buddhist teachings, displaying merits for patrons, and generating religious meanings. The growing importance of *xiang*, or images in the sixth century, on the micro level, is already pinpointed above by the emphasis, highlight, and iconization of individual images that carry specific meanings, such as the Parinirvāṇa scene. *Xiang*, or "image," is usually distinguished from other forms in terms of their roles primarily as objects of worship (Wong 2004, introduction). The veneration of images in China has also developed over the course of centuries. The establishment of image making or the image cult and its correlation to the dissemination of Buddhism in China since the second century has been a key issue in debate in the field of art history as well as religious studies. Minku Kim, in his dissertation on the very topic, has combed through the historiography of the topic and provided a definition of Buddhist image worship in a Chinese context (Kim 2011, 2014). Following his scholarship, the essential factor defining image worship is related to whether there is any ritualistic Buddhist significance displayed. Furthermore, recent scholarship in Buddhist studies also shows an intentional construction of the image worship as a distinctly Buddhist practice since the late fifth century (Greene 2018). It was a gradual process for the image worship to be integrated into medieval Chinese society.

7. Concluding Remarks

This paper discusses the phenomenon of making stacked pagodas in sixth-century northern China. On the one hand, stacked pagodas were complementary in form and placement to real pagoda buildings. On the other hand, they feature their own peculiarity in terms of material, making, and venerating. Images on stacked pagodas have provided a wonderful outlet for the unbridled imagination of the artisan. Although there is a variety of delicate details to defy an attempt to impose a systematic classificatory scheme of arranging images on stacked pagodas, a survey of surviving images reveals a specific preference for certain images and allows further study of the context of their correlation with the broader artistic milieu. More importantly, an investigation of the relation between carved images and inscribed texts against the historical context of dedicating Buddhist statues and constructing pagodas sheds light on the making and perception of pagodas in non-built form in sixth century. In addition to the definition of sacred space with architectural forms, this case study pinpoints the quintessential role played by the adornment of images in generating sacredness. **Funding:** This research received no external funding. The APC was

covered by research funding from the Center for Global Asia, New York University Shanghai.

Conflicts of Interest: The authors declare no conflict of interest.

Notes

[1]   Regarding the region's history of Buddhism and the role played by ethnic groups, see Hou Xudong (Hou 2008); Wei and Wu (2009); Song (2009); Gansu Beishikusi Wenwu Baohu Yanjiusuo (2013); Zheng et al. (2014); Wang (2015). Eastern Gansu and the bordering Shaanxi has formed an important center since the fourth century, where Buddhism flourished during the rule of the non-Han Chinese dynasties, including the Former Qin 前秦 (351–394), Later Qin 後秦 (384–417), and Xiongnu Da Xia 匈奴大夏 (407–431) before the Northern Wei's capture of Chang'an. For the area's history and its non-Han culture, see (Ma 1985).

[2]   The discussion on the initial appearance of stūpas is vast. See (Hawkes and Shimada 2009; Fogelin 2012, 2015).

[3]   Each of "stūpa" and "pagoda" has a broader reference and a more contested history of use in various research contexts. The terminological ambiguity has persisted through the field over the past century. To avoid ambiguity, this study uses pagoda to refer to multistory structures whereas stūpa refers to all other forms. The last section in this introduction will provide a more

detailed discussion of the terminology. For related discussion, also see Miu (2012), Miller (2014), Mukai (2020). Meanwhile, a second-century relief carving of a multistory structure from Sichuan is considered by scholars as one among the earliest depictions of pagoda. See Xie (1987).

[4] They are usually named "stūpa" due to the hemispherical dome.

[5] There is a number of studies on the topic. See, among other sources, (Wang 1999; Yin 2000; Abe 2002).

[6] It has been preserved in the Chongfu 崇福 Monastery in northwest Shanxi, but was originally commissioned at Pingcheng, the capital city of Northern Wei. Its pedestal and tower body were taken to Japan during World War II and later returned to Taiwan after the war, while the ornamental top was preserved by a local person in Shuozhou. See (Shi 1980; Wang 2011b).

[7] At the Nannieshui Museum, a stone piece with a roof imitating wooden structure is placed on top of a stacked pagoda. Yet it remains uncertain how universal this practice is, without any more similar pieces located.

[8] For instance, the inscription found on a stone block from Xiejiamiao in eastern Gansu reads, "永熙三年太歲在寅八月十四日弟子/縣張生德為忘息大奴敬/造石佛圖三劫願上生天上/諸佛下生人間口王長壽若/三速令解脱善願從心" (On the fourteenth day of the eighth month, the third year of the Yongxi era [the year of the Tiger], Zhang Shengde from the county of … dedicated a three-story stone pagoda for his deceased Danu for his ascension into the heaven … Buddhas descend to the mundane world … longevity … achieve emancipation from … to follow what the heart desires …) The stacked pagoda that this stone block belongs to dates 534, Northern Wei. There is no archaeological report, but a general overview, on the group of statues found in the Xiejiamiao site ever published. Current studies weigh on the ethnic group of the donors for this group. See (Wang 2015, 2016; Zhang 2000, pp. 108–12).

[9] The difference between the three words is generally attributed to the distinctive strategies of translation used between the third and the eighth century. Both *futu* 浮屠 and *futu* 浮圖 are used to refer to the Buddha in early Chinese historical texts, denoting an interchangeable relationship between the Buddha and the sacred structure. The usage is considered a result of phonetic confusion caused by the transliteration of Sanskrit phrases. In *Hou Han shu* 後漢書, *futu* is described as miraculous images that appear together with Laozi 老子, the indigenous saint who later became a quasi-deity of Daoism. Additionally, both *futu* and Laozi are housed in *ci* 祠 or *miao* 廟, which both refer to a ritual shrine in Chinese. See (*Hou Han shu* 1984, p. 16). For recent studies of the three terms' literal meanings, see (Greene 2018).

[10] Miniature pagodas and pagoda reliefs have been examined as evidence for the study of early Chinese architectural history in almost all major studies in the field. These studies contributed to many aspects, but their discussion on these miniatures and pagoda images is in the form of an overview. For major works, see (Liang 1961, 1962; Ledderose 1980; Seckel 1980; Sun 1984; Xiao 1989; Wang 2011a; Steinhardt 2011, 2014, 2019). In recent years, we have also seen discussions focusing on miniatures; however, the discussion is still confined in the scope of developing a typological system based on their structural traits (Wang 2006; Tang 2016; Xu 2016). Further studies in light of the study of Chinese miniatures in other forms and recent theoretical discussion of miniaturization is much needed. See (Ledderose 1983, 2000; Stein 1990; Steward 1993; Selbitschka 2005; Guo 2010; Hong 2015; Wu 2015; Luo 2016; Graves 2018; Martin and Langin-Hooper 2018; Davy and Dixon 2019; Elsner 2020).

[11] To employ the term "hybridity," some clarification should be made in response to recent scholarly discourse on the topic in the field of archaeology. Scholars have rectified the perception of hybridity by examining the issue of receptivity. See Stockhammer (2013), Andreeva (2018). To summarize within the scope of this footnote, this research confines the definition of hybridity within the scope of specific styles and motifs that have been developed and transmitted in northern China, as displayed by major Buddhist artworks. It agrees with the strand of scholarship that challenges the narrative of uninterrupted transmission of dominant styles and motifs in provincial areas. Rather, this study showcases the complexity about the way how styles and motif of various origins were combined unevenly in subordinate regions in sixth-century northern China.

[12] From the east to the west, five of the seven counties of Pingliang are Lingtai 靈台, Jingchuan 涇川, Chongxin 崇信, Huating 華亭, and Zhuanglang 莊浪, all of which boast several Buddhist cave-temples as well. See (Gansu Sheng Wenwu Gongzuodui, and Qingyang Bei Shiku Wenwu Baoguansuo 1987; Zhang 1994; Cheng 1998, p. 41; Dong 2008).

[13] See footnote 1.

[14] A total number of 209 cave-temples are carved out of the cliffs, dated from the Later Qin and Western Qin of the Sixteen States to the Tang dynasty. See (Yan 1984; Wei 2005; Gansu Sheng Wenwu Gongzuodui 1987; Steinhardt 2014, pp. 90–92).

[15] The 503 stone block's inscription reads, "景明四年太/歲在癸未/太陰在/巳大將軍/在午白虎/在寅青龍/在子四月癸馬" (In the fourth month of the fourth year of the Jingming era, the year at Guiwei and the lunar cycle at Si, the Great General is in the year of the Horse, the White Tiger in the year of the Tiger, the Dragon in the year of the Rat, the fourth month … Kui … Horse …) See (Zhang 2000, pp. 98–104; Gansu Sheng Wenwu Ju 2014, pp. 58–59).

[16] Its inscription reads, "延昌三/年十五日/亥/涇州郡" (The third year of the Yanchang era … the fifth day … the year of the Pig … Jingzhou Prefecture …). See (Zhang 2000, p. 101).

[17] Its inscription reads, "神龜元 … 孫亡 … 天上亡 … " (The first year of the Shengui era … Sun … deceased … deceased in the heaven). See (Zhang 2000, p. 102).

[18] The earliest identified Twin Buddhas motif is found in Cave 169 of the Binglingsi 炳靈寺 Cave-temples in Gansu. Cave 169 was commissioned during the Western Qin 西秦 Dynasty (385–431 CE) in the early fifth century. In the scene, two Buddhas are sitting side by side below a niche, above which three *chattra*-like elements protruding upwards. It was thus considered the earliest

representation of the Twin Buddhas concept and of the "Jeweled Stūpa" chapter from the *Lotus Sūtra*. Yet the motif was not depicted frequently until the 470s. See, among many other sources, (Davidson 1954; Wong 2004, chp. 8; Wang 2005, chp. 1; Hurvitz 2009; Williams 2009).

[19] Huating has been an important regional economic hub along the Silk Road. The pit's location matches the historical site of a temple of the Northern dynasties. There is no archaeological report, but a general overview, on the group of statues found in the Xiejiamiao site ever published. Current studies weigh in on the ethnic group of the donors for this group. See (Zhang 2000, pp. 108–12; Wang 2016).

[20] The inscription of the 516 piece reads, "熙平元年／太歲／在申／為張／何迴／張雙／清信士供養河門／大小／張永／奴／／河門大小／者得" (On the first year of the Xiping era, the year of Shen, this . . . is dedicated to Zhang Hejiong, Zhang Shuang, men of pure faith . . . Zhang Yongnu of Hemen . . . to follow what the heart desires . . . from Hemen . . .) See (Wang 2016, Figure 1). That of the 558 piece reads "二年歲次戊寅六月癸寅朔十七日己丑清／信弟子路夫功曹南／中敬造石像一區願三涂地／願一切生花三得成佛道所／所願心／佛弟子安家大小常住三" (On the seventeenth day [*jichou*], of the sixth month [*guiyin*], the second year [*wuyin*], Lu Weifu, a man of pure faith . . . dedicated a stone image . . . Samadhi . . . Wish all the beings achieve the Buddhahood . . . to follow what the heart desires . . . the Buddhist disciple . . .) See (Zhang 2000, p. 110; Gansu Sheng Wenwu Ju 2014, pp. 55–56).

[21] Su Bai has discussed this feature briefly. In addition to examples from Longmen and Gongxian, a statue from the White Horse Temple of Luoyang also features the Buddha's right foot in the manner. It is on display in the Museum of Fine Arts, Boston. See (Su 1996, pp. 153–76).

[22] For more Zhuanglang pieces that are not examined in this section, see (Zhang 2000, p. 107; Gansu Sheng Wenwu Ju 2014, pp. 48–49, 51–52; E and Yang 2014a, 2014b).

[23] One of the five stone blocks was found in the first half of the twentieth century while the other four were discovered in 1974. See (Cheng and Ding 1997; Zhang 2000, pp. 113–24; Wang 2004; E 2011; Gansu Sheng Wenwu Ju 2014, pp. 49–50).

[24] See (Ding 2016, pp. 68–69). Another piece, brought to the US by the expedition of Warner in 1923, is currently in the repository at the Fogg Museum, Harvard. On Warner's expedition, see (Warner 1926; Jayne 1929; Liu 2000). Jayne's work was translated into Chinese by Liang Xuping. See also (Wang and Mrozowski 1990; Ding 2016, p. 74).

[25] See footnote 2.

[26] A record of the original excavation is helpful in exploring the original purpose of these hoarding pits. For instance, the excavation of hoarding pits located in Qingzhou, Shandong, shows that the statues and steles had been deposited in the pit in several layers, with the well-preserved items in the center, and fragments in the surrounding area. See (Nickel 2002, p. 35).

[27] The period of Eastern Wei is not specified here due to its short life. Works produced during the Eastern Wei are usually grouped with the Northern Wei or the Northern Qi based on stylistic affinity.

[28] On the origin of the Lushui hu, see (Tang 1955; Zhou 1963, pp. 156–57; Wang 1985; Zhao 1986; Wang 1997; Hou 2008; Liu 2008, pp. 9–11).

[29] This study does not agree with any absolute reconstruction of the way in which each side of the five stone blocks is aligned. However, for convenience of discussion, I refer to each side based on the way the pagoda is currently displayed.

[30] The Gandhāran tradition depicts the horse at center with the prince standing aside in ordinary royal dress, while the prince of northern China is featured dominantly in the conventional look of a pensive figure, who has one leg pendant and the other raised and brought across to rest upon the knee of the pendant leg, and with one arm raised towards the face. The Great Departure scene appears first in reliefs at Yungang in the 490s among fifteen other narrative episodes of the Buddha's life story. There are sixteen scenes depicting episodes related to the Buddha's birth on the central pillar of the Yungang Cave 6, and another set of sixteen on the lower register of the interior walls in the main chamber.

[31] The motif is noticeably absent from any other stone works in the fifth century, indicating intentional neglect of it. For a comprehensive discussion of the Parinirvāṇa scene in Chinese Buddhist art tradition, see (Lee 2010, introduction, pp. 38–42, Figure 1.15).

[32] The Chinese examples are not depicted in association with any other events of the Buddha's life, departing from the South Asian and Central Asian traditions, in which the Parinirvāṇa image is always represented together with other episodes of the Buddha's life story.

[33] John S. Strong has translated the text's surviving Sanskrit version to English. See (Strong 1983).

[34] According to Victor Meir, the original text that the *Xianyu jing* was translated from has a Central Asian origin. *Xianyu jing* 賢愚經 (The Sutra of the Wise and the Foolish), trans., by Hui Jue 慧 et al. T no. 4: 202.368c. See (Junjirō 1901; Mair 1993).

[35] Caves 5–11, 5–38, 25, 28, 29, 33, 33–34, and 34. For an overview of the story depicted in Yungang cave-temples, see (Yi 2017, chps. 5 and 6). For an example, see (Yungang Shiku Wenwu Baoguansuo 1991, Figure 197).

[36] Hu Wenhe has provided a comprehensive discussion of the story. See (Hu 2005; Yi 2017, chp. 3). Regarding the narrative scenes carved on figured steles, Li Jingjie proposed a different identification of the scene. He argues for a representation of the Dīpaṃkara *Jātaka* instead of the offering dust story, according to several surviving inscriptions that point out the connection between the Buddha Dīpaṃkara and the children (儒童). See (Li 1996).

37 There are a series of research examining the pensive Buddha image in the fifth and the sixth century. This essay will not go into details. See (Rei 1975; Leidy 1990; Lee 1993; Hsu 2002).

38 The study of Nannieshui materials with Gephi was undertaken by the author in the workshop "Social Network Analysis in Buddhist Studies," organized by "From the Ground Up" project in August 2018, National Singapore University. A more detailed discussion is forthcoming. For more discussion about using Gephi in the study of Buddhism, see (Bingenheimer 2020).

39 The Cao Tiandu Pagoda is also known for its turbulent history of displacement. Its base and body were looted from China and brought to Japan; later it was returned to Taiwan, now in preservation at the National Museum of History in Taipei. The *chattra* top was saved by a local person during the war and returned to Chongfu Monastery in 1953.

40 The county borders with the Pingcheng area and, throughout history, has been included in the Northern Shanxi cultural sphere. The Yu county belongs to the Kingdom Dai 代 in the fourth century, the precedent of Northern Wei. See (Yuxian Difangzhi Bianzuan Weiyuanhui 1995; Huang 2015).

41 This claim is further affirmed by the absence of the Han mode of dresses that developed and entered the scene of the Pingcheng Buddhist art in the late 480s. The Han mode is a new dress style that features Sinicized traits, such as loose robes and wide girdles (*baoyi bodai* 褒衣博帶). It echoes the Sinicization reform in clothing promoted by Emperor Xiaowen 孝文 during the Taihe 太和 era (486–495). During the Taihe period, the "Era of Supreme Harmony," Emperor Xiaowen and his court instituted a series of reforms that integrate intensively historical Chinese administrative institutions, rituals, urban design, etc. One of the defining features of this process is *Hanhua*, "becoming like the Han," revealing the very nature of refashioning the Xianbei Northern Wei regime as an imperial Chinese dynasty. See (Bachhofer 1946, p. 66; Okada and Ishimatsu 1993, pp. 181–203; Abe 2002, p. 89).

42 The pagoda is preserved and on display in the recently founded Museum of Northern dynasties. It is said to be discovered at a local construction site. Yet no archaeological report is available at this moment.

43 It was first mentioned in the initial report on the discovery of the monastery in 1954. Yet it was reported stolen in 2000. See (Li 2008b).

44 For a proposed chronology of the three stone pagodas under discussion, see (Zhao 2020).

45 The Twin Buddhas motif signifies that more than one Buddha can exist at the same time in the cosmos. This is a new Mahāyāna theme, as early Buddhists believe there is only one Buddha in each age. See (Liu 1958; He 1992; Mizuno and Nagahiro 1951–1956, vols. 8 and 9, pp. 73–75).

46 For a detailed examination of narrative scenes on the miniature pagodas, see (Zhao 2020).

47 The Cao Tianhu pagoda was excavated in present-day Jiuquan 酒泉, Gansu province. The inscription on its pedestal records that it was commissioned in 499 by a local person named Cao Tianhu. Jiuquan is part of the east-west corridor of the Hexi region. The name exhibits a resemblance to Cao Tiandu of Pingcheng. Yet no further evidence shows connections between the two. Except for Chen Bingying's description, the Cao Tianhu Pagoda has not been studied beyond a brief report. See (Chen 1988).

48 A similar arrangement is also found on a pagoda fragment preserved in the Palace Museum in Beijing. See (Li 1986).

49 The conclusion is based on the study of three types of texts on the Buddha's austerity. For a detailed examination, see (Zhao 2020). For a topical study of the three types of texts, see (Rhi 2006).

50 However, the epigraphic inscriptions of the pagodas of 514 and 518 are severely worn, leaving the donors' identities unrecognizable.

51 Another factor to be considered in the study of this statue is the ethnic background of the local laity. The surnames of most of the donors may indicate their "Hu" identity. Yet the scarcity of textual records is not sufficient to support further discussion. See (Wang 2016).

52 There are two on the upper register and three on the lower. From left to right, the upper two are "deceased father Bu Waitong" 亡父卜外通 and "deceased mother Le Baozhu" 亡母樂保朱, while the lower three are "general Bu yong" 上將卜永, "deceased brother Bu An" 亡雄卜安, and "deceased sister Bu Yonghe" 亡妹卜永禾.

53 For a discussion of the procession scene on steles from Shanxi, see (Wong 2004, chp. 5). Such a unique way nevertheless reminds us of a parallel tradition in Chinese funerary art, which depicts exactly the deceased, or the owners of the tomb, in ox carts or on horses. Appearing as early as the Eastern Zhou, and continued in later periods, a funerary procession was usually depicted on side walls in tombs, representing the escorting of the "soul carriage" of the dead from his mundane life to the otherworldly abode. The tradition continued to flourish in the following centuries in tombs located in various regions in northern China. One finds exactly the same juxtaposition of horses and an ox chariot on the Zhuanglang Pagoda as well as the two other steles from eastern Gansu. In the Central Binyang Cave at Longmen, and the Gongxian cave-temples in northern Henan, the procession of the emperor and empress still astounds visitors with magnificent craftsmanship. See (Wu 2010, pp. 60–70).

54 With attendants flanking or not, donor images are usually separated from each other by cartouches of inscriptions. Kate Lingley has written extensively on donors of Buddhist art in the sixth century. For instance, see (Lingley 2006, 2010).

55 For instance, the Wei Wenlang 魏文朗 Stele of 424, one of the earliest surviving Buddhist steles from Shaanxi, features a donor figure riding on a horse with an attendant and an ox cart following. See (Li 2008a, p. 33). Another stele of 546 from Pingliang depicts two registers of horse and cart riders on the lower part of its façade. See (Zhang 2000, pp. 172–74).

[56] See the Quan Daonu 權道奴 Stele dated to 563, the third year of the Baoding era, Northern Zhou, and the Wang Lingwei 王令猥 Stele from Zhangjiachuan. See (Zhang 2000, pp. 205–6, 222–23).

[57] *Yizi* 邑子, villagers who were members of *yiyi*, also perform charitable works for the benefit of the entire community. For more on the history of *yiyi* during the Northern dynasties, see (Twitchett 1957; Michihata 1967; Tanigawa 1985; Hou 1998, 2005, 2007; Gernet 1995, pp. 259–77).

[58] For instance, Stanley Abe refers to the Shaanxi piece (see Figure 8) as a square stele. See (Abe 2001).

[59] For full inscription, see note 7.

[60] This is found on the Quan 權 Pagoda from Qin'an 秦安. It was discovered by locals in Wujiachuan 吳家川 in 1941, according to notes written on a rubbing of its pedestal. Without any other records of its discovery, it remains in debate whether other parts of the pagoda were unearthed at the same site. This study follows the theory of Wen Jing and Wei Wenbin, who argue that while the upper stone block of the pagoda is the original part dated 536, Western Wei, the lower two are from a separate set due to their display of a typical Northern Zhou style. This theory also draws evidence from the incompatibility between the three blocks and the three eaves. Meanwhile, the names of donors indicate its provenance in situ, since the clan of Quan 權, the surname of most donors, is among the most prominent families in the region since the fifth century. See (Wen and Wei 2012; Zhang 2000, p. 213).

[61] "於本鄉南北舊宅，上為二圣造三級浮圖各一區." See (Yen 2008, no. 1).

[62] The inscription mentions the construction date of the second year of Xiaochang era, 526. "以寺內有五級浮圖一區，建自永昌，後因兵劫 … " (… for the reason that there was a five-story pagoda in the monastery. Built in the Yongchang era, and because of warfare …) See (Yen 2008, no. 25).

[63] The usage of "shi" (stone) highlights the choice of medium and material.

[64] A number of scholars have examined the rich corpus of domed stūpa imagery of the sixth century. Among many, see (Tsiang 2000; Su 2006, 2010).

References

Primary Sources in Chinese

Guoqu xianzai yinguo jing 過去現在因果經 (Sūtra on Past and Present Causes and Effects). Translated by Guṇabhadra 求那跋陀羅 (394–468). T. no. 189.

Hou Han shu 後漢書 (The History of the Later Han). Fan Ye 范曄 (398–445). Taipei: Taiwan shangwu yinshuguan, 1984.

Luoyang qielan ji jiaoshi 洛阳伽蓝记校释 (A Record of Buddhist Monasteries of Luoyang). Yang Xianzhi 楊衒之. Commented by Zhou Zumo 周祖谟. Beijing: Zhonghua shuju, 1963.

Miaofa lianhua jing 妙法蓮華經 (Skt. Saddharma-puṇḍarīka-sūtra). Translated by Kumārajīva 鳩摩羅什 (343/4–413). T. 262.

Puyao jing 普曜經 (General Radiance Sūtra; Skt. *Lalitavistara*). Translated by Dharamarakṣa 竺法護 (239–316). T. no. 186.

Shui jing zhu 水經註 (Commentary on the Water Classic). Li Daoyuan 酈道元 (~527). Beijing: Zhonghua shuju, 2007.

Taizi ruiying benqi jing 太子瑞應本起經 (Sūtra on the Life of the Prince in Accordance with Good Omens). Translated by Zhi Qian 支謙 (222–252). T. no. 185.

Wei shu 魏書. Wei Shou 魏收 (505–572) et al. Beijing: Zhonghua shuju, 1974.

Weimojie suoshuo jing 維摩詰所説經 (Skt. Vimalakīrti-Nirdeśa-Sūtra). Translated by Kumārajīva 鳩摩羅什 (343/4–413). T. 14, 0475.

Xu gaoseng zhuan 續高僧傳 (Supplement to the Biographies of Eminent Monks). Daoxuan 道宣 (596–667). T. 2060, 50.

Zheng fahua jing 正法華經 (Lotus Sūtra). Translated by Dharamarakṣa 竺法護 (239–316). T. no. 263.

Secondary Sources

Abe, Stanley K. 2001. Provenance, Patronage, and Desire: Northern Wei Sculpture from Shaanxi Province. *Ars Orientalis* 31: 1–30.

Abe, Stanley K. 2002. *Ordinary Images*. Chicago: University of Chicago Press.

Andreeva, Petya. 2018. Fantastic Beasts of the Eurasian Steppes: Toward A Revisionist Approach to Animal-Style Art. Ph.D. dissertation, University of Pennsylvania, Philadelphia, PA, USA.

Bachhofer, Ludwig. 1946. *A Short History of Chinese Art*. New York: Pantheon.

Behrendt, Kurt. 2003. *The Buddhist Architecture of Gandhāra*. Leiden: Brill.

Behrendt, Kurt, ed. 2007. *The Art of Gandhara in the Metropolitan Museum of Art*. New Haven and London: Yale University Press.

Bingenheimer, Marcus. 2020. On the Use of Historical Social Network Analysis in the Study of Chinese Buddhism: The Case of Dao'an, Huiyuan, and Kumārajīva. *Journal of the Japanese Association of Digital Humanities* 5: 84–131. [CrossRef]

Brancaccio, Pia, and Kurt Behrendt, eds. 2006. *Gandharan Buddhism: Archaeology, Art, Texts*. Vancouver: UBC Press.

Cao, Xuexia. 2011. Nannieshui shike zaoxiang de minjian tese: Qiantan shike yishu zai qinxian de chuancheng yu fazhan 南涅水石刻造像的民間特色：淺談石刻藝術在沁縣的傳承與發展 (Features of the Nannieshui Stone Statues: On the Inheritance and Development of Stone Sculptures at County Qin). *Wenwu Shijie* 4: 30–42.

Caswell, James O. 1975. The 'Thousand-Buddha' Pattern in Caves XIX and XVI at Yün-Kang. *Ars Orientalis* 10: 35–54.

Chen, Bingying. 1988. Bei Wei Cao Tianhu zao fang shita 北魏曹天護造方石塔 (Square stone pagoda commissioned by Cao Tianhu in Northern Wei). *Wenwu* 3: 83–85, 93.

Chen, Yikai. 1996. Lvelun Bei Wei shiqi Yungang shiku, Longmen shiku fudiao taxing" 略論北魏時期雲岡石窟龍門石窟浮雕塔型 (On Typology of Northern Wei Pagodas in Relief in Yungang and Longmen Cave-temples). In *Longmen shiku yiqian wubai zhounian guoji xueshu taolunhui lunwenji* 龍門石窟一千五百週年國際學術研討會論文集 *(Proceeding of the International Symposium upon the Fifty Hundred Years Anniversary of the Long-men Cave-Temple)*. Beiijng: Wenwu Chubanshe, pp. 220–51.

Cheng, Xiaozhong. 1998. Zhuanglang Yunyasi shiku neirong zonglu 莊浪雲崖寺石窟內容總錄 (Zhuanglang Yunyasi Cave Temples). *Dunhuang yanjiu* 敦煌研究 1: 41.

Cheng, Xiaozhong, and Guangxue Ding. 1997. Zhuanglang xian chutu Bei Wei shi zaoxiang ta 莊浪縣出土北魏石造像塔 (Northern Wei Stone Pagodas Discovered in Zhuanglang). *Dunhuangxue jikan* 敦煌學輯刊 2: 134.

Cheng, Xiaozhong, and Fuxue Yang. 2003. *Zhuanglang shiku* 莊浪石窟 *(Zhuanglang Cave-Temples)*. Lanzhou: Gansu Wenhua Chubanshe.

Davidson, J. Leroy. 1954. *The Lotus Sutra in Chinese Art: A Study in Buddhist Art to the Year 1000*. New Haven: Yale University Press.

Davy, Jack, and Charlotte Dixon, eds. 2019. *Worlds in Miniature: Contemplating Miniaturisation in Global Material Culture*. London: University College London Press.

Ding, Detian. 2016. Jingchuan Fojiao Kaogu Yanjiu. Ph.D. dissertation, Lanzhou University, Lanzhou, China.

Dingxian County Museum 定縣博物館. 1972. Hebei Dingxian faxian liangzuo Songdai Taji 河北定縣發現兩座宋代塔基 (Two Pagoda Foundations of the Song Dynasty Discovered at County Ding, Hebei). *Wenwu* 文物 8: 39–51.

Dong, Guangqiang. 2008. Zhuanglang Yunyasi deng shiku de diaocha jianbao 莊浪雲崖寺等石窟的調查簡報 (Report on Zhuanglang Yunyasi Cave Temples). *Dunhuang Yanjiu* 敦煌研究 3: 44–54.

Du, Yusheng. 1981. Beiwei Yongningsi taji fajue jianbao 北魏永寧寺塔基發掘簡報. *Kaogu* 考古 3: 223–24.

E, Yunan. 2011. Gansu sheng bowuguan cang Bushi shita tuxiang diaocha yanjiu" 甘肅省博物館藏卜氏石塔圖像調查研究 (On Images of the Bu Pagoda in the Gansu Provincial Museum). *Dunhuangxue Jikan* 敦煌學輯刊 4: 67–78.

E, Yunan, and Fuxue Yang. 2014a. Gansu sheng bowuguan shoucang de yijian weikan Beichao canta 甘肅省博物館收藏的一件未刊北朝殘塔 (An Unpublished Piece of Fragmentary Stupa in Gansu Provincial Museum). *Dunhuang Yanjiu* 敦煌研究 4: 17–22.

E, Yunan, and Fuxue Yang. 2014b. Qin'an Xi Wei shita quansuo 秦安西魏石塔权索 (On Western Wei Stone Pagodas from Qin'an). *Xinjiang Shifan Daxue Xuebao* 新疆师范大学学报 1: 87–96.

Elsner, Jaś. 2020. *Figurines: Figuration and the Sense of Scale*. Oxford: Oxford University Press.

Fogelin, Lars. 2012. Material Practice and the Metamorphosis of a Sign: Early Buddhist Stupas and the Origin of Mahayana Buddhism. *Asian Perspectives* 51: 278–310. [CrossRef]

Fogelin, Lars. 2015. *An Archaeological History of Indian Buddhism*. Oxford: Oxford University Press.

Gansu Beishikusi Wenwu Baohu Yanjiusuo 甘肅北石窟寺文物保護研究所, ed. 2013. *Qingyang Beishikusi neirong zonglu* 慶陽北石窟寺內容總錄 *(On the North Cave Temple of Qingyang)*. 2 vols. Beijing: wenwu chubanshe.

Gansu Sheng Wenwu Gongzuodui 甘肅省文物工作隊, ed. 1987. *Hexi Shiku* 河西石窟 *(Hexi Cave-Temples)*. Beijing: Wenwu Chubanshe.

Gansu Sheng Wenwu Gongzuodui 甘肅省文物工作隊, and Qingyang Bei Shiku Wenwu Baoguansuo 慶陽北石窟文物保管所. 1987. *Longdong Shiku* 隴東石窟 *(Eastern Gansu Cave Temples)*. Beijing: Wenwu Chubanshe.

Gansu Sheng Wenwu Ju 甘肅文物局, ed. 2014. *Gansu Gu Ta Yanjiu* 甘肅古塔研究 *(Gansu Pagodas)*; Beijing: Kexue Chuban She.

Gao, Meng. 2012. Tuta yu lifo 圖塔與禮佛 (Decorated Pagodas and Buddhist Worship). Ph.D. dissertation, Zhongyang meishu xueyuan 中央美術學院, Beijing, China.

Gernet, Jacques. 1995. *Buddhism in Chinese Society: An Economic History from the Fifth to Tenth Centuries*. Translated by Franciscus Verellen. New York: Columbia University Press.

Graves, Margaret. 2018. *Arts of Allusion: Object, Ornament, and Architecture in Medieval Islam*. Oxford: Oxford University Press.

Greene, Eric M. 2018. The 'Religion of Images'? Buddhist Image Worship in the Early Medieval Chinese Imagination. *Journal of the American Oriental Society* 138: 455–84. [CrossRef]

Guo, Yong. 1959. Shanxi Qinxian Faxian le Yipi Shike Zaoxiang 山西沁縣發現了一批石刻造像 (A Group of Stone Statues Discovered in the County of Qin, Shanxi). *Wenwu* 文物 3: 54–5.

Guo, Tongde. 1979. Shanxi Qinxian Nannieshui de Beiwei Shike Zaoxiang 山西沁縣南涅水的北魏石刻造像 (Northern Wei Stone Statues from Nannieshui, County of Qin, Shan-Xi). *Wenwu* 文物 3: 91–2.

Guo, Qinghua. 2010. *The Mingqi Pottery Buildings of Han Dynasty China, 206 BC–AD 220*. Portland: Sussex Academic Press.

Hawkes, Jason, and Akira Shimada, eds. 2009. *Buddhist Stūpas in South Asia*. New Delhi: Oxford University Press.

He, Shizhe. 1992. Guanyu Shiliuguo Beichao shiqi de sanshi Fo yu san Fo zaoxiang zhuwenti: 1 關於十六國北朝時期的三世佛與三佛造像諸問題（一）(On Buddhas of the Three Ages and images of three Buddhas during the period of the Sixteen Kingdoms and Northern Dynasties). *Dunhuang Yanjiu* 敦煌研究 4: 1–20.

Hebei Sheng Wenhuaju Wenwu Gongzuodui 河北省文化局文物工作隊. 1966. Hebei Dingxian chutu de Beiwei shihan 河北定縣出土的北魏石函 (Northern Wei Stone Chamber Discovered from County Ding, Hebei). *Kaogu* 考古 5: 252–59.

Hong, Jeehee. 2015. Mechanism of Life for the Netherworld: Transformations of Mingqi in Middle-Period China. *Journal of Chinese Religion* 2: 161–93. [CrossRef]

Hou, Xudong. 1998. *Wu Liu Shiji Beifang Minzhong Fojiao Xinyang* 五六世紀北方民衆佛教信仰 *(Buddhist Worship in the North of the Fifth and the Sixth Centuries)*. Beijing: Zhongguo Shehui Kexue Chubanshe.

Hou, Xudong. 2005. *Beichao cunmin de shenghuo shijie: Chaoting, zhouxian yu cunli* 北朝村民的生活世界：朝廷，州縣與村裡 *(On Hamlets in the Northern Dynasties)*. Beijing: Shangwu Yinshuguan.

Hou, Xudong. 2007. On Hamlets (Cun 村) in the Northern Dynasties. *Early Medieval China* 1: 99–141.

Hou, Xudong. 2008. Bei Wei jingnei huzu zhengce chutan: Cong 'Da Dai chijie Binzhou cishi Shangongsi bei' shuoqi 北魏境內胡族政策初探：從"大代持節豳州刺史山公寺碑"説起 (On the Barbarian Policy in the Northern Wei Dynasty). *Zhongguo Shehui Kexue* 中國社會科學 5: 168–82.

Hsu, Eileen Hsiang-Ling. 2002. Visualization Meditation and the Siwei Icon in Chinese Buddhist Sculpture. *Artibus Asiae* 1: 5–32. [CrossRef]

Hu, Wenhe. 2005. Yungang shiku mouxie ticai neirong he zaoxing fengge de yuanliu tansuo 雲岡石窟某些題材內容和造型風格的源流探索 (A survey of the themes and stylistic features of stories depicted in the Yungang Grotto). In *2005 nian Yungang guoji xueshu yantaohui lunwenji 2005年雲岡國際學術研討會論文集 (Proceedings for the 2005 International Conference on Yungang)*. Beijing: Wenwu Chubanshe, pp. 17–37.

Huang, Shaoxiong. 2015. *Daiguo Daijun tongzhi* 代國代郡通志 *(History of Dai)*. Beijing: Zhonghua shuju.

Hurvitz, Leon. 2009. *Scripture of the Lotus Blossom of the Fine Dharma (the Lotus Sutra)*. New York: Columbia University Press.

Ishimatsu, Hinako 石松日奈子. 2005. *Hokugi bukkyō zōzōshi no kenkyū* 北魏教造像史研究 *(Studies on the History of Northern Wei Buddhist Sculpture)*. Kunitachi-shi: Buryukke.

Jayne, Horace F. 1929. *The Buddhist Caves of Ching Ho Valley*, Eastern Art: Part. 1.

Junjirō, Takakusu. 1901. Tales of the Wise Man and the Fool, in Tibetan and Chinese. *Journal of the Royal Asiatic Society of Great Britain and Ireland* 3: 447–60.

Seishi, Karashima. 2017. Stūpas Described in the Chinese Translations of the Vinayas. In *Annual Report of the International Research Institute for Advanced Buddhology at Soka University for the Academic Year 2017*. Tokyo: Soka University, pp. 439–70.

Kim, Minku. 2011. The Genesis of Image Worship: Epigraphic Evidence for Early Buddhist Art in China. Ph.D. dissertation, University of California Los Angeles, Los Angeles.

Kim, Minku. 2014. Claims of Buddhist Relics in the Eastern Han Tomb Murals at Horinger: Issues in the Historiography of the Introduction of Buddhism to China. *Ars Orientalis* 44: 134–54. [CrossRef]

Ledderose, Lothar. 1980. Chinese Prototypes of the Pagoda. In *Stūpa: Its Religious, Historical and Architectural Significance*. Edited by Anna L. Dallapiccola. Wiesbaden: Steiner, pp. 238–48.

Ledderose, Lothar. 1983. The Earthly Paradise: Religious Elements in Chinese Landscape Art. In *Theories of the Arts in China*. Edited by Susan Bush and Christian F. Murch. Princeton: Princeton University Press, pp. 165–80.

Ledderose, Lothar. 2000. *Ten Thousand Things: Module and Mass Production in Chinese Art*. Princeton: Princeton University Press.

Lee, Junghee. 1993. The Origin and Development of the Pensive Bodhisattva Images of Asia. *Artibus Asiae* 53: 311–53. [CrossRef]

Lee, Sonya. 2010. *Surviving Nirvana: Death of the Buddha in Chinese Visual Culture*. Hong Kong: Hong Kong University Press.

Leidy, Denise P. 1990. The Ssu-wei Figure in Sixth Century A.D. Chinese Buddhist Sculpture. *Archives of Asian Art* 43: 21–37.

Li, Huaiyao. 1986. Bajian gudai tong shi zaoxiang 八件古代銅、石造像 (Eight pieces of bronze and stone sculptures). *Gugong Bowuyuan Yuankan* 故宮博物院院刊 4: 30–31.

Li, Jingjie. 1996. Zaoxiangbei Fo bensheng benxing gushi diaoke 造像碑佛本生本行故事雕刻 (Narrative scenes from the Buddha's life story and birth stories on figured steles). *Gugong Bowuyuan Yuankan* 故宮博物院院刊 4: 66–83.

Li, Song. 2008a. *Fojiao meishu quanji 8: Shaanxi fojiao yishu* 佛教美術全集 8：陝西佛教藝術 *(Buddhist Art Series 8: Shaanxi Buddhist Art)*. Beijing: Wenwu Chubanshe.

Li, Yuqun. 2008b. Wutaishan Nanchansi jiucang Bei Wei jingang baozuo shi ta 五台山南禪寺舊藏北魏金剛寶座石塔 (The Vajra-based Stone Pagoda Preserved in the *Nanchan* Monastery on Mount Wutai). *Wenwu* 文物 4: 82–99.

Liang, Sicheng. 1961. Zhongguo di Fojiao jianzhu 中國的佛教建築 (Chinese Buddhist Architecture). *Qinghua Daxue Xuebao* 清華大學學報 2: 51–74.

Liang, Sicheng. 1962. Mantan fota 漫谈佛塔. *Guangming Ribao* 光明日报, May 26.

Lin, Wei-Cheng. 2016. Performing Center in a Vertical Rise: Multilevel Pagodas in China's Medieval Period. *Ars Orientalis* 46: 100–35. [CrossRef]

Lingley, Kate. 2006. The Multivalent Donor: Zhang Yuanfei at Shuiyusi. *Archives of Asian Art* 56: 11–30. [CrossRef]

Lingley, Kate. 2010. Patron and Community in Eastern Wei Shanxi: The Gaomiaoshan Cave Temple *Yi*-Society. *Asia Major* 1: 127–72.

Liu, Huida. 1958. Bei Wei shiku zhong de sanfo 北魏石窟中的三佛 (Three Buddhas in Northern Wei cave-temples). *Kaogu Xuebao* 考古學報 4: 91–101.

Liu, Jinbao. 2000. Huaerna jiqi Dunhuang kaochatuan lunshu 華爾納及其敦煌考察團論述 (On Warner and his Dunhuang Expedition). *Zhongguo Bianjiang Shidi Yanjiu* 中國邊疆史地研究 1: 83–92.

Liu, Shufen. 2005. Bei Wei shiqi de Hedong Shu Xue 北魏時期的河東蜀薛 (The Xue Clan of Shu during the Northern Wei in Hedong Prefecture). In *Jiazu Yu Shehui* 家族與社會. Edited by Huang Kuanzhong and Liu Zenggui. Beijing: Zhongguo dabaike Quanshu Chubanshe, pp. 259–81.

Liu, Shufen. 2008. *Zhonggu de Fojiao yu shehui* 中古的佛教與社會 *(Medieval Chinese Buddhism and Society)*. Shanghai: Shanghai Guji Chubanshe.

Luo, Di. 2016. A Grain of Sand: *Yingzao Fashi* and the Miniaturization of Chinese Architecture. Ph.D. dissertation, University of Southern California, Los Angeles, CA, USA.

Ma, Changshou. 1985. *Beiming suo jian Qian Qin zhi Sui chu de Guanzhong buzu* 碑銘所見前秦至隋初的關中部族 *(Guanzhong Clans form the Former Qin to early Sui Dynasty in Inscription)*. Beijing: Zhonghua Shuju.

Mair, Victor H. 1993. The Linguistic and Textual Antecedents of the Sūtra of the Wise and the Foolish. In *Sino-Platonic Papers*. Philadelphia: University of Pennsylvania.

Martin, S. Rebecca, and Stephanie M. Langin-Hooper, eds. 2018. *The Tiny and the Fragmented: Miniature, Broken, or Otherwise Incomplete Objects in the Ancient World*. Oxford: Oxford University Press.

Michihata, Ryōshū. 1967. *Chūgoku Bukkyō to Shakai Fukushi Jigyō* 中國仏教と社會福祉事業 *(Chinese Buddhism and the Social Wellfare)*. Kyoto: Hōzōkan.

Miller, Tracy. 2014. Perfecting the Mountain: On the Morphology of Towering Temples in East Asia. *Journal of Chinese Architecture History* 10: 419–49.

Miu, Zhe. 2012. Chongfang louge 重訪樓閣 (Revisiting Towers and Pavilions). *Meishushi Yanjiu Jikan* 美術史研究集刊 33: 2–111.

Mizuno, Seiichi, and Toshio Nagahiro. 1951–1956. *Unkō Sekkutsu* 雲岡石窟 *(Yungang Cave-Temples)*. 16 vols. Kyoto: Kyoto Daigaku Jimbun Kagaku Kenkyūsho.

Mukai, Yūsuke. 2020. *Chūgoku Shoki Buttō no Kenkyū* 中国初期仏塔の研究 *(On Early Chinese Buddhist Pagodas)*. Kyōto: Rinsen Shoten.

Nickel, Lucas. 2002. *Return of the Buddha: The Qingzhou Discoveries*. London: Royal Academy of Arts.

Okada, Ken, and Hinako Ishimatsu. 1993. Chūgoku Nanbokuchaō jidai no nyorai zōchakui no kenkyūs. *Bijutsu Kenkyu* 356: 181–203.

Rei, Sasaguchi. 1975. The Image of the Contemplating Bodhisattva in Chinese Buddhist Sculpture. Ph.D. dissertation, Harvard University, Cambridge, MA, USA.

Rhi, Juhyung. 2006. Some Textual Parallels for Gandhāran Art: Fasting Buddhas, *Lalitavistara*, and *Karuṇāpuṇḍarīka*. *Journal of the International Association of Buddhist Studies* 1: 125–53.

Seckel, Dietrich. 1980. Stūpa Elements Surviving in East Asian Pagodas. In *Stūpa: Its Religious, Historical and Architectural Significance*. Edited by Anna L. Dallapiccola. Wiesbaden: Steiner, pp. 249–59.

Selbitschka, Armin. 2005. Miniature Tomb Figurines and Models in Pre-imperial and Early Imperial China: Origins, Development and Significance. *World Archaeology* 1: 20–44. [CrossRef]

Shanxi Sheng Kaogu Yanjiusuo 山西省考古研究所, ed. 1994. *Shanxi Kaogu Sishi Nian* 山西考古四十年 *(Forty Years of the Archaeological Research in Shanxi)*. Taiyuan: Shanxi Renmin Chubanshe.

Shi, Shuqing. 1980. Beiwei Cao Tiandu zao qianfo shita 北魏曹天度造千佛石塔 (Thousand-Buddha stone pagoda commissioned by Cao Tiandu of Northern Wei). *Wenwu* 文物 1: 68–71.

Song, Wenyu, ed. 2009. *Beishikusi Lunwenji* 北石窟寺論文集 *(Collected Papers on the North Cave Temple)*. Qingyang: Gansu Beishikusi Wenwu Baohu Yanjiusuo.

Stein, Rolf A. 1990. *The World in Miniature: Container Gardens and Dwellings in Far Eastern Religious Thought*. Translated by Phyllis Brooks. Stanford: Stanford University Press.

Steinhardt, Nancy S. 2011. The Sixth Century in East Asian Architecture. *Ars Orientalis* 41: 27–71.

Steinhardt, Nancy S. 2014. *Chinese Architecture in an Age of Turmoil, 200–600*. Honolulu: University of Hawai'i Press.

Steinhardt, Nancy S. 2019. *Chinese Architecture: A History*. Princeton: Princeton University Press.

Steward, Susan. 1993. *On Longing: Narratives of the Miniature, the Gigantic, the Souvenir, the Collection*. Durham and London: Duke University Press.

Stockhammer, Philip W. 2013. From Hybridity to Entanglement, from Essentialism to Practice in Archaeology and Culture Mixture. *Archaeological Review from Cambridge* 28: 11–28.

Strong, John S. 1983. *The Legend of King Asoka: A Study and Translation of the Aśokāvadāna*. Princeton: Princeton University Press.

Su, Bai. 1996. Luoyang diqu Beichao shiku de chubu kaocha 洛陽地區北朝石窟的初步考察 (A Preliminary Study of Cave-temples of the Northern Dynasties near Luoyang). In *Zhongguo shikusi yanjiu* 中國石窟寺研究. Beijing: Wenwu Chubanshe, pp. 153–76.

Su, Xuanshu. 2006. Dong Wei Bei Qi baotawen yanjiu 東魏北齊寶塔紋研究 (On the Treasure *Stūpa* Pattern in Eastern Wei and Northern Qi). *Yishushi yanjiu* 藝術史研究 8: 323–64.

Su, Xuanshu. 2010. Gudai Dongya zhuguo danceng fangta yanjiu 古代東亞諸國單層佛塔研究 (Study of the One-story Cubical *Stūpas* in East Asian). *Wenwu* 文物 11: 71–80.

Su, Bai. 2011. Donghan Weijin Nanbeichao Fosi buju chutan 东汉魏晋南北朝佛寺布局初探 (A Preliminary Study on the Layout of Buddhist Temples in the Eastern Han, Wei-Jin era, and Six Dynasties). In *Weijin Nanbeichao Tangsong kaogu Wengao Jicong* 魏晉南北朝唐宋考古文稿集成 *(On Archaeology of the Wei, Jin, Northern and Southern Dynasties, Tang, and Song)*. Beijing: Wenwu Chubanshe, pp. 230–47.

Sun, Ji. 1984. Zhongguo zaoqi gaoceng fota zaoxing de yuanyuan wenti 中国早期高层佛塔造型的渊源问题 (Issues on the Origin of Early Chinese Multistory Pagoda's Structure). *Zhongguo Lishi Bowuguan Guankan* 中國歷史博物館館刊 6: 41–47, 130.

Tang, Changru. 1955. Wei Jin Za hu kao 魏晉雜胡考 (Barbarians of Wei and Jin). In *Wei Jin Nan Bei Chao Shi Luncong* 魏晉南北朝史論叢. Beijing: Sanlian Shudian, pp. 403–14.

Tang, Zhongming. 2016. Zhongyuan diqu Beichao Fo ta yanjiu 中原地北朝佛塔研究 (Study on Buddhist Pagodas of the Northern Dynasties in the Central Plain). *Kaogu* 考古 11: 95–103.

Tanigawa, Michio. 1985. *Medieval Chinese Society and the Local "Community"*. Translated by Joshua A. Fogel. Berkeley: University of California Press.

Tsiang, Katherine. 2000. Miraculous Flying *Stupas* in Qingzhou Sculpture. *Orientations* 31: 45–53.

Twitchett, Denis. 1957. The Monasteries and China's Economy in Medieval Times. *BSOAS* 3: 526–49. [CrossRef]

Wang, Zonghui. 1985. Handai Lushui Hu De Zuming Yu Judi Wenti 漢代盧水胡的族名與居地問題 (On the name and territory of Lushui hu during the Han dynasty). *Xibei shidi* 西北史地 1: 25–34.

Wang, Qing. 1997. Ye lun Lushui hu yiji Yuezhi hu de juchu he zuyuan 也論盧水胡以及月氏胡的居處和族源 (On Habitat and Origin of the Lushui hu and Rouzhi hu). *Xibei Shidi* 西北史地 2: 25–30.

Wang, Eugene Y. 1999. What do Trigrams have to do with Buddhas? The Northern Liang Stupas as a Hybrid Spatial Model. *RES: Anthropology and Aesthetics* 1: 70–91. [CrossRef]

Wang, Xiaohong. 2004. Gansu sheng bowuguan cang liangjian Beichao fojiao shike 甘肅省博物館藏兩件北朝佛教石刻 (Two Pieces of Buddhist Stone Sculptures of the Northern Dynasties in Gansu Provincial Museum). *Sichou Zhi Lu* 絲綢之路 1: 31–33.

Wang, Eugene Y. 2005. *Shaping the Lotus Sutra: Buddhist Visual Culture in Medieval China*. Seattle: University of Washington Press.

Wang, Yuanlin. 2006. Bei Wei zhong xiao xing zaoxiang shi ta de xingzhi yu neirong 北魏中小型造像石塔的形製與内容 (The typology and content of images on Northern Wei stone pagodas of the medium and small sizes). In *2005 Nian Yungang Guoji Xueshu Yantaohui Lunwen Ji* 2005 年雲岡國際學術研討會論文集研究卷. Beijing: Wenwu Chubanshe, pp. 547–73.

Wang, Guixiang. 2011a. Lve lun Zhongguo gudai gaoceng jianzhu de fazhan 略論中國古代高層建築的發展 (On the Development of Multi-level Buildings in Classical China). In *Zhongguo Gudai Mugou Jianzhu Bili Yu Chidu Yanjiu* 中國古代木構建築比例與尺度研究 *(Study on the Ratio of Classical Chinese Wooden Buildings)*. Beijing: Zhongguo Jianzhu Gongye Chubanshe.

Wang, Miaozhen. 2011b. Bei Wei Cao Tiandu jiuceng shita xiyi 北魏曹天度九層石塔析疑 (On the nine-story Cao Tiandu pagoda of Northern Wei). *Heihe Xuekan* 黑河學刊 2: 79–80.

Wang, Huaiyou. 2015. Pingliang chutu Bei Wei fojiao shike zaoxiang tanxi 平涼出土北魏佛教石刻造像探析 (Northern Wei Buddhist Stone Statues Excavated at Pingliang). *Sichou Zhi Lu* 絲綢之路 4: 29–31.

Wang, Huaiyou. 2016. Gansu Huating xian chutu Beichao fojiao shike zaoxiang gongyangren zushu kao 甘肅華亭縣出土北朝佛教石刻造像供養人族屬考 (On the Ethnicity of Donors of Buddhist Stone Statues Excavated in Huating, Gansu). *Dunhuangxue jikan* 敦煌學輯刊 2: 133–145.

Wang, Jiqing, and Susan Elizabeth Mrozowski. 1990. Dunhuang shoucang de Dunhuang yu Zhongya yishupin 敦煌收藏的敦煌與中亞藝術品. *Dunhuangxue Jikan* 1: 116–28.

Warner, Longdon. 1926. *The Long Old Road in China*. Garden City: Doubleday, Page & Campany.

Wei, Wenbin. 2005. 20 shiji zao zhongqi Gansu shiku de kaocha yu yanjiu zongshu 20世紀早中期甘肅石窟的考察與研究綜述 (On the Investigationa and Study on Gansu Cave-temples in the early- and mid-20th Century). *Dunhuangxue Jikan* 敦煌學輯刊 1: 72–81.

Wei, Wenbin, and Hong Wu. 2009. *Gansu Fojiao Shiku Kaogu Lunji* 甘肅佛教石窟考古論集 *(Archaeology of the Buddhist Cave-temples of Gansu)*. Beijing: Minzu Chubanshe.

Wen, Jing, and Wenbin Wei. 2012. Gansu guancang fojiao zaoxiang diaocha yu yanjiu zhiyi 甘肅館藏佛教造像調查與研究之一 (Investigation and Research on Buddhist Statues from Gansu Collections, part 1). *Dunhuang Yanjiu* 敦煌研究 4: 34–44.

Williams, Paul. 2009. *Mahāyāna Buddhism: The Doctrinal Foundations*. New York: Routledge.

Wong, Dorothy C. 2004. *Chinese Steles: Pre-Buddhist and Buddhist Use of a Symbolic Form*. Honolulu: University of Hawaii Press.

Wu, Hung. 2010. *The Art of the Yellow Springs: Understanding Chinese Tombs*. London: Reaktion Books.

Wu, Hung. 2015. The Invisible Miniature: Framing the Soul in Chinese Art and Architecture. *Art History* 2: 286–303.

Xia, Nai. 1966. Hebei Dingxian taji shelihan zhong Bosi Sashan chao yinbi 河北定縣塔基舍利函中波斯薩珊朝銀幣 (Persian Silver Coins from the Relic Chamber Excavated from the Pagoda Foundation at County Ding, Hebei). *Kaogu* 考古 5: 267–70.

Xiao, Mo. 1989. *Dunhuang Jianzhu Yanjiu* 敦煌建築研究 *(Dunhuang Architecture)*. Beijing: Wenwu Chubanshe.

Xie, Zhicheng. 1987. Sichuan Handai huaxiangzhuan shang de fota xingxiang 四川代像上的佛塔形象 (Pagoda Image on Sichuan Pictorial Bricks of the Han Dynasty). *Sichuan Wenwu* 四川文物 4: 62–4.

Xu, Pingfang. 1994. Zhongguo sheli taji kaoshu 中国舍利塔基考述 (On Chinese Pagoda Foundations with Relics). *Chuantong Wenhua Yu Xiandaixing* 傳統文化與現代性 4: 59–74.

Xu, Yitao. 2016. Gongyuan wu zhi shisan shiji Zhongguo zhuanshi Fo ta tabi Zhuangshi leixing fenqi yanjiu 公元5至13世紀中國磚石佛塔塔壁裝飾類型分期研究 (Study on the Categorization of Decorations on Exterior of Chinese Brick and Stone Pagodas). *Gugong Bowuyuan Yuankan* 故宮博物院院刊 2: 6–15.

Yagi, Haruo. 2004. *Chūgoku Bukkyō bijutsu to Kan minzokuka: Hokugi jidai kōki o chūshin to shite* 中美術と漢民族化—北魏時代後期を中心として *(Sinicization of Chinese Buddhist art in late Northern Wei)*. Kyōto: Hōzōkan.

Yan, Wenru. 1984. *Maijishan Shiku* 麥積山石窟. Lanzhou: Gansu Renmin Chubanshe.

Yen, Chuan-Ying. 2008. *Beichao Fojiao Shike Tapian Baipin* 北朝佛教石刻拓片百品 *(Rubbings of Buddhist Stone Carvings from the Northern Dynasties)*. Taipei: Zhongyang Yanjiuyuan.

Yi, Lidu. 2017. *Yungang: Art, History, Archaeology, Liturgy*. London and New York: Routledge.

Yin, Guangming. 2000. *Bei Liang Shita Yanjiu* 北涼石塔研究 *(On Northern Liang Stone Stūpas)*. Xinzhu: Chueh-Feng Buddhist Art & Cultural Foundation.

Yungang Shiku Wenwu Baoguansuo 雲岡石窟文物保管所. 1991. *Zhongguo Shiku: Yungang Shiku* 中國石窟：雲岡石窟 *(Chinese Cave-Temples: Yungang)*. Beijing: Wenwu Chubanshe, vol. 2.

Yuxian Difangzhi Bianzuan Weiyuanhui 蔚縣地方志編纂委員會. 1995. *Yuxian Zhi* 蔚縣志 *(Records of the County Yu)*. Beijing: Zhongguo Sanxia Chubanshe.

Zhang, Baoxi, ed. 1994. *Gansu Shiku Yishu: Diaosu Bian* 甘肅石窟藝術：雕塑編 *(Gansu Cave-Temple Art: Sculpture)*. Lanzhou: Gansu Meishu Chubanshe.

Zhang, Baoxi, ed. 2000. *Gansu Fojiao Shike Zaoxiang* 甘肅佛教石刻造像 *(Gansu Buddhist Stone Statues)*. Lanzhou: Gansu renmin Meishu Chubanshe.

Zhang, Mingyuan. 2005. *Shanxi Shike Zaoxiang Yishu Jicui* 山西石刻造像藝術集萃 *(Selected Stone Sculptures from Shanxi)*. Taiyuan: Shanxi Kexue Jishu Chubanshe.

Zhao, Yongfu. 1986. Guanyu Lushui hu zu de zuyuan ji qianyi 關於盧水胡的族源及遷移 (On the Origin and Migration of the Lushui hu). *Xibei Shidi* 西北史地 4: 43–53.

Zhao, Jinchao. 2020. Reconfiguring the Buddha's Realm: Pictorial Programs on Fifth-century Chinese Miniature Pagodas. In *Tradition, Transmission, and Transformation: Perspectives on East Asian Buddhist Art*. Edited by Dorothy C. Wong. Delaware: Vernon Press.

Zheng, Binglin, Yingchun Wei, and Qingshan Zhao, eds. 2014. *Longdong Hexi Shiku Yanjiu Wenji* 隴東河西石窟研究文集 *(Collected Papers on the Hexi Cave-Temples of Eastern Gansu)*. 2 vols. Lanzhou: Gansu Wenhua Chubanshe.

Zhongguo Shehui Kexueyuan Kaogu Yanjiusuo 中国社会科学院考古研究所. 1996. *Bei Wei Luoyang Yongningsi: 1979–1994 nian kaogu fajue baogao* 北魏洛阳永宁寺: 1979–1994年考古发掘报告 *(Northern Wei Yong-ning Monastery in Luoyang: A Report on Archaeological Excavations from 1979 to 1994)*. Beijing: Zhongguo Dabaike Quanshu Chubanshe.

Zhongguo Shehui Kexueyuan Kaogu Yanjiusuo 中国社会科学院考古研究所, and Hebei Sheng Wenwu Yanjiusuo Yecheng Kaogudui 河北省文物研究所鄴城考古隊. 2010. Hebei Linzhangxian Yecheng yizhi Zhaopengcheng beichao fosi yizhi de kantan yu fajue 河北臨漳縣鄴城遺址趙彭城北朝佛寺遺址的勘探與發掘 (Excavations at the Monastic Sites of the Northern Dynasties at Zhanpengcheng site, Yecheng, in present-day Linzhang, Hebei). *Kaogu* 考古 7: 31–42.

Zhongguo Shehui Kexueyuan Kaogu Yanjiusuo 中国社会科学院考古研究所, and Hebei Sheng Wenwu Yanjiusuo 河北省文物研究所. 2013. Hebei Yecheng yizhi Zhaopengcheng Beichao fosi yu Beiwuzhuang fojiao zaoxiang maicangkeng 河北鄴城遺址趙彭城北朝佛寺與北吳莊佛教造像埋藏坑 (The Zhaopengcheng Buddhist Temple and the Beiwuzhuang Buddhist Sculptures Hoarding Pit from the site of Yecheng, Hebei). *Kaogu* 考古 7: 49–68.

Zhou, Yiliang. 1963. Beichao de minzu wenti yu minzu zhengce 北朝的民族問題與民族政策 (Ethnicity and Regulations of the Northern Dynasties). In *Wei Jin Nan Beichao Shi Lunji* 魏晉南北朝史論集. Beijing: Zhonghua Shuju, pp. 117–76.

Zürcher, Erik. 1972. *The Buddhist Conquest of China: The Spread and Adaptation of Buddhism in Early Medieval China*. Leiden: Brill.

Article

Struggle on the Axis: The Advance and Retreat of Buddhist Influences in the Political Axis of Capitals in Medieval China (220–907)

Yifeng Xie

Department of History, Yuelu Academy, Hunan University, Changsha 410082, China; xieyifeng@hnu.edu.cn

Abstract: Buddhist influences on the sacred axis of the capital during Medieval China (220–907) underwent a process of starting with little impact during the era of Eastern Han, Caowei, and Western Jin (220–317) to a more prominent influence from the late Southern and Northern Dynasties (386–589) to early Tang (618–907), peaked during the reign of Wu Zetian (690–705), and roughly returned to the layout patterns from the late Southern and Northern Dynasties to early Tang after the death of Wu Zetian. As maintained below, the process appears complex in terms of the interaction between Buddhism and political space throughout early Medieval China. There are roughly two modes of integration and interaction between Buddhist buildings and ritual buildings with Buddhist influences and the political axis of the capital: the first mode can be regarded as a typical mode after its establishment in the late Northern Wei Dynasty. This mode exhibits major Buddhist influences, particularly regarding the huge scale of monasteries and pagodas, and the location of high-rise pagodas as landmarks flanking the political axis of the capital. The second mode should be regarded as an atypical mode occurring during the late period of Emperor Wu of the Liang (464–549, r. 502–549), the period of Northern Qi (550–577), and the reign of Wu Zetian. At this point, Buddhist buildings and imperial ritual buildings with Buddhist characteristics and symbolic meanings were placed directly on the political axis of the capital, close to or located at the core of the palace. This practice was a sign that the influence of Buddhism in the political culture and ideology of the entire empire during these eras of Emperor Wu of the Liang, the Northern Qi, and the reign of Wu Zetian had reached their culmination. Architecture reflected the most intuitive embodiment of an external visual form in presenting the most symbolic image of power. With the decline of political enthusiasm for advocating Buddhism, Buddhist and related buildings no longer occupied the political axis of the capital. Various forces majeure such as natural fires, demolition, and reconstruction by subsequent rulers also led to the demise of Buddhist influence on the pollical axis of capital architecture in subsequent eras.

Keywords: capitals; the political axis; Buddhist influences; Emperor Wu of the Liang; Wu Zetian

Citation: Xie, Yifeng. 2021. Struggle on the Axis: The Advance and Retreat of Buddhist Influences in the Political Axis of Capitals in Medieval China (220–907). *Religions* 12: 984. https://doi.org/10.3390/rel12110984

Academic Editors: Shuishan Yu and Aibin Yan

Received: 26 August 2021
Accepted: 4 November 2021
Published: 10 November 2021

1. Introduction

The spatial presentation and interaction of political authority and religious forces has always been a significant topic in the layout of capitals during Medieval China (220–907). The period started from the end of the Eastern Han (25–220) to a stage during what is labeled a transition during the Tang–Song periods, as defined by Naitō Konan and Miyazaki Ichisada (Liu 1992, pp. 10–18, 153–241). In terms of the political traditions of imperial China, political status is not only related to bureaucratic hierarchy, but also to the spatial distance from the center of absolute power. According to Liang Sicheng (1901–1972) "palaces, government offices, monasteries and even residences composed by multiple buildings are usually arranged in an absolutely neat and symmetrical layout [. . .] The most important thing is the establishment of a main symmetry axis" (Liang 2011, p. 8). During the Qin and Han Dynasties (221 BCE–220 CE), the spatial layout of the capital was organized primarily in a multi-palace system and by locating a single palace structure high on a platform to suggest the magnificence and power. With the construction of capitals at Yecheng and Luoyang, the emperors of Caowei (220–265) adopted the single-palace system and the central axis location of the capital by connecting the most important political and ritual spaces to a whole. The Altar of Heaven and the Taiji Palace, which symbolizes the highest imperial power, occupied the north and south ends of this power axis in Luoyang of Caowei (Sagawa 2016, pp. 107–32).

The political axis of this capital was composed of the most important imperial power and ritual buildings, such as the imperial palace, palace gates, imperial avenue, city gates, and the Altar of Heaven. The only building located in front of the imperial palace main hall was the Altar of Heaven, the most important ritual building involving sacrifices to heaven. The ideal capital plan of Medieval China was the symmetrical structure, the central axis of power. All kinds of government offices, and even the Ancestral Shrine and *Sheji* (the Altar of Land and Grain) sat symmetrically on both sides of this political axis. It should be noted that this political axis was sometimes not located in the absolute center of the entire capital city (such as Luoyang in the Sui and Tang Dynasties). Therefore the more precise term "political axis" is used rather than "central axis."

In studying the axis of the existing capitals, most scholars in the past have focused on Beijing during the Yuan, Ming, and Qing Dynasties (1271–1911). In recent years, Chen Jing traced the origin of the development of the "central axis" of the ancient capital (Chen 2020). Zeng Lei focused his viewpoint on the spatial expression of political ideals represented by gates, axes, and avenues in Qin and Han (Zeng 2020). Scholars such as Chen Jianjun and Wang Tao paid special attention to the development of the capital axis in Luoyang from Wei to Tang (220–907) (Chen et al. 2020; Wang et al. 2015). Other research on Buddhist architecture and political influences includes studies by Liang (1998), Sickman and Soper (1956), Fu (1998, 2017), and Steinhardt (1997; 2004, pp. 228–54; 2019). Yet few scholars have systematically studied the connection and interaction between Buddhist architecture in connection with the capital's political axis.

The earliest Buddhist monastery in Chinese history was established in Luoyang, the capital of the Eastern Han Dynasty (25–220). Based on limited available historical records, Buddhist monasteries in Eastern Han and Jin were mostly located far away from this axis of power, even outside the core area of the capital. For instance, the world-famous Baima (White Horse) Monastery in Luoyang was located in the western suburbs of Luoyang in Han and Jin, not in the heart of the city. The location of the forty-two monasteries in Luoyang during the Western Jin Dynasty, as recorded in the preface of *Luoyang Qielan Ji*, is very unclear (Yang 2018, p. 1). These monasteries, however, were not so-called imperial monasteries, and thus they were less likely to be located on both sides of the political axis to demonstrate imperial power.

Although Buddhism had been favored by some rulers, its intersection with political power was still limited. Was Buddhism to challenge the political and ritual landscape of capitals in the traditional capital plan? Important turning points in the advance and retreat of the Buddhist influences in the sacred political axis of ancient capitals will be analyzed below.

2. The Initial Establishment of the Relation between Buddhist Monasteries and the Political Axis of Capitals

The clarion call for Buddhist monasteries to march on the political axis was sounded during the Northern Wei period (386–534). After the capital, Pingcheng (nowadays Datong), was established in the early Northern Wei, the relation between Buddhism and imperial power entered a new stage. The rulers of Northern Wei first developed the political philosophical concept that "the emperor is the Buddha," which firmly tied politics and religion together (Luo 2021, p. 244). The rulers built large Buddhist monasteries (pagodas), recruited monks, and built grottoes (such as Tanyao Five Caves in Yungang Grottos) in Pingcheng in the name of the imperial family, and brought Buddhism under the supervision of political power. Different from the White Horse Monastery outside Luoyang of the Eastern Han and the 42 monasteries in Luoyang during the Yongjia period (307–311) of the

Western Jin, the pagodas and Buddhist monasteries in Pingcheng had become an organic component of the capital landscape.

When Emperor Daowu (371–409, r. 386–409) established the capital in Pingcheng, he "started to make the five-story Buddha pagoda, the Grdhrakūta Mountain and the Hall of Miru Mountain, and decorated them. Additionally, he built lecture halls, meditation halls and seats of monks. These things were prepared very well" (Wei 2017, p. 3292). During the reign of Emperor Taiwu (408–452, r. 424–451), Buddhism in Northern Wei was severely destroyed, but it did not totally fail. After Emperor Xianwen (454–476, r. 465–471) ascended to the throne in the sixth year of the Heping period (465), he re-supported Buddhism, and successively established important Buddhist buildings such as Yongning Monastery, Tiangong Monastery, and the three-story stone pagoda in Pingcheng (Wei 2017, p. 3300). Although archaeologists have not yet clarified the specific location of the seven-story Yongning Monastery Pagoda or carried out archaeological excavations on this building, according to the description of existing documents, the height of the pagoda of Yongning Monastery in Pingcheng reached more than 300 *chi* (if the total length of 1 *chi* is 27.868 cm, it should be above 83.7 m), totaling seven stories (Wei 2017, p. 3300). The total height and volume of this multi-story pagoda was a record for this time. It was a civil-wood mixed structure containing a huge core in its center. Archaeological excavations have revealed the location of Siyuan Pagoda in the Yonggu Mausoleum on the mountain in the northeast of Pingcheng (Datongshi Bowuguan 2007, pp. 4–26). Yet the specific location of the most important Buddhist architectural landscapes in Pingcheng, such as the Yongning Monastery Pagoda, are still unclear and thus the relations between such Buddhist monasteries and pagodas regarding the axis of the capital is also unclear. An exception may be the continued use of the title of Yongning Monastery Pagoda in Luoyang (Yang 2018, p. 11), the capital of the late period of the Northern Wei (494–534). It is most likely that the distribution of large Buddhist monasteries and pagodas in Pingcheng was no longer in remote suburbs or scattered in the city, but had become an important part of the organic development of the city as a whole.

That the locations of Buddhist monasteries, especially the imperial monasteries, were close to the central axis of power in the capital in the late Northern Wei is demonstrated by the arrangement of Yongning Monastery, Jingming Monastery, and Qin Taishang Gong Double Monasteries in Luoyang, flanking the sides of Bronze Camel (*Tongtuo*) Avenue, the central axis of power in the capital of the Northern Wei.[1] The five-story double towers of Qin Taishang Gong Monastery, seven-story Jingming Monastery Pagoda, and nine-story Yongning Monastery Pagoda from south to north formed one arithmetically balanced sequence of stories of Buddhist buildings.[2] The arithmetically balanced sequence gradually increased as the distance decreased between these pagodas and Taiji Hall, the center of absolute power. In terms of the south-to-north spatial layout of Luoyang in the Northern Wei, it followed the following distribution: (1) *Siyi Li*, *Siyi Fang* (the area where foreign envoys and immigrants reside), and the *Sitong* Market to the south of the Luo River; (2) a regular block (*Lifang*) district, a large area of residents in the south of Xuanyang Gate; (3) the Ancestral Shrine, the Altar of Land and Grain, and the districts of official buildings on both sides of *Tongtuo* Street, the axis of the previous city of Han and Jin (the so-called inner city); and (4) the palace in the north part of the inner city. Luoyang in the Northern Wei showed a unidirectional layered tandem structure on the axis from the palace in the north to the Altar of Heaven in the south. With the exception of the Altar of Heaven in the south end of this axis, the distribution unfolded from barbarians to Chinese, from the outside to the inside, and from low status to high status. However, the monasteries established by Buddhism from the Western Regions (*xiyu*) were not totally located in *Siyi Fang* and *Siyi Li* to the south of the Luo River because of their foreign religious status, but were located in the *Lifang* district and the official institutes district to the north of the Luo River. In other words, compared to the White Horse Monastery located in the western suburbs of the capital in the Eastern Han, the grand monasteries with imperial backgrounds such as Yongning Monastery and Jingming Monastery at this time had already approached the inner area of the palace to symbolize the highest imperial power. Yongning Monastery was the only imperial Buddhist monastery in the city that was planned according to the "Capital Planning" of Emperor Xiaowen (467–499, r. 471–499), who moved to Luoyang (Wei 2017, p. 3306). Jingming Monastery was built by his son, Emperor Xuanwu (483–515, r. 499–515), with a slightly lower status in being located farther away from the palace (Yang 2018, p. 133). In addition, the Qin Taishang Gong Double Monasteries built by Empress Ling (?–528) and her sister Hu Xuanhui (?–556)[3] for their father, Hu Guozhen (438–518), were located in the south of Jingming Monastery and north of the Luo River, farther away from the palace.

Luoyang's Buddhist architectural landscape in the late period of the Northern Wei created a new pattern completely different from the Eastern Han, Wei, and Jin periods both in terms of its relation with the political axis and its interaction with the power center. With the construction of remarkably high pagodas and grand imperial monasteries (such as Jingming Monastery and Yongning Monastery) occupying one *fang* (the standard block in the capital of the Northern Wei) and a half *fang* on the both sides of the central axis of the capital as new power monuments, the urban landscape of Luoyang had undergone fundamental changes. The importance of Buddhist architecture in the construction of the political landscape of the capital and the display of power had been unprecedentedly strengthened, dwarfing the declining ritual buildings in the southern suburbs.

A detailed evolution of the spatial distribution of Luoyang monasteries in the late Northern Wei Dynasty is provided in four stages, according to the research results of Hu Manli's M. A. dissertation, "Buddhist Monasteries and Urban Space in Luoyang of Northern Wei (495–534): based on *Luoyang Qielan Ji*"[4]:

The first stage was from the eighteenth year of the Taihe period of Emperor Xiaowen (494) to the first year of the Zhengshi period of Emperor Xuanwu (504). There were 12 Buddhist monasteries built by the records of *Luoyang Qielan Ji*, two in the inner city and two in the east area of the inner city, five in the south area of the inner city, three in the west area of the inner city, and none in the north. There were relatively more Buddhist monasteries built in the south of the city. Yongning Monastery on the west side of the central axis of the Official Institutes District and Jingming Monastery on the east side of the central axis of *Lifang* District, located in the south area of the inner city, had been built at this time, but the pagodas of these two monasteries had not yet been built. The three monasteries of Baode, Longhua, and Zhuisheng, built in *Quanxue Li* in the south area of the inner city, had approached or even invaded the traditional space of Confucian ritual architecture, reflecting the deep involvement of Buddhist architecture in the political core area of the capital. In terms of pagoda construction, there were three pagodas with three or more stories built in Luoyang at that time by the records in *Luoyang Qielan Ji*, one each in the inner city, the east area of the inner city, and the west area of the inner city. The tallest pagoda in Yaoguang Monastery[5] was located to the northwest of the inner city, close to Jinyong Fortresses. Judging from the location of the commanding heights of the capital, according to the information provided by *Shuijing Zhu, Taiping Yulan, Henan Zhi*, and other documents, the inherent commanding heights of Luoyang in the Han and Jin Dynasties should be located in the Jinyong Fortresses and *Baichilou* (a one-hundred-*chi*-high pavilion)[6] area, located to the northwest of the inner city. The construction of the five-story pagoda of Yaoguang Monastery at the beginning of Luoyang as the new capital of the Northern Wei did not change this basic pattern. Therefore, in the early years of the reign of Emperor Xuanwu, the grand Buddhist monasteries in the Luoyang of the Northern Wei (such as Yongning Monastery and Jingming Monastery) had been lined up on the both sides of the political axis. However, because the pagodas of Yongning Monastery and Jingming Monastery were not built, in the three-dimensional space, the remarkably high pagodas still had not much influence or disturbance on the political and cultural landscape along the central axis of the capital. The progressive city gates and palace gates were still the main buildings of this axis.

The second stage was the ten years from the first year of the Zhengshi period to the third year of the Yanchang period (514), mainly during the reign of Emperor Xuanwu. There was no significant change in the Buddhist landscape near Luoyang's political axis. Yongning Monastery and Jingming Monastery were still the remarkable cores of the Buddhist landscape in Luoyang. The commanding height of the city was still the five-story pagoda of Yaoguang Monastery to the south of Jinyong Fortresses and to the northwest of the inner city. Compared with the first ten years after moving the capital to Luoyang, there was no significant transformation. With the development of the outer city area, a series of emerging Buddhist monasteries, such as Pingdeng Monastery, Zhengshi Monastery, Yongming Monastery, and Dajue Monastery, had indeed appeared in the east and west of the city. However, these monasteries were far away from the political axis of Luoyang and did not remarkably disturb it.

The third stage was from the third year of the Yanchang period to the fifth year of the Zhengguang period (524) of Emperor Xiaoming (510–528, r. 515–528), mainly during the reign of Empress Ling. This period was a stage of vigorous development for Buddhist monasteries in Luoyang, and it was also a peak for the construction of high and super-high pagodas. According to my statistics, there were only two large-scale pagodas with nine or seven stories in Luoyang, the new capital of the Northern Wei, all of which were built during the reign of Empress Ling. Among the eight five-story medium-sized pagodas, there were also two pagodas in Qin Taishang Jun Monastery and Qin Taishang Gong West Monastery built by Empress Ling, and Hutong Monastery and Qin Taishang Gong East Monastery were built by her aunt and younger sister. These four monasteries already accounted for 50% of the total number of monasteries with five-story pagodas. If Chongjue Monastery and Rongjue Monastery, which were roughly built in the same period or later, related to Yuan Yi (487–520), were added to this list, then the number of monasteries with five-story pagodas actually accounted for three-quarters of the total at that time. Only the pagoda of Yaoguang Monastery (Yang 2018, p. 48), built in the time of Emperor Xuanwu, and the five-story pagoda built in Pingdeng Monastery (Yang 2018, p. 109)[7] after Emperor Xiaowu (Yuan Xiu, 510–535, r. 532–535) ascended the throne, were not established in the period of Empress Ling. Additionally, I prefer not to analyze the three-story pagodas recorded in *Luoyang Qielan Ji* because of their small volume and the difficulty in determining their construction date clearly. In summary, most of the large and medium-sized pagodas with more than five stories in Luoyang in the Northern Wei were built during the period of Empress Ling (see Table 1).[8]

Table 1. Statistics of the monasteries with pagodas above three stories in Luoyang of the Northern Wei.

Number of Stories of Pagodas	Title of Monasteries	Quantity
9	Yongning Monastery	1
7	Jingming Monastery	1
5	Yaoguang Monastery, Hutong Monastery, Qin Taishang Jun Monastery, Pingdeng Monastery, Qin Taishang Gong West Monastery, Qin Taishang Gong East Monastery, Chongjue Monastery, and Rongjue Monastery	8
3	Changjiu Monastery, Mingxuan Nuns Monastery, Lingying (Taikang) Monastery, Wangdianyu Monastery, and Baoguang Monastery	5

The distribution of Buddhist monasteries in the two-dimensional space made the east side of the political axis from the north bank of the Luo River to Xuanyang Gate almost entirely occupied by Buddhist monasteries. This area was originally the location of "Three

Yong" buildings (*Lingtai, Mingtang* (The Bright Hall) and *Piyong*), traditional Confucian ritual buildings in the southern suburbs during the Eastern Han Dynasty. At that time, the *Lingtai* and *Piyong* had long been abandoned, and the *Mingtang* was rebuilt after the coup by Yuan Cha (Wei 2017, p. 283). The decline of the ritual architecture area in the southern suburb of the capital at that time was obvious, and could not compete with the architecture of increasingly glorious imperial Buddhist monasteries. Looking north from the south bank of the Luo River, the visual landscape that greeted the viewer was no longer ritual architecture but the magnificent three pagodas of Qin Taishang Gong Double Monasteries and Jingming Monastery in a triangular arrangement (Xie). The latter had become a new visual center in the southern area of the capital and even on both sides of the central city's political axis.

The fourth stage dated from the fifth year of the Zhengguang period (524) to the third year of the Yongxi period (534), which was the year when Yongning Monastery was burned and of the division of the Eastern and Western Wei. Additionally, it was also the final stage of Luoyang as the capital of the Northern Wei. Monasteries inside and outside the inner city continued to grow, particularly after the massacre in 528, and as more nobles died, tributary monasteries were built. However, the new Wei court seemed unable to build giant, super-high pagodas, such as the pagodas of Yongning Monastery and Jingming Monastery. Buddhist monasteries were no longer part of a unified city plan (see Figure 1).

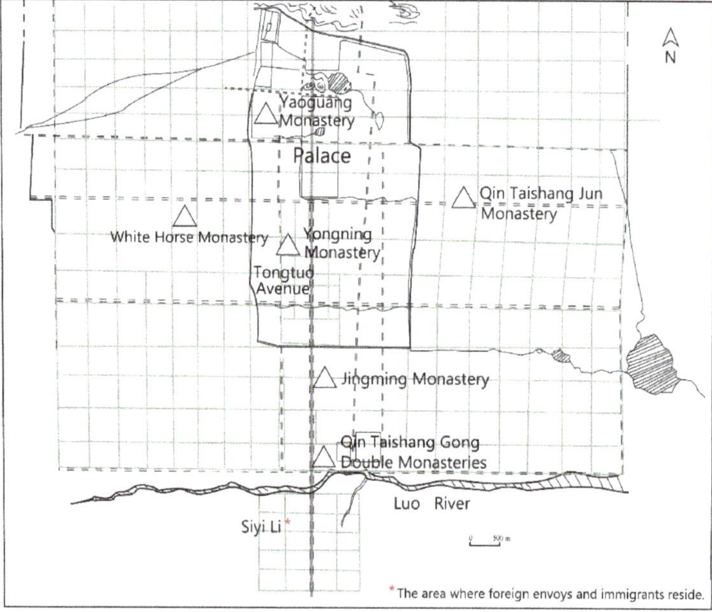

Figure 1. Schematic diagram of the locations of the main Buddhist monasteries in Luoyang in the Northern Wei.[9]

Nonetheless, changes in the number of Buddhist monasteries were minimal. With the division of the Eastern Wei (534–550) and Western Wei (535–556) and the transfer of the capital of the empire to Yecheng and Chang'an in 534, I use Yecheng, the capital of the Eastern Wei and Northern Qi (550–577), to exemplify changes in the dynamics between political expediency and Buddhist influences. In recent years, the site of Zhaopengcheng Buddhist Monastery and Da Zhuangyan Monastery was unearthed by the Yecheng Team of the Institute of Archaeology of the Chinese Academy of Social Sciences. The two were located on the east side of Zhuming Gate Avenue along the central axis of Yecheng.[10]

The relative position of the Zhaopengcheng Buddhist Monastery was almost exactly the same as that of Jingming Monastery in Luoyang during the Northern Wei. The pagodas of Zhaopengcheng Monastery (Da Zongchi Monastery)[11] and Da Zhuangyan Monastery may have had seven stories, not reaching the height of the nine-story Yongning Monastery pagoda, but important landmarks in the three-dimensional landscape of Yecheng. The relative position and landscape continued the construction logic and visual effects of the Buddhist pagodas of Yongning Monastery and Jingming Monastery in Luoyang of the Northern Wei (see Figure 2).

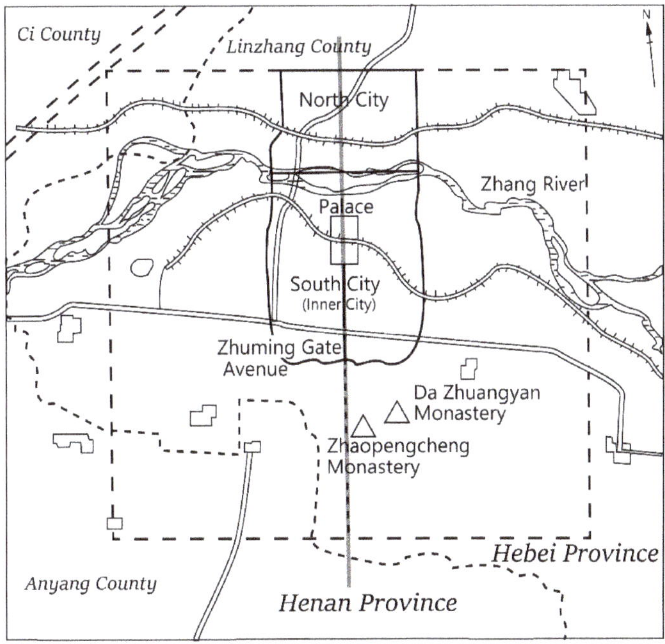

Figure 2. Schematic diagram of the locations of the main imperial Buddhist monasteries in Ye.[12]

An excellent case of Buddhist influences occupying the axis of the capital concerns the performance activity on the eighth day of the fourth lunar month (the birthday of Buddha in Chinese tradition). A few days before the birthday, Buddha statues from the entire city left their monasteries and gathered on the square of Jingming Monastery in the south of Luoyang, surrounded by guards of honors and believers. On that special day, the parade line of statues departed from Jingming Monastery and went north until they received the emperor's flowers and respect in front of Changhe Gate, the main gate of the palace (Yang 2018, pp. 133–34). During this period, the division caused by the well-ordered *Lifang* system of the capital in Medieval China was briefly broken, as Buddhist influences maintained the most important and political symbolic axis from Taiji Hall, Changhe Gate to Xuanyang Gate, on this special day. The political axis of the capital profoundly reflected the integration of Buddhism with imperial power. Yet there were limitations: Firstly, there was no permanent Buddhist architecture on the central axis, and the emergence of Buddhist influences (the parade of Buddhist statues) on it was still temporary. Secondly, the parade of Buddhist statues stopped in front of Changhe Gate, without entering the core area of the imperial palace.

As represented by Luoyang, the capital of the late period of the Northern Wei, Buddhist monasteries and pagodas, whether in two or three dimensions, effectively interacted with the political axis of the capital in an orderly distribution on both sides of it, showing a trend of continuous growth in influence on the capital landscape. In the performance activities on the eighth day of the fourth lunar month, Buddhist influence even temporarily occupied the political axis of the capital itself.

3. The Development of Buddhist Architecture in the Capital of the Southern Dynasties (420–589) and the Breakthrough of Tongtai Monastery

Buddhism during the Southern Dynasties integrated with the political culture of the capital. Due to the large number of monasteries in the capital, Jiankang (nowadays Nanjing), the title of "four hundred and eighty monasteries of the Southern Dynasties" was coined. According to the historical records, Jiankang City during the Eastern Jin and the Southern Dynasties was roughly composed of three areas: the palace, the inner capital city, and the outer city. From south to north, the southern part of the city consumed a broad space and was composed of a southern suburban ritual architecture area, a civilian residential area, commercial districts, and residential areas of princes and nobles on both sides of the Huai River (i.e., the Qinhuai River); to the north were the district of the official institutes, the palace, the Hualin Imperial Garden, and other imperial gardens. The important monasteries during the Eastern Jin, Changgan Monastery (originally Jianchu Monastery) and Waguan Monastery, were arranged on the east and west sides of the capital city axis, namely, the *Changgan Li* area in the broad sense, defined by Xu Zhiqiang (Seo 2019, Figure 75; Zhang 2021, pp. 340–54). Changgan Monastery had an especially sacred status due to the bone relics of the Buddha unearthed there and thus was recognized as one of the most significant monasteries of Jiankang Buddhism in the Eastern Jin to the early Southern Dynasties periods. This was also where the Ashoka Pagoda was located.[13] The relative relation between its location and the axis of the capital city was also similar to the relation between Jingming Monastery and the political axis of Luoyang during the Northern Wei. However, because the city planning of Jiankang from Eastern Jin to the Southern Dynasties was not as meticulous as Luoyang in the late period of the Northern Wei, it is difficult to make clear whether the plan's relation with the central axis concept was accidental or the result of prior planning.

The imperial monastery in Jiankang during the Southern Dynasties was Da Zhuangyan Monastery, built by Liu Jun, Emperor Xiaowu of Song (430–464, r. 453–464) (Xiao 2017, p. 1010). The title of this monastery was also witnessed in Ye of Northern Qi and Chang'an in Sui and Tang (581–907). The prefix *da* ("great") was added to the monastery title in demonstrating its important and prominent status. During Emperor Xiaowu's reign, the court of Song had already built a seven-story pagoda as the highest pagoda in Jiankang, earlier than the huge pagoda of Yongning Monastery with the same number of stories. In addition, its geographical location was not far from the political axis of Jiankang. It was located on the west side of the Ancestral Shrine on the north bank of the Huai River, and was adjacent to the district of official institutes on the north side.[14] Judging from this location, to compare with the locations of Changgan and Waguan monasteries in *Changgan Li* to the south, the location of the Da Zhuangyan Monastery was closer to the core area of imperial power and the political axis of the capital (see Figure 3).

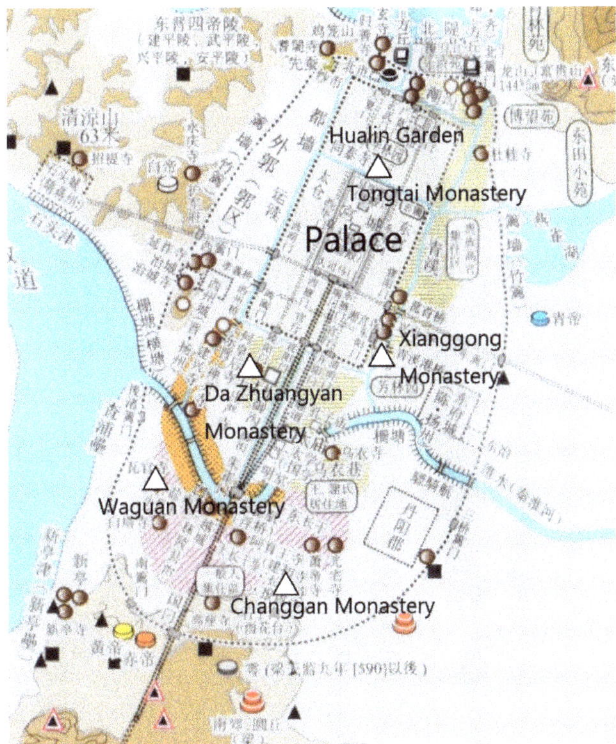

Figure 3. Schematic diagram of the main monasteries in Jiankang during the Southern Dynasties.[15]

Later, Liu Yu, Emperor Ming of Song (439–472, r. 466–472), who seized victory in the struggle for court power, desired to surpass his half-brother by building a ten-story super-high pagoda in Xianggong Monastery, his former residence on the east side of the palace in Jiankang. However, technical limitations prevented his grand plan from being fully realized. In the end, he could only take the expedient method to build two five-story pagodas to compose a conceptual ten-story pagoda. In the process of building the pagoda, Lu Yuan once bluntly persuaded him to stop the project. However, in Liu Yu's view, the construction of the pagoda had great merits and its own important meanings and functions. It was not a "face-saving project" that wasted labor and money (Xiao 2017, p. 1010; Li 1975, p. 1710). Because the location of the pagoda was the old residence of Emperor Ming of the Song Dynasty, it was located to the east of the palace, and a certain distance from the political axis of the capital was maintained.

During the Liang Dynasty (502–557), Xiao Yan, Emperor Wu of the Liang (464–549, r. 502–549), named the "Bodhisattva Emperor," greatly admired Buddhism. According to *Liangshu* and *Xu Gaoseng Zhuan*, Emperor Wu once built the famous Da'aijing Monastery in Mount Zhong, in the northeastern suburb of Jiankang (Yao 1973, p. 159; Daoxuan 2014, p. 9). According to *Jiankang Shilu*, the Da'aijing Monastery was built in the first year of the Putong period (520). Based on a different text, the "*Da'aijing Si Chaxia Ming* (Inscription under the spire of Pagoda of Da'aijing Monastery)," written by Xiao Gang, Emperor Jianwen (503–551, r. 549–551), a seven-story pagoda was built in Da'aijing Monastery in the third year of the Putong period (522) (Li 1966, p. 4149a). Although the Da'aijing Monastery was built by Emperor Wu with a seven-story pagoda, and Emperor Jianwen himself wrote the inscription under the spire of its pagoda, it was located on Mount Zhong (now Zijing Mountain in Nanjing) to the northeast of Jiankang. There was a considerable distance between this monastery and the political axis of the capital and the palace. Thus, by the middle period of Emperor Wu, distance between the Buddhist architecture of the Southern Dynasties and the political axis of the capital still existed (e.g., Da Zhuangyan Monastery, Changgan Monastery, etc.).

In the first year of the Datong period (527), after completing the construction of the seven-story pagoda of Da'aijing Monastery in Mount Zhong, Emperor Wu also started construction on Tongtai Monastery in Hualin Garden to the north of the palace in Jiankang. According to the "Biography of Shi Baochang" in *Xu Gaoseng Zhuan*,

> "In the first year of the Datong period, the Datong Gate was opened in the north of *Taicheng* (in the palace), and Tongtai Monastery was established. The pavilions and halls in it were established to follow the imperial standard of palace; and the nine-story pagoda was remarkably high, [sic, as if] to float on the cloud. There were many hills, trees, gardens and ponds located in it. On the sixth day of third lunar month, the emperor came to this monastery to tribute the Buddha ritually. This had become a routine and standard" (Daoxuan 2014, p. 10).

Tongtai Monastery had a competitive relationship with Yongning Monastery in *Weishu* and *Luoyang Qielan Ji* on three levels (Wei 2017, p. 3306; Yang 2018, pp. 11–13). Firstly, the number of stories of the Tongtai Monastery pagoda was also nine, which was the same as the pagoda of Yongning Monastery in Luoyang; secondly, the pavilions and halls followed the imperial standard of the palace; and thirdly, Emperor Wu, as Empress Ling and Emperor Xiaoming did in the Northern Wei, visited Tongtai Monastery after its completion. According to *Nanshi* and *Jiankang Shilu*, the title of Tongtai Monastery was the inverted word "Datong" in Chinese (Li 1975, p. 205; Xu 1986, p. 477); thus, it was directly related to the new reigning title of Emperor Wu and had distinctly political attributes. In the following two decades,[16] Tongtai Monastery became famous for the four times of ordination rituals of Emperor Wu of the Liang, and became the most important imperial monastery in the Liang Dynasty. Every time Emperor Wu of the Liang went to Tongtai Monastery to give up his secular life, he was accompanied not only by large-scale Buddhist activities such as Pañcavārṣika Assemblies (Chen 2006, pp. 43–103), but also by steps for amnesty and a change in a new reigning title.

The Tongtai Monastery was in the Hualin Garden adjacent to the Datong Gate in the south, which was the north gate of the palace of the Liang. Although it has not been clearly excavated, based on historical records and the hypothesis of Seo Tatsuhiko and others, Tongtai Monastery was located on the political axis of Jiankang, the capital of the Liang (Seo 2019, Figure 75). This pagoda was similar in number of stories to the pagoda of Yongning Monastery in the Northern Wei, yet it was located closer to the core of power of the capital. Although the location of Yongning Monastery had moved to the area of institutions next to the palace, it was still located on the side of the political axis of Luoyang in the Northern Wei; Tongtai Monastery had completely and permanently occupied the northern part of the political axis of Jiankang, the capital of the Liang, located in the imperial garden behind the palace, as the representative and symbolic axis of power-sharing with Taiji Hall, the highest power symbol of imperial authority. From the perspective of the capital landscape, the pagoda of Tongtai Monastery was like a "background version" of the palace. Its huge volume and shocking height reflected supremacy amidst imperial monasteries, on the same power axis landscape as Taiji Hall in front of it.

On one night in 546, as Emperor Wu explained Buddhist sutras and held Buddhist rituals at Tongtai Monastery, a fire broke out, and the pagoda of Tongtai Monastery burned to the ground (Yao 1973, p. 90). Nonetheless, Emperor Wu did not abandon his grand ambition to build a higher and larger landmark building on the most important political axis. The construction plan for a twelve-story pagoda was put on the agenda. This remarkably high pagoda was not completed in time, ending because of Hou Jing's rebellion and Emperor Wu's tragic palace confinement (Xu 1986, p. 478). For Emperor Wu, the positions of Tongtai Monastery and its pagoda were comparable to Taiji Hall in symbolizing supreme power. They served as permanent fixtures on the political axis of the capital and as one of the most important political landscapes.

This model may also have affected the construction and site selection of *Queli* (Cakra) Buddhist Monastery in Ye, the capital of the Northern Qi. The word *Queli* is related to the giant skyscraper built during the Kaniṣka period of the Kushan Empire (127–150), the Kaniṣka Stupa, which was the so-called "Queli Futu," as recorded in the biography of Faxian and the biographies of Song Yun and Daorong (and quoted in the *Luoyang Qielan Ji* and *Datang Xiyu Ji*; Faxian 2008, p. 33; Yang 2018, pp. 228–29; Xuanzang and Bianji 2000, pp. 60–61). In September 1908 and November 1910, Dr. D. B. Spooner led a team to excavate the ruins of Cakri Stupa in Shaqikiteri, outside Peshawar (Sun and He 2018, p. 433). According to Sun Yinggang, "The so-called *Queli Futu* means the stupa of Cakravartin. Queli means Cakri, which means the Wheel Treasure" (Sun 2015a, p. 126; Sun 2019, pp. 30–44). According to his argument, the so-called *Queli Futu* thus signifies the stupa of Cakravarti-raja.

The title of *Queli Foyuan* (Cakra Buddhist Monastery), located in Hualin Garden (Li 1972, p. 173; Sima 1956, p. 5281), the imperial Garden of Yecheng in the Northern Qi, directly copied the word *Queli*, meaning "Cakravarti-raja." Different from the location of the pagoda of Yongning Monastery in the Northern Wei, this Buddhist monastery was also located in Hualin Garden[17] to the north of the palace in Ye, and thus on the imperial and political axis. The location of Queli Buddhist Monastery was like the location of Tongtai Monastery in the Liang, and different from the location pattern of Yongning Monastery and Jingming Monastery in the Northern Wei. Whether the appearance of this form at the end of the Northern Dynasties was directly affected by the model of Tongtai Monastery in the Liang is still hypothetical. However, judging from the close contacts and interactions between the northern regime and the southern regime from the late period of the Northern Wei to the Eastern Wei and Northern Qi periods, there is the possibility that the builders of the Northern Qi were influenced by the Southern Liang.

In Daxing City (nowadays Xi'an), newly built in Sui (581–618), the largest monasteries to occupy one *fang*, such as Da Zhuangyan Monastery and Da Zongchi Monastery, stood side by side at the southwest corner of the outer city. Two pagodas standing opposite each other with a height of three hundred *chi* had also become the most important religious fixtures in the southern part of the capital (Wei 2006, pp. 69–70; Song 1991, p. 141; Su 1997, pp. 29–33). However, the two highest-ranking imperial monasteries were far away from the political axis of Daxing City and the main hall of the imperial palace (Taiji Hall), different from the relations amidst the monasteries of the Northern and Southern Dynasties. Da Xingshan Monastery and Jianfu Monastery in Chang'an during the Tang, however, in the capital planning of Chang'an in Sui and Tang, were imperial Buddhist monasteries close to the central axis.[18] Buddhist architecture did not follow the footsteps of Emperor Wu of the Liang and Queli Buddhist Monastery in the Northern Qi in occupying the political axis of the capital for a long time or even permanently. The basic pattern composed by the main halls of the palace and multiple main city gates, palace gates, and imperial avenues on the axis of the capital remained unchanged. The location of Buddhist buildings was still only close to this axis, not on it (see Figure 4).

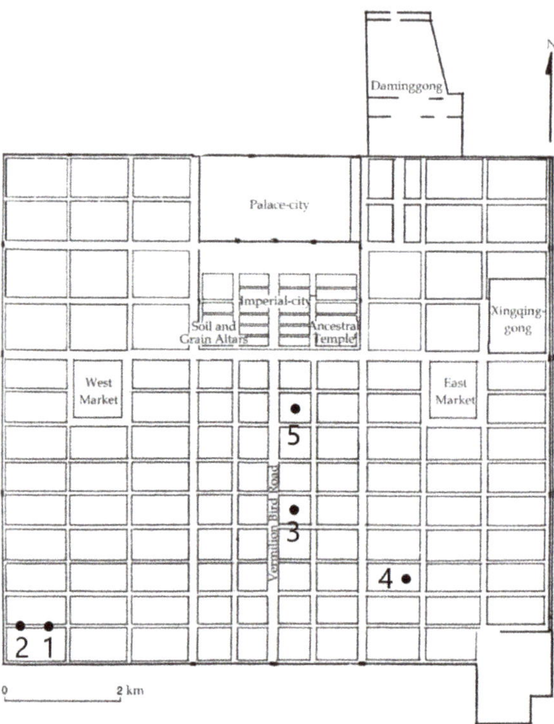

Figure 4. The locations of the main imperial monasteries in the plan of Sui-Tang Daxing-Chang'an, 581–907.[19] 1—Da Zhuangyan Monastery; 2—Da Zongchi Monastery; 3—Da Xingshan Monastery; 4—Da Ci'en Monastery; 5—Da Jianfu Monastery.

4. The Pinnacle of the Buddhist Political Landscape: The Reshaping of the Luoyang Axis Landscape during Wu Zetian's Era (684–705)

Another dramatic change was the construction of a series of imperial ritual buildings with Buddhist elements, such as Mingtang (Bright Hall), Tiantang (Hall of Heaven), and Tianshu (Heaven Pillar) during the Wu Zetian era. According to the records of *Zizhi Tongjian*, the plans to establish a Mingtang were often discussed during the reigns of emperors Taizong (598–649, r. 627–649) and Gaozong (628–683, r. 649–683). However, the Confucian scholars involved were indecisive, and came to no conclusion. The Mingtang was not completed until much later. After Gaozong died and Wu Zetian (624–705, r. 690–705) came to power, Confucian scholars following traditional Confucianism believed that the Mingtang should be built within three *li* (almost 452 m) and seven *li* southeast of the capital. Empress Wu considered that the distant location was too far away from the palace, so in 688, Qianyuan Hall, the main hall of Luoyang Palace in the Tang, which was the former site of Qianyang Hall in the eastern capital of the Sui, was destroyed and replaced by the Mingtang, with the monk Xue Huaiyi (662–694) as the overseer of the project (Sima 1956, p. 6447). Controversy ensued between the Confucian group and the pro-Buddhist forces represented by Xue Huaiyi. Obviously, Wu Zetian did not act according to the suggestions of Confucian scholars, but raised a new plan, trying to build a unique new *mingtang* with Buddhist elements.

In 1988, Wang Yan et al. published a brief report on the excavation of Wu Zetian's Mingtang site (Zhongguo Shehui Kexueyuan Kaogu Yanjiusuo Luoyang Tangchengdui 1988, pp. 227–30, Figures 3 and 4), the location of which has been disputed (see Yu and Li 1993, pp. 90–98; Xin 1989, pp. 149–57; Wang 1993, pp. 949–51; Yang 1994a, pp. 154–61; Yang

1994b, pp. 94–97). Since the beginning of the Tang period, especially during the Gaozong period, important court ceremonies were held at Qianyuan Hall of Luoyang Palace. Wu's purpose for building the Mingtang here was not to build a ceremonial building, a Mingtang in the style of Wang Mang (BCE 45–CE 23) or the Eastern Han (Liu 2009, pp. 490–94; Zhongguo Shehui Kexueyuan Kaogu Yanjiusuo 2010, pp. 80–125) (single-story, multiple eaves, located in the southern suburbs), but to create a huge hall located in the center of the palace—the geometric center of the palace, the imperial city (Fu 2001, p. 21; Wang 2020, p. 137), and the center of the universe (see Figure 5).

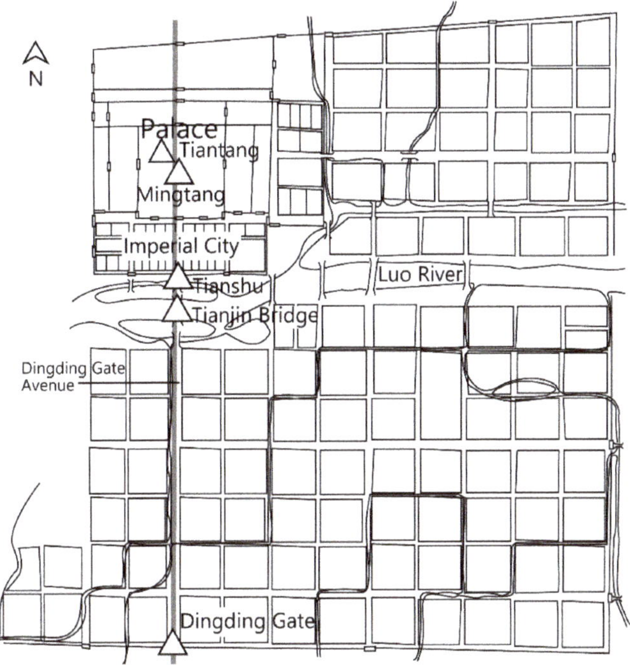

Figure 5. Schematic diagram of the locations of the main ritual buildings showing Buddhist influences in Luoyang.[20]

Buddhist characteristics of Wu's Mingtang concern five aspects: the person who presided over the construction, the external form, the decorative elements, the internal display, and the general function.

Firstly, the person who presided over the construction, as mentioned above, was Wu's favorite monk, Xue Huaiyi. However, only six years after this Mingtang was built, it was reduced to ashes in the unprecedented tragic fire in 695. According to *Jiu Tangshu* and *Zizhi Tongjian*, the fire was caused by Xue's grievances and subsequent arson when he discovered Shen Nanqiu, an imperial physician who was patronized by Wu Zetian. Wu was ashamed and kept silent about the reason for the fire (Liu 1975, p. 4743; Sima 1956, p. 6499). In an earlier record, *Chaoye Qianzai* by Zhang Zhuo, it is simply stated that Gongde Hall (the Hall of Heaven) was on fire, which extended to the Mingtang, without saying that it was caused by Xue's arson (Zhang 1979, pp. 115–16). To be sure, Wu did not intend to hold Xue Huaiyi accountable, and instead continued to appoint him as the person in charge of rebuilding the Mingtang and Heaven Hall (Sima 1956, p. 6499). In summary, since the principal concept behind Wu's two Mingtang construction projects was nominally Buddhist, it was natural for Xue to inject more Buddhist influences into the construction process.

Secondly, in terms of the form of the Mingtang, the existing archaeological reports only show its planar structure and scale, and the complex three-dimensional form requires further research. According to Wang Guixiang, due to the strong Buddhist ideological atmosphere and emphasis of belief in the seven-treasures pagodas and multi-treasure pagodas during the reign of Wu Zetian, she desired to create permanent residences of Sakyamuni Tathagata and Prabhutaratna. These structures would serve as metaphors or manifestations of her profound belief in Buddhism. To realize her wish to become a Cakravartin-raja, it was not necessary to build a Mingtang according to the Confucian classics, but to reproduce the wonderful treasure pavilions revealed in the Buddhist classics in order to fulfill her wish to truly become a Cakravartin-raja (Wang 2011, p. 385). Wang Guixiang found that "the closest structure and spatial form to the building of Wu's Mingtang is the existing three-story wood-like multi-treasure stone pagoda in front of the main hall of Bulguksa Monastery in Gyeongju, South Korea" (Wang 2011, p. 387). Specifically, this Silla pagoda, which was built nearly a hundred years after the date of Wu's Mingtang, had the following characteristics in common: (1) Two of them were three-story pavilion-style buildings, (2) the first story was square with four sloping roofs, (3) both the second and third stories of them were polygonal planes, and (4) the center of these two buildings had a thick central column (Wang 2011, p. 390). From the Southern and Northern Dynasties (386–589) to the Tang Dynasty, the central column had always been a unique structural form in pagodas. The function of this pillar was structural but symbolized the Buddhist universe column (Wang 2006, p. 138).

According to Wang Guixiang, the height of the Mingtang was designed to reach "42 *ren*," an explanatory number that has the symbolic significance of Buddhist space. If one *ren* was converted to seven *chi*, the height of the Mingtang was exactly 294 *chi* (Wang 2011, p. 407). Following Wang's hypothesis for the height and shape of the Mingtang, the shape was different from the previous single-story double eaves Mingtangs as a three-story pavilion with a square and polygonal shape.

Thirdly is the discussion about the decoration of the Mingtang. In this respect, the most representative ones are the decorations of nine dragons and one fire pearl on top of the Mingtang. According to the "Mingtang System" (in Vol. 11 of *Tang Huiyao*), Wu's Mingtang "had three stories: the lower story was the symbol of four seasons, each side with a particular color; the middle story followed the principle of zodiac, with a round cover, and a top plate with nine dragons holding it; the upper story followed the principle of twenty four solar terms, and it also had a round cover" (Wang 1955, p. 277; Sima 1956, pp. 6454–55). Sun Yinggang maintains that Wu Zetian decorated the Mingtang with nine dragons, and there were several golden dragons and a fire pearl at the top, which actually reflects the narrative theme of the spit of water by nine dragons, a popular motif in Buddhist visual culture. As early as the Northern Dynasties, the spit of water emitted by nine dragons had already been used for political purposes. The spit of water by nine dragons above Wu's Mingtang was actually an empowerment ritual. Empowerment was a necessary ceremony for ascending to the throne of Cakravartin-raja (Sun 2015b, p. 46). In this context, the theme of nine dragons on top of the second story of the Mingtang was not just a symbol of the emperor's supreme authority but also a visual representation of the initiation ceremony of the Buddhist Cakravartin-raja.

In the first year of Zhengsheng (695), the old Mingtang was completely destroyed by fire, and Wu ordered a new Mingtang to be built. The scale of the new Mingtang and the original Mingtang was similar, but the decoration changed, and its name changed from *Wanxiang Shengong* to Tongtian Palace. According to *Tongdian* and *Jiu Tangshu*, the top of the original Mingtang was decorated with a golden bird called *yuezhuo*, and the top of the newly built Mingtang was decorated with a precious phoenix (Du 1988, p. 1228; Liu 1975, p. 867). The phoenix was believed to be a symbol of a female, and the phoenix on top of the Mingtang thus symbolized one female, Wu Zetian as an emperor. Nonetheless, within a short period of time, the phoenix at the top of the new Mingtang was replaced by a Buddhist fire pearl (Du 1988, p. 1228; Liu 1975, p. 867). The use of a fire pearl as the top decoration on a pagoda, or on the ridge of a pointed roof, was especially common in Buddhist buildings from the Southern and Northern Dynasties to Sui and Tang (Wang 2011, p. 407). Therefore, Wu used a fire pearl as the top decoration of the newly built Mingtang, named it Tongtian Palace, and changed her reigning title to "Wansui Tongtian" (Long Live Communication with Heaven), which further reinforced the prominence of Buddhist influences on the new Mingtang.

Fourthly are the exhibits in Wu's Mingtang. According to *Zizhi Tongjian*, in 692, "Wu Chengsi, the king of Wei, and five thousand other people, suggested to add the title of the Golden Wheel Holy Emperor (to Wu Zetian). On the day of *yiwei*, the empress came to the Vientiane Shrine, to accept the title and hold amnesty throughout empire. (The court) made seven treasures include the Golden Wheel. They were displayed in the court for every imperial ceremony" (Sima 1956, p. 6492). Additionally, *Xin Tangshu* and the annotations of Hu Sanxing in *Zizhi Tongjian* offer more details on the specific names of the abovementioned seven treasures. According to these records, the seven treasures displayed in the Mingtang were the golden wheel treasure (*cakra*), white elephant treasure (*hasti*), female treasure (*stri*), horse treasure (*asva*), jewelry (*mani*), general treasure (*parinayaka*), and financial officer treasure (*grhapati*) (Ouyang and Song 1975, p. 3482; Sima 1956, p. 6492). The so-called seven treasures are the most important systematic symbol of the birth of the Buddhist Cakravartin-raja. When a grand imperial ceremony was held, the display of the Seven Treasures of Cakravartin-raja in the Mingtang was undoubtedly the use of intuitive visual symbols to maximize the identity of Wu's Buddhist Cakravartin-raja. As a result, the seven treasures appeared in Wu's Mingtang. This practice added undeniable prominent Buddhist characteristics to Wu's Mingtang.

After the old Mingtang was burned in 694, the nine cauldrons and Chinese zodiac signs to correspond to the twelve earth branches, which are more characteristic of Chinese political and cultural traditions,[21] were displayed in the newly built Mingtang. Eight of these *ding* cauldrons were arranged in the eight directions, and each one was 10 *chi* in width (Sima 1956, p. 6499). The practice of setting up nine cauldrons in eight directions with simulated imagery of local mountains, rivers, and products of their corresponding prefectures on these cauldrons was a continuation of the tradition of "making *ding* to symbolize the 10,000 things," as described in the ancient literature. Furthermore, this practice also demonstrated Wu's emphasis on traditional political and cultural symbols. According to Sun Yinggang, "The change of the core ritual implements of the Mingtang from the seven treasures of Cakravartin-raja to the nine cauldrons of a Chinese emperor may be a vivid reflection of the fusion and conflict between two different views of kingship and ideology" (Sun 2015b, p. 47).[22] Wu also composed a song sung in harmony about moving the cauldrons by herself—a reenactment almost reflecting traditional imperial power. In addition, she also wished to build the Great Instrument, a precision instrument for timekeeping (Liu 1975, p. 868). The nine cauldrons and twelve gods symbolized position and space, whereas the great instrument symbolized time, and Wu's hope was to build the Mingtang as a real cosmic clock (Forte 1988b, pp. 95–139). Although there is no currently available direct evidence for the display of seven treasures in the newly built Mingtang, these traditional Chinese political–cultural characteristics represented by the nine cauldrons in the new Mingtang in a later period created tension with the fire-pearl characteristics of Buddhist culture. The Mingtang may be regarded as complex ritual architecture for an ideal stage illustrating the unity of Wu's politics and Buddhism.

Fifthly, I wish to discuss the general function of the Mingtang. As Kaneko Shūichi pointed out, the Mingtang was the de facto main hall of Wu Zetian's regime (Kaneko 2017, pp. 189–214). The structure was ostensibly a Buddhist hall, where Buddhist ceremonies were often held. Based on the remnants of the Capricorn fish unearthed in the Tang stratum on the west side of the Mingtang, located on the axis of Yingtian Gate of Luoyang Palace, Zhang Naizhu believed that the purpose of Wu's construction was to imitate the model of King Ashoka in India and set up a Dharma hall to hold Pañcavārṣika Assemblies with vivid Buddhist influences inside the palace (Zhang 2002, pp. 205–24; Shang and Pei 2017, pp. 1–7, 24).

When Wu Zetian built the Mingtang and the Hall of Heaven for the second time in 695, she accepted the title of *Cishi Yuegu*, "Golden Wheel Holy Emperor"; changed her reigning title to Zhengsheng; and identified Maitreya Buddha with the sovereignty of her emperorship and the divine power. Xue Huaiyi hosted an unprecedentedly grand Pañcavārṣika Assembly in the new Mingtang (Sima 1956, pp. 6497–98). For this grand Buddhist ritual, the Mingtang was decorated with majestic Buddhist objects such as fabrics, Buddha statues, and Buddhist paintings, and was transformed into a Buddhist ritual hall. According to *Zizhi Tongjian*, "on the *yiwei* day (of the first year of the Zhengsheng period [695]), a Pañcavārṣika Assembly was held in the Mingtang . . . (they painted) a large Buddhist portrait whose head measured 200 *chi*. It was said that Huaiyi stabbed his knees and used his blood to paint it. On the *bingshen* day, the portrait was displayed in the south of Tianjin Bridge" (Sima 1956, p. 6498). At the peak of Wu's worship of Buddhism, the Mingtang was not only the place for general administrative business, but also the place where Buddhist Pañcavārṣika Assemblies were held. This function was like that of Tongtai Monastery of Emperor Wu of the Liang. Compared with the pagoda of Tongtai Monastery and Queli Buddhist Monastery of the Northern Qi, its position went further, being in a location directly at the center of the power space of the capital. It is worthy of special attention that this huge Buddha portrait was later set up south of Tianjin Bridge. Tianjin Bridge was located over the Luo River on the political axis of Luoyang. The display of the giant Buddha portrait in this place reveals the profound impact of Buddhism on the political axis of the capital during Wu Zetian's reign.

The Hall of Heaven was built just after the Mingtang, according to the biography of Xue Huaiyi in *Jiu Tangshu*: "the great hall of the Mingtang had three stories, with a total height of 300 *chi*. The Hall of Heaven was built in the north of the Mingtang, and its bottom area was inferior to that of the Mingtang" (Liu 1975, p. 4742). According to this record, the place of the Hall of Heaven should have been to the north of the Mingtang, and its building area should have been smaller than the Mingtang. Its vertical height should have been above the height of the Mingtang, more than 294 *chi*. The description of the Mingtang in *Tongdian* notes, "Firstly, the Mingtang was established, and then the five-story Hall of Heaven was built behind the Mingtang. The third story, could overlook the Mingtang[. . .]" (Du 1988, p. 1228). The size of the statue was described in an extremely exaggerated manner in the record of *Chaoye Qianzai*. It showed that the Buddha statue was "900 *chi* high, with a nose like a thousand-*hu* boat, and the nose can accommodate dozens of people sitting side by side" (Zhang 1979, p. 115). Worship of large-scale Buddha images is documented by a huge lacquer Buddha statue said to have been installed in the Hall of Heaven; its third story was close to the top of the Mingtang, and the total height should have been much higher than the Mingtang. If inferred from the record of the Mingtang's height of 294 *chi*, the total height of heaven should have been above 150 m, which is comparable to the pagoda of Yongning Monastery in the Northern Wei.

Perhaps it was precisely because the base of the Hall of Heaven was not as large as the Mingtang, and its height was far above the Mingtang, that the wind resistance of this high-rise building was quite problematic. According to *Jiu Tangshu*, "at that time, Wu Zetian made the Hall of Heaven behind the Mingtang, with a statue of Buddha, more than a hundred *chi* high. The construction began, and then it was overthrown by strong winds. Soon after, the building was reconstructed again and its work had not been completed finally. On the *bingyin* night of the first lunar month, 695, the Buddhist hall burned and affected the Mingtang. Until the morning of the next day, the two halls were burned altogether" (Ouyang and Song 1975, p. 865). According to this record, the original location of the Hall of Heaven was built on the site of Daye Hall of the Sui. It may not have been completed when burned together with the Mingtang. Seo Tatsuhiko still believes that "This was probably the only example of the construction of a Buddhist hall or pavilion in the center of the imperial palace in Chinese history" (Seo 2019, p. 188). In Luo Shiping's view, "the purpose of Wu's newly built the Hall of Heaven in Luoyang was more than praying for merit, but had the function to advertise the power of emperor from the heaven. With the help of the prophecy of Maitreya Buddha to be Wu Zetian, it is possible to see Wu Zetian and Maitreya Buddha as the same. The Buddha statue in the Hall of Heaven should be regarded as a monument built by Wu Zetian for her enthronement" (Luo 2021, p. 252).

After the fire, Wu Zetian did not abandon her plan to rebuild the Hall of Heaven. According to the note in the paragraph of Mingtang in Du You's *Tongdian*,

> "To rebuild the Hall of Heaven, the scale was lower and narrower than the original one. A huge Buddhist statue was displayed in the Hall of Heaven and a great instrument was also installed After the Hall of Heaven was burned, and the voice of the bell (the great instrument) disappeared. In the reign of Zhongzong (656–710, r. 684, 705–710), he wished to follow and complete the plan of Wu Zetian, and cut the statue to make it short, and build Shengshan Monastery Pavilion to display it" (Du 1988, p. 254).

Although the newly created Hall of Heaven was used to install the Buddhist statue, its size had been reduced. After the fire, Emperor Zhongzong, the third son of Wu Zetian, wanted to continue his mother's plan, so he cut off the giant Buddha statue in the middle, reducing its height, and then built a new pavilion to install it. According to Xu Song, in the Qing Dynasty (1644–1911), although the Hall of Heaven had not been rebuilt, Foguang Monastery was built in its location (Xu 2019, p. 341). After the fire, Wu adjusted the furnishings in the Mingtang, replacing the seven treasures, which symbolized the Buddhist Cakravartin-raja, with nine cauldrons, which symbolized nine Chinese prefectures, marking a retreat of Buddhist influences. Changing the precious phoenix at the top of the Mingtang to the fire pearl, and replacing the name of the Hall of Heaven with the title of Foguang Monastery, Buddhist influences continued to occupy the core area of the palace and political axis during the later period of Wu Zetian's reign.

As for the verification of archaeological materials, in the report on the excavation of the site of the Hall of Heaven in Luoyang during the Sui and Tang, archaeologists directly identified the base of the circular building on the northwest side of the Mingtang site in Luoyang as the Hall of Heaven of Wu Zetian (Luoyangshi Wenwu Kaogu Yanjiuyuan 2016, p. 114). However, this round building to be identified as the Hall of Heaven did not appear on the site of the Daye Hall of the Sui, located in the north of Qianyuan Hall (Mingtang) in the Tang. Before a comprehensive archaeological excavation takes place to the northeast of the current round building site, the location of Daye Hall in the Sui, there seems to be another possibility: that the cleaned-up site may not be the site of the Hall of Heaven by Wu Zetian, but rather the site of Foguang Monastery that was later rebuilt, and even the site of Shengshan Monastery Pavilion built by Zhongzong. If so, the Hall of Heaven, or Foguang Monastery, which had the attributes of a Buddhist hall, may possibly have been removed from the political axis after being destroyed and repositioned to the northwest of the Mingtang. As for the time of this change, it is unknown whether it was at the time of the construction of Foguang Monastery or the time when Shengshan Monastery Pavilion was re-established by Zhongzong.

Another possibility is to turn to admitting the views of archaeologists, who believe it is incorrect that the Hall of Heaven was built to the north of the Mingtang and the location of the Daye Hall, based on historical materials. Why the Hall of Heaven was placed northwest of the Mingtang may be relate to the position of *Qian*, a hexagram corresponding to heaven in the Houtian Eight Diagrams. This explanation corroborates to the judgment of archaeologists that the site of the Hall of Heaven is "the only round building in the palace buildings in ancient China" (Luoyangshi Wenwu Kaogu Yanjiuyuan 2016, p. 114). According to *Jiu Tangshu*, when the Hall of Heaven caught fire, cloudless thunder appeared in the northwest, the so-called *Qian* position in the Houtian Eight Diagrams (Liu 1975, p. 865). This argument also provides a certain degree of logical support for establishing a connection between the Hall of Heaven, heaven, and the *Qian* position in the northwest. However, there is still an obvious contradiction between it and the argument of establishing a hall of heaven on the site of Daye Hall, to the north of the Mingtang according to historical records.

Despite the above disputes, it is certain that "the two buildings were both located in the center of the palace, and the relations between Buddhist architecture and the central building as the political arena were closely integrated, supporting and complementary to each other. The close connection between the Mingtang and the Hall of Heaven reflected the ritualization process of Buddhist architecture around 694 in the reign of Wu Zetian" (Yang 2013, p. 390).

Outside of the palace where the Mingtang and the Hall of Heaven were located, the most noteworthy political landscape on the political axis of Luoyang during the Wu period was Tianshu (Heaven Pillar) of the Great Zhou Dynasty. This huge structure to the north of the Luo River took only eighth months, from 694 to its completion in 695. The purpose of building Tianshu was to commemorate the successes and merits of Wu Zhou (Zhou Dynasty of Emperor Wu) and praise Wu Zetian's achievements. According to Liu Su's record, "In the third year of the Changshou period (694), Wu Zetian collected more than 500,000 *jin* of copper, 3.3 million *jin* iron, and 27,000,000 coins from within the whole empire, to establish an eight-sided bronze pillar in the front of the Dingding Gate. It was ninety *chi* high and twelve *chi* in diameter. The title was 'The Heaven Pillar to record the successes and merits of Great Zhou', which records the achievements of the revolution and derogates the imperial virtue of the Li family. Under the Tianshu, there was an iron mountain, held by bronze dragons, surrounded by lions and unicorns. There was a cloud canopy, decorated by round dragons to hold a fire pearl. The pearl was 10 *chi* high and 30 *chi* circumference. Its golden color was shining brightly to compare with the light of sun and moon" (Liu 1984, p. 126).

Luo Xianglin discussed the relationship between Nestorian Arohan and Tianshu; Li Song started with the analysis of its external form and compared it to stone pillars with Buddhist inscriptions and tomb pillars in the Southern Dynasties. Li maintained that the column was popular at that time. Forte analyzed the construction and abandonment of Tianshu;Zhang Naizhu argued that the construction of Tianshu was influenced by the monumental architectural culture of the Western Regions, such as the Ashoka Stone Pillars and Trajan's Column.[23] In recent years, decorative patterns such as lions and the fire pearl on Tianshu were analyzed by Chen Huaiyu and Peng Lihua (Chen 2012, p. 308; Peng 2020, pp. 31–50; 2021, pp. 26–50). The Tianshu monument appears similar to the form of the Ashoka Pillars (Zhang 1994, pp. 44–46), decorated with the fire pearl and stone lions, and surrounded by the so-called Iron Mountain, which symbolizes the *cakravda* in the Buddhist world, or its symbolic meaning as a pillar to connect earth and heaven. These Buddhist symbols amplified Wu's great ambitions as Cakravartin-raja, the lord of the world.

To sum up, in the early and middle periods of Wu's reign, the Mingtang, an example of Confucian ritual architecture with obvious Buddhist elements, and Tianshu, which was closely related to the nature and form of the Ashoka Pillars, were located in the geometric center of the palace and the small island to the north of Tianjin Bridge and south of Duan Gate, to occupy the most prominent positions on the political axis of the sacred city, Luoyang, symbolizing the center of the world. The Hall of Heaven, built by Xue Huaiyi to install a giant Maitreya statue, was possibly located in the foundation of the Daye Hall to the north of the Mingtang. After being destroyed by fire, it was renamed Foguang Monastery. The other possibility is that the Hall of Heaven was always located northwest of the Mingtang, in corresponding to the position of *Qian*, symbolizing heaven in the Houtian Eight Diagrams, without its position ever having been moved. In the process of displaying the seven treasures symbolizing the power of Cakravartin-raja, welcoming the Buddha bone relics,[24] and holding an unprecedented scale of the Pañcavārṣika Assembly, the Mingtang, as a traditional Confucian ritual space, played an extremely important political role. If the political axis of Luoyang, the sacred capital of Wu Zetian, was extended to the south, its southern endpoint would have been the Longmen Grottoes on the west bank of the Yi River. In the capital of Wu, Buddhist space and buildings were no longer limited to "giving up the main axis and occupying the two compartments," but composed the primary political axis and cosmological axis in the capital. The most symbolic and representative so-called "Seven Heaven Architectures" on the political axis of Luoyang during the reign of Wu Zetian included Tiantang (Hall of Heaven), Tiangong (Mingtang), Tianmen (Yingtian Gate), Tianshu (Heaven Axis), Tianjin (Tianjin bridge), Tianjie (Heaven Avenue), and Tianque (Longmen Grottoes in Yique Valley), and buildings (or grottoes) with clear Buddhist influences made up four of these seven. Wu's construction of the Mingtang and the Hall of Heaven on the political axis of Luoyang, the sacred capital, had a certain logical connection with the actions of Emperor Wu of the Liang. As Chen Jinhua said, all three key components of Wudi's palace chapel—the Chongyun Hall, the Sanxiu Pavilion, and the astronomical edifice (called Cengcheng Cengchengguan, Chuanzhenlou, or Tongtianguan)—have their counterparts in Wu Zetian's Mingtang complex (Chen 2006, p. 92).

5. The Return to the Model of the Southern and Northern Dynasties: The Relation between Architecture of Buddhist Influences and the Political Axis of the Capital in the Post-Wu Zetian Era

Too much water drowned the miller. The fires of the Mingtang and the Hall of Heaven in the middle and late periods of Wu Zetian's reign foretell the decline of Buddhist influences in the political axis of the capital. Wu did not completely abandon the construction of large-scale Buddhist buildings with political symbolism. After she moved the ruling center back to Chang'an, she created a stage of seven treasures in Guangzhai Monastery located on the south axis of Daming Palace.[25] The direct sponsors of the statues on this stage were mostly high-ranking officials and monks, rather than Wu herself (Yang 2013, pp. 363–71). Wu insisted in her later years on building a giant Buddha statue in Bai Sima Ban, located far away, 30 *li* northeast of Luoyang, but she was opposed by Di Renjie (630–700), Li Qiao (645–714), Zhang Tinggui (663–741), and others, and ultimately failed (Matsumoto 1934, pp. 13–49; Hida 2010, pp. 130–36). Evidently, the direct occupation of the political axis of the capital axis by Buddhist buildings only existed in the period when Wu officially proclaimed herself the emperor.

This rebuilt Mingtang was still used until Xuanzong's reign (712–756). In 717, Emperor Xuanzong (685–762, r. 712–756) issued an edict: "Nowadays the Mingtang is located next to the palace. It is not correct or adequately respectful to compare with the ritual requirements. If it did not follow the Confucian principle, what is the model of this building?" (Ouyang and Song 1975, p. 178). Emperor Xuanzong thought that the location of the Mingtang was too close to the palace and that the height was too high. Emperor Xuanzong ordered ritual masters, high-ranking officers, to discuss this topic together, and later renamed the Mingtang the Qianyuan Hall, the original name of the building before the Mingtang was built. In 722, Xuanzong, without explanation, "re-named Qianyuan Hall as Mingtang again" (Ouyang and Song 1975, p. 184). In 739, "the upper story of the Mingtang in the eastern capital (Luoyang) was destroyed, and the lower two stories were rebuilt as Qianyuan Hall" (Ouyang and Song 1975, p. 212). According to the paragraph of the Mingtang system in *Tang Huiyao*, Xuanzong's initial plan was to demolish the Mingtang entirely, in ordering an imperial master craftsman to go to the eastern capital to carry out the demolition. In the end, the emperor simply suggested demolishing the top story of the building and reduced its height to two-thirds of the original height. The center wood pillar was removed, and an octagonal building was placed on the middle story between the first and second stories, with eight dragons rising to hold a fire pearl. The size of the fire pearl was also smaller than the previous one (Wang 1955, p. 281).

This approach was obviously a degradation in terms of building regulations. A similar practice was seen in Qianyang Hall of Luoyang in the Sui and Qianyuan Hall in the same location in the early Tang. The location of the two halls was the location of Wu's Mingtang, which was the main hall of the imperial palace. According to Wang Guixiang's textual research, Qianyang Hall in the Sui had three eaves, whereas Qianyuan Hall in the Tang had only two eaves, and the height difference between the two was exactly 50 *chi* (Wang 2012,

p. 129). The degradation of the main hall of Luoyang Palace in the early Tang was obviously related to the fact that the capital was set in Chang'an during this period, and Luoyang once lost its status as the eastern capital. In the later period of Emperor Gaozong of the Tang Dynasty, especially after Wu Zetian established the Great Zhou Dynasty and established Luoyang as the actual capital (the so-called sacred capital), the status of Luoyang rose sharply, and the magnificence of the Mingtang was an intuitive manifestation of its political status. After Emperor Xuanzong re-established Chang'an as the only central political center, the status of Luoyang was reduced again. After 739, the demolition plan of the Mingtang and its final reconstruction and ultimate degradation became a logical and inevitable result.

As for the demolition of the center pillar of the Mingtang, the reduction of nine dragons into eight dragons, and the reduction of the size of the fire pearl, they all clearly reflect the reduction of Buddhist influences, in addition to the abovementioned downgrading requirements. As mentioned earlier, the major features of Wu's Mingtang differentiated from any previous Mingtang buildings in having a multi-story pavilion and a huge core pillar. According to Forte's research, the Mingtang, which was built for the last time and reconstructed by Emperor Xuanzong, was also called Tongtian Palace. The building had three stories and the top story was a pagoda (Forte 1988b, p. 174). If Forte's analysis is valid, the third story of the Mingtang was eventually demolished by Xuanzong, which was the pagoda part. Emperor Xuanzong's purpose was to remove the material representation of the connection between Buddhism and political culture in Wu Zetian's reign.

As for the final outcome of this Mingtang (or Qianyuan Hall), built during Wu Zetian's reign and later rebuilt by Xuanzong, it was completely burned down during the Anshi Rebellion (755–763). According to *Fengshi Wenjianji*, Shi Siming (703–761) was killed by his son, Shi Chaoyi (?–763), in 761. The Mingtang and the Shengshan Monastery Pavilion, dedicated to Maitreya, were also burned down (Feng 2005, p. 35). Up to this period, the Mingtang and the Hall of Heaven had completely disappeared and withdrawn from the political axis of Luoyang.

Outside the palace, the other magnificent building in Luoyang during the Wu Zetian period, the Tianshu, was also a target for elimination by Xuanzong. According to *Datang Xinyu*, "in the Kaiyuan period, the emperor ordered to destroy Tianshu, and the soldiers took more than one month to melt it" (Liu 1984, p. 126).

From this point of view, a series of measures such as the reconstruction of the Mingtang and the melting down of Tianshu after Xuanzong's accession to the throne marked his fundamental reshaping of the landscape of the political axis of Luoyang, with a plan to remove Buddhist influences from the political axis as much as possible. Marked by the aforementioned dramatic transition, the positions of Buddhist buildings in Chang'an and Luoyang after the reign of Xuanzong had once again returned to the model of being to the side (especially to the east side) of the central axis, as it was during the Northern and Southern Dynasties.

In the Song (960–1276), Liao (907–1125), and Jin (1115–1234), the relations between significant Buddhist monasteries and the political axis was as follows: The great imperial monastery in the capital of the Northern Song, Da Xiangguo Monastery, also followed this principle—located in a similar position, but not necessarily the result of pre-planning because it was inherited from previous dynasties. Lin'an (nowadays Hangzhou), the capital of the Southern Song, was particularly special because of its palace sitting in the south. However, the schematic plan shows that only a few imperial or state monasteries, such as Bao'en Guangxiao Monastery, were located on the two sides of the imperial avenue to the north of Chaotian Gate, which was not a completely centralized political axis. The influence of Buddhism on the political axis of the capital was weaker than during the Northern Song. The so-called five capitals of the Liao did not actually operate with a centralized political center. The central capital of the Jin (nowadays Beijing) was expanded following the southern capital of the Liao: The monasteries of Da Kaitai, Da Haotian, Da Wan'an Chan, Tianwang (nowadays Tianning), Faguang, Lingquan Chan, Shousheng, Shifang Wanfo Xinghua, and others gathered on both sides of the political axis of the capital city from the gates of Tongxuan and Gongchen through the palace to Xuanyang and Fengyi Gate in the south.[26] The monasteries of Da Kaitai, Da Haotian, Tianwang, Lingquan, and (Zhaoti) Shousheng had existed on both sides of the south capital of the Liao and were not newly built during the Jin (Meng 2019, Figures 3–6). The Buddhist buildings on both sides of the political axis of the capitals of the Song, Liao, and Jin, although different in size and scale, maintained the basic model of the Northern Wei and the Xuanzong period of the Tang.

6. Conclusions

During the early Imperial Period of China (the Qin and Han dynasties), the capital adopted a multi-palace system without a political central axis. When Buddhism entered China, Buddhist sites were located in the western suburbs of Luoyang during the Eastern Han Dynasty, and thus were not organically included in the central planning of the capital. Due to limited historical records, the distribution of dozens of Buddhist monasteries in the capital during the Western Jin cannot be documented. The political axis and organic integration of monasteries at the new capital of Pingcheng became prominent during the early Northern Wei Dynasty. In the late period of the Northern Wei, the distribution pattern within Luoyang began to focus Buddhist monuments to the left and right sides of the central axis in the northwest corner of the inner city, forming a political landscape from south to north, with a gradual increase in the number of pagoda stories. Additionally, the parade of Buddha statues on Buddha's birthday incentivized Buddhism to occupy the political axis of the capital.

During the Eastern Jin through to the early period of Emperor Wu of the Liang, significant monasteries were located on both sides of the political axis of the capital, or to the east of the palace, whereas Da'aijing Monastery, with a seven-story pagoda, was located on Mount Zhong, to the northeast of the capital. A crucial breakthrough occurred in the later reign of Emperor Wu of the Liang. The position of the imperial Buddhist monastery (Tongtai Monastery and its nine-story pagoda) was positioned in the imperial garden on the north end of the palace, occupying the political axis of the capital itself. This situation was also seen in the *Queli* Buddhist Monastery, which was also located in the imperial garden on the political axis during the Northern Qi.

During the Sui Dynasty and the early Tang dynasties, the imperial Buddhist monasteries did not continue the position of Tongtai Monastery, built in the late period of Emperor Wu of the Liang, but continued the tradition of the Northern Dynasties, arranged on both sides of the political axis of the capital or other areas. During the reign of Wu Zetian, Buddhism and imperial political culture were fused, as is especially marked in the development of Cakravatin-raja thought and its related Buddhist ideology. The Mingtang of Confucian ritual architecture of traditional Chinese political culture was combined with Buddhist decor and design concepts. The seven treasures of Cakravatin-raja were placed in the Mingtang, symbolizing the legitimacy and authority of the Buddhist monarch. More significantly, Wu Zetian also built a higher five-story building (Tiantang) at an important location to the north or northwest of the Mingtang, inside of which was with a monumental Buddha statue. Tianshu, which was quite similar to the Ashoka Pillars, was also placed as an important landmark monument to the north of Tianjin Bridge, which was part of the political axis of Luoyang. In this particular period, the influence of Buddhism not only occupied the political axis of the capital for Emperor Wu of the Liang, but also became rooted into the core area of the palace, highly integrated with imperial power. When the Mingtang and the Hall of Heaven were burned down in fires, Wu Zetian returned to Chang'an. Especially after Emperor Xuanzong of the Tang held power, the second Mingtang, built by Wu, was reconstructed; the metal Tianshu was melted; and the Buddhist influences in the political axis of Luoyang gradually decayed.

In summary, there are two modes of integration and interaction between Buddhism or architecture with Buddhist influences and the political axis of the capital city: First, it can be regarded as typical after the late period of the Northern Wei to locate landmark Buddhist buildings with huge scale, important political influence, or high multi-story pagodas on both sides of the political axis of the capital. This typical mode was considerably stable, lasting from the Northern Wei to the Tang and even extending its influence to the capitals during the Song and Liao. The second mode was atypical, formed in the later period of Emperor Wu of the Liang and developed in the Northern Qi and the reign of Wu Zetian, placing Buddhist buildings or imperial ritual buildings with remarkably Buddhist influences and symbolic meanings directly on the political axis of the capital. This atypical mode was the product of the high integration and close interaction between Buddhism and political culture in the above three periods but considerably related to the emperor's personal political and cultural orientation.

Buddhism in Medieval China was not only a foreign religious belief, ideology, and culture, but also a new political and cultural factor that penetrated the political system and affected capital planning at the time. Buddhist monasteries (Appendix A) and pagodas, as well as certain imperial ritual architecture with Buddhist elements, were no longer just testimonies of the Buddha's actions and the emperor's personal Buddhist beliefs, but had also become a material carrier, visual representation, and cultural landscape with a high degree of politically powerful symbolism, which constituted new expressions of political power by Chinese rulers.

Funding: This research was funded by the National Social Science Foundation Youth Project "Studies on the relations between religion and politics in the 'moving frontiers' from the 7th to 13th century" (project approval number: 18CZJ001).

Institutional Review Board Statement: Ethical review and approval were waived for this study, due to that this research does not include research on identifiable human materials and data.

Informed Consent Statement: Not applicable.

Data Availability Statement: The study did not report any data.

Acknowledgments: In the procedure to collect historical and archaeological materials and write this paper, Hu Manli and other students were very helpful and inspired me in some points. On the revision of figures, Li Qiyuan and Shengyulan Design Service Co. Ltd. did much work. The main ideas and main content of this paper were presented at the online conference "Luoyang Architecture and City in Northern Wei: Technology Dissemination and Knowledge Integration in the 5th and 6th Centuries," hosted by Harvard Chinese Art Media Lab on 22 July 2021, and I obtained the comments and suggestions of Yu Shuishan, an associate professor at the College of Arts, Media and Design of Northeastern University. Absolutely, I also need to thank the three anonymous reviewers for the corrections and suggestions.

Conflicts of Interest: The author declares no conflict of interest.

Appendix A

Table A1. List of Buddhist monasteries to be discussed.

Title of Monastery	Location	Current Location	Period
Baima Monastery	Luoyang	Luoyang	Eastern Han
Yongning Monastery	Pingcheng	Datong	Northern Wei
Tiangong Monastery	Pingcheng	Datong	Northern Wei
Yongning Monastery	Luoyang	Luoyang	Northern Wei
Jingming Monastery	Luoyang	Luoyang	Northern Wei
Qin Taishang Gong East Monastery	Luoyang	Luoyang	Northern Wei
Yaoguang Monastery	Luoyang	Luoyang	Northern Wei
Pingdeng Monastery	Luoyang	Luoyang	Northern Wei
Zhengshi Monastery	Luoyang	Luoyang	Northern Wei

Table A1. *Cont.*

Title of Monastery	Location	Current Location	Period
Yongming Monastery	Luoyang	Luoyang	Northern Wei
Dajue Monastery Luoyang	Luoyang	Luoyang	Northern Wei
Qin Taishang Jun Monastery	Luoyang	Luoyang	Northern Wei
Qin Taishang Gong West Monastery	Luoyang	Luoyang	Northern Wei
Hutong Monastery	Luoyang	Luoyang	Northern Wei
Chongjue Monastery	Luoyang	Luoyang	Northern Wei
Rongjue Monastery	Luoyang	Luoyang	Northern Wei
Changjiu Monastery	Luoyang	Luoyang	Northern Wei
Mingxuan Nuns Monastery	Luoyang	Luoyang	Northern Wei
Lingying (Taikang) Monastery	Luoyang	Luoyang	Northern Wei
Wangdianyu Monastery	Luoyang	Luoyang	Northern Wei
Baoguang Monastery	Luoyang	Luoyang	Northern Wei
Dazongchi Monastery	Yecheng	Linzhang	Northern Qi
Dazhuangyan Monastery	Yecheng	Linzhang	Northern Qi
Queli Buddhist Monastery	Yecheng	Linzhang	Northern Qi
Changgan Monastery (originally Jianchu Monastery)	Jiankang	Nanjing	Eastern Jin
Waguan Monastery	Jiankang	Nanjing	Eastern Jin
Da Zhuangyan Monastery	Jiankang	Nanjing	Liu Song
Xianggong Monastery	Jiankang	Nanjing	Liu Song
Da'aijing Monastery	Jiankang	Nanjing	Liang
Tongtai Monastery	Jiankang	Nanjing	Liang
Da Zhuangyan Monastery	Daxing (Chang'an)	Xi'an	Sui and Tang
Da Zongchi Monastery	Daxing (Chang'an)	Xi'an	Sui and Tang
Da Xingshan Monastery	Daxing (Chang'an)	Xi'an	Sui and Tang
Jianfu Monastery	Chang'an	Xi'an	Tang
Da Ci'en Monastery	Chang'an	Xi'an	Tang
Guangzhai (Qibaotai) Monastery	Chang'an	Xi'an	Tang
Foguang Monastery	Luoyang	Luoyang	Tang
Shengshan Monastery	Luoyang	Luoyang	Tang
Bulguksa Monastery	Gyeongju	Gyeongju	Unify Silla, Korea
Da Xiangguo Monastery	Bianjing	Kaifeng	Northern Song
Bao'en Guangxiao Monastery	Lin'an	Hangzhou	Southern Song
Da Kaitai Monastery	Nanjing	Beijing	Liao
Da Haotian Monastery	Nanjing	Beijing	Liao
Da Wan'an Chan Monastery	Nanjing	Beijing	Liao
Tianwang Monastery	Nanjing	Beijing	Liao
Faguang Monastery	Nanjing	Beijing	Liao
Lingquan Chan Monastery	Nanjing	Beijing	Liao
Shousheng Monastery	Nanjing	Beijing	Liao
Shifang Wanfo Xinghua Monastery	Nanjing	Beijing	Liao

Notes

[1] On the locations of these monasteries, see (Yang 2018, pp. 11, 133, 140).

[2] On the story of pagodas in these monasteries, see (Yang 2018, pp. 11, 133, 140).

[3] On the name of Hu Xuanhui and her other information, see "Wei Gu Shichijie Shizhong Piaoji Dajiangjun Lingjun Jiangjun Shangshuling Jizhou Cishi Jingzhao Wangfei zhi Ming", in (Datong Beichao Yishu Yanjiuyuan 2016, p. 131).

[4] Some contents and more details in the analysis of four stages, see (Hu 2021).

[5] On the location of Yaoguang Monastery and stories of its pagoda, see (Yang 2018, pp. 48–49).

[6] On the descriptions of *Baichilou*, see (Li 2007, p. 393; Li 1960, pp. 859b, 873b; Xu 1994, p. 64).

[7] According to the record in *Weishu*, Emperor Xiaowu ascended the throne in the fourth lunar month of the second year of the Zhongxing period or the first year of the Taichang period (532). Therefore, the five-story pagoda in Pingdeng Monastery should be built in this year or a little bit later, see (Wei 2017, p. 332).

[8] On the materials in the Table 1, see (Yang 2018; Wang 2016, pp. 154–59).

[9] The diagram is taken from the work of Yang Yong in the attached map at the end of *Luoyang Qielan Ji* (Yang 2018, Figure 1), and modified by (Zhou 2018, p. 145).

[10] On the archaeological excavation reports of these two monasteries, see (Zhongguo Shehui Kexueyuan Kaogu Yanjiusuo and Hebeisheng Wenwu Yanjiusuo Yecheng Kaogudui 2010, pp. 31–42, Figures 6–9; 2013a, pp. 49–68; 2013b, pp. 25–35; 2016, pp. 563–92, Figures 1–8).

[11] Guo Jiqiao argued that Zhaopengcheng Monastery was Dao Zongchi Monastery, a imperial Buddhist monastery in Northern Qi, see (Geng and Zhang 2017, p. 9).

[12] On the base map of this schematic diagram, see (Zhongguo Shehui Kexueyuan Kaogu Yanjiusuo and Hebeisheng Wenwu Yanjiusuo Yecheng Kaogudui 2016, p. 564, Figure 1).

[13] Many historical materials record that Liu Sahe discovered the bone relic of Buddha in Changgan Monastery in the Ningkang period of Eastern Jin (373–375), see (Yao 1973, p. 791; Li 1975, pp. 1954–55; Xu 1986, p. 672).

[14] On its particular location, see (Seo 2019, Figure 75).

[15] On the base map of this schematic diagram, see (Seo 2019, Figure 75).

[16] On the related records, see (Yao 1973, pp. 71, 73, 90, 92). On the related researches, see (Yan 2010, pp. 250–319).

[17] On the precise location of Hualin Garden in Ye, see (Jia 2013, pp. 125–28).

[18] On the locations of Zhuangyan Monastery, Zongchi Monastery, Da Xingshan Monastery and Da Ci'en Monastery, see (Su 2009, p. 28).

[19] On the base map of this schematic diagram, see (Steinhardt 2019, p. 105).

[20] On the base map of this schematic diagram, see (Fu 2009, p. 351, Figure 3-1-11).

[21] On the legend of nine cauldrons and its monumentality, see (Wu 1995, pp. 1–15).

[22] On the meanings of Wu Zetian's nine cauldrons, see (Forte 1988a, pp. 85–96).

[23] See (Luo 1966, pp. 57–70; Li 1985, pp. 41–45; Forte 1988b, pp. 235–45; Zhang 1994, pp. 44–46). Besides these discussions, the related researches also include the paper by Norman Harry Rothschild, see (Rothschild 2008, pp. 199–234).

[24] On the records of welcoming the Buddha bone relics, see (Zanning 1987, p. 332; Takakusu 1934, pp. 283c–284a).

[25] For the most representative researches on the stage of seven treasures, see (Yen 1986, 1987, pp. 1–16; Yan 1987, pp. 41–47; 1998, pp. 829–42; Yang 2013).

[26] On the above-mentioned schematic plans of Eastern Capital of Northern Song, Lin'an of Southern Song and the middle capital of Jin, see (Meng 2019, Figures 1-2, 2-2 and 4-3).

References

Chen, Huaiyu. 2012. *Dongwu yu Zhonggu Zhengzhi Zongjiao Zhixu*. Shanghai: Shanghai Guji Chubanshe.

Chen, Jianjun, Zhou Hua, and Hu Xiao. 2020. Caowei Luoyanggong Taijidian Qijian Weizhi Zaitan: Jianlun Weijin Luoyangcheng de Zhongzhouxian. *Xuchang Xueyuan Xuebao* 4: 12–18.

Chen, Jing. 2020. Gudu "Zhongzhouxian" Suyuan ji Fazhan Bianhua: Jianlun Gudu Beijing "Zhongzhouxian" Lishi Yiyi. *Zhongguo Wenhua Yichan* 6: 85–96.

Chen, Jinhua. 2006. Pañcavārṣika Assemblies in Liang Wudi's Buddhist Palace Chapel. *Harvard Journal of Asiatic Studies* 66: 43–103.

Daoxuan. 2014. *Xu Gaoseng Zhuan*. Beijing: Zhonghua Shuju.

Datong Beichao Yishu Yanjiuyuan, ed. 2016. *Beichao Yishu Yanjiuyuan Cangpin Tulu*. Beijing: Wenwu Chubanshe.

Datongshi Bowuguan. 2007. Datong Beiwei Fangshan Siyuan Fosi Yizhi Fajue Baogao. *Wenwu* 4: 4–26.

Du, You. 1988. *Tongdian*. Beijing: Zhonghua Shuju.

Faxian. 2008. *Faxian Zhuan Jiaozhu*. Beijing: Zhonghua Shuju.

Feng, Yan. 2005. *Fengshi Wenjianji Jiaozhu*. Beijing: Zhonghua Shuju.

Forte, Antonino, ed. 1988a. *Tang China and Beyond: Studies on East Asia from the Seventh to the Tenth Century*. Kyoto: Istituto Italiano di Cultura, Scuola di Studi Sull'asia Orientale.

Forte, Antonino. 1988b. *Mingtang and Buddhist Utopias in the History of Astronomical Clock: The Tower, Statue and Armillary Sphere Constructed by Empress Wu*. Roma: Istituto Italiano per il Medio ed Estremo Oriente Paris, Ecole Française d'Extrême-Orient.

Fu, Xi'nian. 1998. *Fu Xi'nian Jianzhu Shi Lunwenji*. Beijing: Wenwu Chubanshe.

Fu, Xi'nian. 2001. *Zhongguo Gudai Chengshi Guihua, Jianzhuqun Buju ji Jianzhu Sheji Fangfa Yanjiu*. Beijing: Zhongguo Jianzhu Gongye Chubanshe, vol. 1.

Fu, Xi'nian. 2017. *Traditional Chinese Architecture: Twelve Essays*. Translated by Nancy S. Steinhardt, and Alexandra Harrer. Princeton: Princeton University Press.

Fu, Xi'nian, ed. 2009. *Zhongguo Gudai Jianzhu Shi*. Beijing: Zhongguo Jianzhu Gongye Chubanshe, vol. 2.

Geng, Jiankuo, and Yunpeng Zhang. 2017. Hebei Yecheng Zhaopengcheng Beichao Fosi bei Renwei 1500 nian qian Beiqi Huangjia Fosi Da Zongchi Si. *Guangming Ribao* 9.

Hida, Romi. 2010. *Fengxiansi Dong Dafo yu Bai Simaban Dafo in Shikusi Yanjiu*. Beijing: Wenwu Chubanshe.

Hu, Manli. 2021. Beiwei Luoyang de Fosi yu Chengshi Kongjian: Yi Luoyang Qielanji wei Zhongxin. Master's dissertation, Hunan University, Changsha, China.

Jia, Jun. 2013. *Weijin Nanbeichao Shiqi Luoyang, Jiankang, Yecheng Sandi Hualinyuan Kao in Jianzhu Shi*. Beijing: Qinghua Daxue Chubanshe, vol. 31, pp. 125–28.

Kaneko, Shūichi. 2017. *Zhongguo Gudai yu Huangdi Jisi*. Translated by Shengzhong Xiao, Sisi Wu, and Caojie Wang. Shanghai: Fudan Daxue Chubanshe.

Li, Baiyao. 1972. *Beiqi Shu*. Beijing: Zhonghua Shuju.

Li, Yanshou. 1975. *Nanshi*. Beijing: Zhonghua Shuju.

Li, Song. 1985. Tianshu: Woguo Gudai Yizhong Jinianbei Yangshi. *Meishu* 4: 41–45.

Li, Daoyuan. 2007. *Shuijing Zhu Xiaozheng*. Beijing: Zhonghua Shuju.

Li, Fang, ed. 1960. *Taiping Yulan*. Beijing: Zhonghua Shuju.

Li, Fang, ed. 1966. *Wenyuan Yinghua*. Beijing: Zhonghua Shuju.

Liang, Sicheng. 1998. *Liang Sicheng Quanji*. Beijing: Zhongguo Jianzhu Gongye Chubanshe.

Liang, Sicheng. 2011. *Zhongguo Jianzhu Shi*. Beijing: Sanlian Shudian.

Liu, Xu. 1975. *Jiu Tangshu*. Beijing: Zhonghua Shuju.

Liu, Su. 1984. *Datang Xinyu*. Beijing: Zhonghua Shuju.

Liu, Junwen, ed. 1992. *Riben Xuezhe Yanjiu Zhongguo Shi Lunzhu Xuanyi*. Beijing: Zhonghua Shuju, vol. 1.

Liu, Xujie, ed. 2009. *Zhongguo Gudai Jianzhu Shi*. Beijing: Zhongguo Jianzhu Gongye Chubanshe, vol. 1.

Luo, Xianglin. 1966. *Tang Yuan Erdai zhi Jingjiao*. Hong Kong: Zhongguo Xueshe.

Luo, Shiping. 2021. *Tuxiang yu Yangshi: Hantang Fojiao Meishu Yanjiu*. Beijing: Wenwu Chubanshe.

Luoyangshi Wenwu Kaogu Yanjiuyuan, ed. 2016. *Suitang Luoyangcheng Tiantang Yizhi Fajue Baogao*. Beijing: Kexue Chubanshe.

Matsumoto, Bunzaburō. 1934. Sokuten Bukō no Shiro Shiba Saka Daizō ni tsuite. *Tōyō Gakuhō* 5: 13–49.

Meng, Fanren. 2019. *Songdai zhi Qingdai Ducheng Xingzhi Buju Yanjiu*. Beijing: Zhongguo Shehui Kexue Chubanshe.

Ouyang, Xiu, and Qi Song. 1975. *Xin Tangshu*. Beijing: Zhonghua Shuju.

Peng, Lihua. 2020. Dazhou Wanguo Songde Tianshu shang de Shizi Zaoxing Lun. In *Tangshi Luncong*. Edited by Wenyu Du. Xi'an: Sanqin Chubanshe, vol. 30, pp. 31–50.

Peng, Lihua. 2021. Dazhou Wanguo Songde Tianshu shang de Huozhu Zhuangshi Lun. In *Tangshi Luncong*. Edited by Wenyu Du. Xi'an: Sanqin Chubanshe, vol. 32, pp. 26–50.

Rothschild, Norman Harry. 2008. The Koguryan Connection: The Quan (Yon) Family in the Establishment of Wu Zhao's Political Authority. *Journal of China Studies* 4: 199–234.

Sagawa, Eiji. 2016. *Chūgoku Kodai Tojō no Sekkei to Shisō: Enkyū Saishi no Rekishiteki Tenkai*. Tōkyō: Bensei Shuppan.

Seo, Tatsuhiko. 2019. *Suitang Changan yu Dongya Bijiao Ducheng Shi*. Translated by Bingbing Gao, Xueni Guo, and Haijing Huang. Xi'an: Xibei Daxue Chubanshe.

Shang, Chunfang, and Xuejie Pei. 2017. Mojieyu Shike yu Wu Zetian de Zhuanlunwang Zhengjiao Sixiang Guanxi Yanjiu. *Luoyang Ligong Xueyuan Xuebao (Shehui Kexue Ban)* 4: 1–7, 24.

Sickman, Laurence C. S., and Alexander Coburn Soper. 1956. *The Art and Architecture of China*. Baltimore: Penguin Books.

Sima, Guang. 1956. *Zizhi Tongjian*. Beijing: Zhonghua Shuju.

Song, Minqiu. 1991. *Chang'an Zhi*. Beijing: Zhonghua Shuju.

Steinhardt, Nancy S. 1997. *Liao Architecture*. Honolulu: University of Hawaii Press.

Steinhardt, Nancy S. 2004. The Tang Architectural Icon and the Politics of Chinese Architectural History. *The Art Bulletin* 96: 228–54. [CrossRef]

Steinhardt, Nancy S. 2019. *Chinese Architecture: A History*. Princeton and Oxford: Princeton University Press.

Su, Bai. 1997. Suidai Fosi Buju. *Kaogu yu Wenwu* 2: 29–33.

Su, Bai. 2009. Shilun Chang'an Fojiao Siyuan de Dengji Wenti. *Wenwu* 1: 27–40.

Sun, Yinggang. 2015a. *Qibao Zhuangyan: Zhuanlunwang Xiaozhuan*. Beijing: Shangwu Yinshuguan.

Sun, Yinggang. 2015b. Wu Zetian de Qibao: Fojiao Zhuanlunwang de Tuxiang, Fuhao jiqi Zhengzhi Yihan. *Shijie Zongjiao Yanjiu* 2: 43–53.

Sun, Yinggang. 2019. Bufa Yanni de Beiqi Huangdi: Zhonggu Randengfo Shouji de Zhengzhi Yihan. *Lishi Yanjiu* 6: 30–44.

Sun, Yinggang, and Ping He. 2018. *Jiantuoluo Wenming Shi*. Beijing: Sanlian Shudian.

Takakusu, Junjirō, ed. 1934. *Taishō Shinshū Daizōkyō*. Tōkyō: Taishō Issaikyō Kankōkai.

Wang, Pu. 1955. *Tang Huiyao*. Beijing: Zhonghua Shuju.

Wang, Yan. 1993. Guanyu Tang Dongdu Wu Zetian Mingtang Yizhi de Jige Wenti. *Kaogu* 10: 949–51.

Wang, Guixiang. 2006. *Dongxifang de Jianzhu Kongjian: Chuantong Zhongguo yu Zhongshiji Xifang Jianzhu de Wenhua Chanshi*. Tianjin: Baihua Wenyi Chubanshe.

Wang, Guixiang. 2011. Tang Luoyanggong Wushi Mingtang de Jiangouxing Fuyuan Yanjiu. In *Zhongguo Jianzhu Shi Huikan*. Beijing: Qinghua Daxue Chubanshe, pp. 369–455.

Wang, Guixiang. 2012. *Gudu Luoyang*. Beijing: Qinghua Daxue Chubanshe.

Wang, Guixiang. 2016. *Zhongguo Hanchuan Fojiao Jianzhu Shi: Fosi de Jianzao, Fenbu yu Siyuan Geju, Jianzhu Leixing Jiqi Bianqian (1st, Chuchuan yu Dingsheng)*. Beijing: Qinghua Daxue Chubanshe.

Wang, Shulin. 2020. *Beisong Xijing Chengshi Kaogu Yanjiu*. Beijing: Wenwu Chubanshe.

Wang, Tao, Ping Ye, Zhilin Fu, and Hang Xu. 2015. Qianxi Suitang Luoyangcheng Chengshi Zhouxian de Fazhan. *Sichuan Jianzhu* 5: 27–28.

Wei, Shu. 2006. *Liangjing Xinji*. Xi'an: Sanqin Chubanshe.

Wei, Shou. 2017. *Weishu (Revision)*. Beijing: Zhonghua Shuju.

Wu, Hung. 1995. *Monumentality in Early Chinese Art and Architecture*. Stanford: Stanford University Press.

Xiao, Zixian. 2017. *Nanqishu (Revision)*. Beijing: Zhonghua Shuju.

Xie, Yifeng. Forthcoming. *Lingtaihou de Futu: Fojiao yu Beiwei Houqi Luoyang Dushi Jingguan de Chongsu*. Shanghai: Shanghai Guji Chubanshe, Unpublished Paper.

Xin, Deyong. 1989. Tang Dongdu "Wu Zetian Mingtang" Yizhi Zhiyi. *Zhongguo Lishi Dili Luncong* 3: 149–57.

Xu, Song. 1986. *Jiankang Shilu*. Beijing: Zhonghua Shuju.

Xu, Song. 1994. *Henan Zhi*. Beijing: Zhonghua Shuju.

Xu, Song. 2019. *Zuixin Zengding Tang Liangjing Chengfang Kao*. Revised and Enlarged by Jianchao Li. Xi'an: Sanqin Chubanshe.

Xuanzang, and Bianji. 2000. *Datang Xiyu Ji*. Beijing: Zhonghua Shuju.

Yan, Juanying. 1987. Wu Zetian yu Tang Chang'an Qibaotai Shidiao Foxiang. *Yishu Xue* 1: 41–47.

Yan, Juanying. 1998. Tang Chang'an Qibaotai Shike de zai Xingsi. In *Yuanwang Ji: Shannxisheng Kaogu Yanjiusuo Huadan Sishi Zhounian Jinian Wenji*. Edited by Shannxisheng Kaogu Yanjiusuo. Xi'an: Shannxi Renmin Meishu Chubanshe, vol. 2, pp. 829–42.

Yan, Shangwen. 2010. *Zhongguo Zhonggu Fojiao Shilun*. Beijing: Zongjiao Wenhua Chubanshe.

Yang, Huanxin. 1994a. Luelun Beisong Xijing Luoyanggong de Jizuo Dianzhi. *Zhongyuan Wenwu* 4: 94–97.

Yang, Huanxin. 1994b. Shitan Tang Dongdu Luoyanggong de Jizuo Zhuyao Dianzhi. In *Hantang yu Bianjiang Kaogu Yanjiu*. Beijing: Kexue Chubanshe, vol. 1.

Yang, Xiaojun. 2013. *Wuzhou Shiqi de Fojiao Zaoxiang: Yi Chang'an Guangzhai Si Qibao Tai de Fudiao Shifo Qunxiang wei Zhongxin*. Beijing: Wenwu Chubanshe.

Yang, Xuanzhi. 2018. *Luoyang Qielan Ji*. Beijing: Zhonghua Shuju.

Yao, Silian. 1973. *Liangshu*. Beijing: Zhonghua Shuju.

Yen, Chuan-Ying. 1986. The Tower of Seven Jewels: The Style, Patronage and Iconography of the T'ang Monument. Ph.D. dissertation, Harvard University, Cambridge, MA, USA.

Yen, Chuan-Ying. 1987. The Tower of Seven Jewels and Empress Wu. *National Palace Museum Bulletin* 22: 1–16.

Yu, Fuwei, and Defang Li. 1993. Tang Dongdu Wu Zetian Mingtang Yizhi Tansuo. In *Zhongguo Gudu Yanjiu (vols. 5 and 6): Zhongguo Gudu Xuehui Diwu, Liu jie Nianhui Lunwenji*. Beijing: Beijing Guji Chubanshe, pp. 90–98.

Zanning. 1987. *Song Gaoseng Zhuan*. Beijing: Zhonghua Shuju.

Zeng, Lei. 2020. *Menque, Zhouxian yu Daolu: Qinhan Zhengzhi Lixiang de Kongjian Biaoda*. Guilin: Guangxi Shifan Daxue Chubanshe.

Zhang, Zhuo. 1979. *Chaoye Qianzai*. Beijing: Zhonghua Shuju.

Zhang, Naizhu. 1994. Wuzhou Wanguo Tianshu yu Xiyu Wenming. *Xibei Shidi* 2: 44–46.

Zhang, Naizhu. 2002. Cong Luoyang Chutu Wenwu kan Wuzhou Zhengzhi de Guoji Wenhua Qingcai. In *Tang Yanjiu*. Beijing: Beijing Daxue Chubanshe, vol. 8.

Zhang, Xuefeng, ed. 2021. *"Duchengquan" yu "Duchengquan Shehui" Yanjiu Wenji: Yi Liuchao Jiankang wei Zhongxin*. Nanjing: Nanjing Daxue Chubanshe.

Zhongguo Shehui Kexueyuan Kaogu Yanjiusuo, ed. 2010. *Hanwei Luoyang Gucheng Nanjiao Lizhi Jianzhu Yizhi 1962–1992 nian Kaogu Fajue Baogao*. Beijing: Wenwu Chubanshe.

Zhongguo Shehui Kexueyuan Kaogu Yanjiusuo, and Hebeisheng Wenwu Yanjiusuo Yecheng Kaogudui. 2010. Hebei Linzhangxian Yecheng Yizhi Zhaopengcheng Beichao Fosi Yizhi de Kantan yu Fajue. *Kaogu* 7: 31–42.

Zhongguo Shehui Kexueyuan Kaogu Yanjiusuo, and Hebeisheng Wenwu Yanjiusuo Yecheng Kaogudui. 2013a. Hebei Yecheng Yizhi Zhaopengcheng Beichao Fosi yu Beiwuzhuang Fojiao Zaoxiang Maicangkeng. *Kaogu* 7: 49–68.

Zhongguo Shehui Kexueyuan Kaogu Yanjiusuo, and Hebeisheng Wenwu Yanjiusuo Yecheng Kaogudui. 2013b. Hebei Linzhangxian Yecheng Yizhi Zhaopengcheng Beichao Fosi 2010–2011 nian de Fajue. *Kaogu* 12: 25–35.

Zhongguo Shehui Kexueyuan Kaogu Yanjiusuo, and Hebeisheng Wenwu Yanjiusuo Yecheng Kaogudui. 2016. Hebei Linzhang Yecheng Yizhi Hetaoyuan Yihao Jianzhu Jizhi Fajue Baogao. *Kaogu Xuebao* 4: 563–91.

Zhongguo Shehui Kexueyuan Kaogu Yanjiusuo Luoyang Tangchengdui. 1988. Tang Dongdu Wu Zetian Mingtang Yizhi Fajue Jianbao. *Wenwu* 3: 227–30.

Zhou, Yin. 2018. Beiwei Wu Ming Shiqi Luoyang Siyuan Buju yu Lifang Guihua. *Shehui Kexue Zhanxian* 10: 143–55.

Article

Memories of Ups and Downs: The Vicissitudes of the Chongshansi in Taiyuan

Jing Wen

College of Architecture & Urban Planning, Tongji University, Shanghai 200092, China; wenj@tongji.edu.cn

Abstract: This article traces the erection of and changes in the Buddhist temple of Chongshansi in Taiyuan through the process of spatial production under the social background of the Ming and Qing Dynasties. It is stated that the founding time of the temple complies with the reorganization policies aimed at Buddhist institutions in the early Ming Dynasty, which confirms the setup of the Prefectural Buddhist Registry as the motivation for erecting the temple. Within the spatial structure of Taiyuan in the Ming Dynasty, its relative position with the Princely Palace of Jin (completed in 1375) and the expanded Taiyuan City is analyzed, revealing how its layout participated in the construction of the ritual path of Taiyuan under the control of the palace. The article concludes with a description of the fall of the temple following the loss of protection from the Jin Principality by tracing back its original form through the remains still evident in the city. The vicissitudes of the physical space of the temple are deeply connected to its role in the political space of the city. The article, thus, presents the changes in the temple throughout history. In positioning the temple back to the power and physical space of the imperial court, as well as the Jin Principality, a new perspective is provided into regional monasteries during the Ming Dynasty.

Keywords: Chongshansi; Buddhist temple; Taiyuan City; Ming Dynasty; imperial clan; Prince of Jin

Citation: Wen, Jing. 2022. Memories of Ups and Downs: The Vicissitudes of the Chongshansi in Taiyuan. *Religions* 13: 785. https://doi.org/10.3390/rel13090785

Academic Editors: Shuishan Yu and Aibin Yan

Received: 5 May 2022
Accepted: 23 August 2022
Published: 26 August 2022

1. Introduction

Chongshansi 崇善寺 is a government-sponsored Buddhist institution erected during the early Ming Dynasty in Taiyuan 太原, Shanxi Province. As one of the few early Ming structures existing to date, this temple offers an extremely rare case for our understanding of architectural monuments of the 14th-century Chinese imperial palace and state temple. Currently, Chongshansi is located in the southeast corner of the old city of Taiyuan, adjacent to Wenmiao 文廟 (the Confucian Temple, now Shanxi Folk Museum) by an alley. It once enjoyed vast land including the current site, as well as the entire area now occupied by the Wenmiao after a devastating fire in the third year of Tongzhi 同治 during the Qing Dynasty (1864). The Great Compassion Hall 大悲殿 that survived the fire is now the main hall of the temple (Figure 1). It is a rare example of a high-ranking official style building in the Hongwu period 洪武 (1368–1398) during the early Ming Dynasty. It enjoys the highest preservation level and among the existing Ming buildings within Shanxi Province, it is the one that has kept its historical style the most. In addition to its architecture, it is also famous for its rich collection of Buddhist sculptures, paintings, and publications, attracting scholarly attention as early as the 1900s[1].

The history of Chongshansi can be traced back to the Hongwu period during the Ming Dynasty. It was a royal temple founded by Zhu Gang 朱棡, the Prince Gong of Jin 晉恭王, and the third son of Emperor Hongwu Zhu Yuanzhang 朱元璋. Throughout the Ming Dynasty, it served as the Prefectural Buddhist Registry "Senggang Si" 僧綱司 of Taiyuan, known as the "family temple" 家廟 of the Jin Principality 晉藩. Throughout history, Chongshansi has experienced rises and declines along with the vicissitudes of the Jin Principality and the political changes of the Ming and Qing Dynasties. Studies on the imperial clan 宗藩 of the Ming Dynasty have mainly focused on the political and economic aspects. Since the 21st century, academic interest in the cultural (Wang 2012) and artistic achievements of the imperial clan (Clunas 2013) has gradually increased. Unfortunately, due to a lack of historical materials, specialized studies on the Jin Principality are rare, let alone Chongshansi, a religious site that embraced an intimate relationship with the Princes of Jin. Although studies on the temple buildings appeared early, progress has been limited. In the research context, studies on the temple have mainly been based on *Taiyuan Chongshansi wenwu tulu* 太原崇善寺文物圖錄 (the catalog of cultural relics of Chongshansi in Taiyuan) (Zhang and An 1987). The architecture and artistic works of Chongshansi introduced to the west by M. S. Weidner (Weidner 2001), for example, came from the *Tulu*. She interpreted their characteristics as an iconographical representation of the empire's authority. However, although the *Tulu* provides a rich documentation, its historicity is not very reliable due to a lack of proper analysis. Discussions on the historical context of the temple, as well as its physical space, are also lacking. To fully understand the historical transformation of the temple, careful studies on the historical facts in the founding phase, as well as the physical traces still left in the surrounding environment, are needed.

Figure 1. The Great Compassion Hall of Chongshansi (photo by Chongshansi, 2006, and provided by Chongshansi).

To fully grasp the history of Chongshansi, this article refers to multiple historical documents, including the official historical records the *Ming Shilu* 明實錄 that document the political life of emperors during the Ming Dynasty, the *Taizu Huangdi Qinlu* 太祖皇帝欽錄 (see Chen 2003) once kept by the Jin Principality and now stored in the Palace Museum in Taipei, and gazetteers of various versions. It also consults the antiques reserved in the temple, including a wooden inscription, *The Founding Story of the Temple* 建寺緣由, stone inscriptions that document the restoration of the temple, and the *Plan of Chongshansi* 崇善寺全圖 that depicts the layout of the temple. Furthermore, the preservation project of the Great Compassion Hall starting in 2019 offered an opportunity to carry out an investigation not only on the literature but also on the cultural relics and the temple buildings. On the basis of the physical and textual materials collected from the on-site survey, this article is aimed at analyzing the historical background of the erection of the temple and its relation to the Jin Principality. Its geographical relationship with the newly built Princely Palace reveals its role as the Buddhist registry for the new Dynasty. The new palace generated a new urban ritual axis. It was the temple's transverse connection with the axis that included it into the ritual spatial system of the Jin Principality within the city of Taiyuan. Unfortunately, with the fall of the Ming Dynasty, the temple no longer occupied a key role in the city. During the Qing Dynasty, the vast land of the temple was gradually abandoned and occupied by other functional buildings. By the time a fire took place in the mid-19th century, the temple had lost most of its land. The ritual lane once connecting the temple to the main road of the city was also interrupted by the emergence of a modern school nearby, which eventually overwhelmed the temple in the quick urbanization following the Qing Dynasty.

2. Historical Background of the Foundation of Chongshansi and Its Official Identity

As Chongshansi is closely related to the Jin Principality, it is necessary to introduce the successive princes first to clarify the historical activities of different Jin generations. The Jin Principality had 13 princes (Figure 2), of which Prince Gong, Prince Ding and his younger brother (a deposed prince), Prince Zhuang, and Prince Jian are connected to our historical survey.

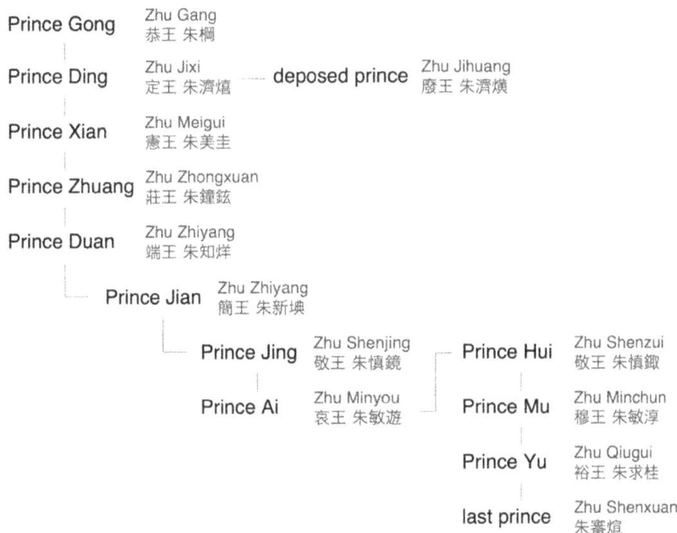

Figure 2. Simplified generations of the Princes of Jin (drawing by author).

Compared to other monasteries in Taiyuan, Chongshansi played a crucial political role during the early Ming Dynasty. Its founding was intimately connected to a series of religious reorganization policies at that time. In earlier China, Buddhist monasteries usually served as prayer sites for the nation. This function was strengthened after religious regulations issued on Buddhist monasteries during the early Ming Dynasty, during which the Buddhist monasteries helped assist and stabilize national institutions. Chongshansi was constructed by the Prince of Jin precisely as a national institution, and it was the location of the prefectural Buddhist registry Senggang Si of Taiyuan.

The historical literature differs with respect to the founding year of Chongshansi: the sixth year (1373), the 14th year (1381), the 16th year (1383), or the 24th year (1391) of Hongwu. The insufficiency of information makes it impossible to confirm which date is correct. However, it should be noted that, in ancient Chinese literature, the "founding" of temples or monasteries does not necessarily refer to the physical completion of the buildings. More often, it refers to the establishment of the institution. The four dates are all connected to the Buddhist consolidation during the early Ming Dynasty, which explains the "founding" of Chongshansi as an institution.

It is difficult to clearly historize the founding of the temple. The stories of the Prince Gong of Jin in the Hongwu period were to some extent falsified after his younger brother Zhu Di 朱棣 took the throne, becoming Emperor Yongle[2] 永樂 (Yang 2015; Yang 2021). Moreover, the internal conflicts within the Jin household (Meng and Zhang 2017) from the Yongle to Xuande 宣德 periods (1403–1435) made it unlikely for the Princes of Jin to document minor issues such as the founding of Chongshansi[3].

The Chongshansi was not newly built during the Ming Dynasty; at its location, there was a predecessor. The *Yuan yitong zhi* 元一統志, a gazetteer compiled around 1286 that survived in the *Yongle Canon* 永樂大典, may provide some useful information. *Yuan yitong zhi* records that a rebuilt temple called Yanshousi 延壽寺 stood 2 li 里 to the east side of Taiyuan City.[4] Chongshansi is 1 km (almost 2 li) away from the east wall of the old city. *The Inscription of the Restoration of Chongshansi* written by Kong Tianyin 孔天胤 in 1563 also marked the excavation of a stone tablet from Yanshousi[5]. In this case, at least around 1286 when the *Yuan yitong zhi* was compiled, there was indeed a temple called Yanshousi at the site where Chongshansi was located.

Furthermore, Chongshansi may have merged before being affirmed as an official temple. Among various sources regarding the founding of the temple, the *Shanxi tongzhi* (comprehensive gazetteer of Shanxi) published in 1475 mentions that the temple was erected as the merging of two monasteries. It says "Chongshansi was set up in the sixth year of Hongwu with the Senggang Si located therein. Meanwhile, two monasteries, Wenshusi 文殊寺 and Anguosi 安國寺, were merged [into Chongshansi]".[6] The *Tulu* treats the date of the sixth year as a miswriting of the 16th year of Hongwu. However, this date should not be easily ignored, for it exactly coincided with the time when Emperor Hongwu decreed the order of the merging of temples and monasteries nationwide. There were several mergers of Buddhist temples and monasteries during the reign of Emperor Hongwu. In July of the fifth year of Hongwu (1372), for example, the monks and properties of Tianxisi 天禧寺 and Nengrensi 能仁寺 in Nanjing were merged into Jiangshansi 蔣山寺[7]. In December of the sixth year of Hongwu, the merger was implemented nationwide, ordering that each prefecture should only keep one temple, with other monasteries being merged and manipulated[8] (Du 2013, pp. 40–48). Although some scholars have pointed out that the policy was not thoroughly carried out (He and Li 2018), this was undoubtedly not the case in Taiyuan, one of the most important political and military centers of north China. It is highly possible that Chongshansi underwent an imperial-decreed merger before it started a great bustle of masons and carpenters under the name of *Chongshan Chansi* 崇善禪寺.

In the restoration inscriptions composed by Prince Zhuang of Jin in 1480[9] and Prince Jian in 1563[10], the founding date of Chongshansi is recorded as the 14th year of Hongwu (1381). This was also the year in which Emperor Hongwu tightened his rule over Buddhist institutions. From 1381 onward, Emperor Hongwu's attitude toward Buddhism changed radically from supportive to discouraging. He ordered the Ministry of Rites to formulate a policy, establishing a bureaucratic structure of Buddhist registries at the national (Senglu Si 僧錄司), prefectural (Senggang Si 僧綱司), sub-prefectural (Sengzheng Si) 僧正司, and county (Senghui Si 僧會司) levels, and this policy was implemented on 24 June 1381. Moreover, he required local monasteries to report to the government their founding members and date to receive an official name plaque from the emperor[11] (Brook 2005, p. 127). In April of the following year, Emperor Hongwu commanded the formal setup of bureaucratic institutions[12]. In May, monasteries all over China were categorized into three types: meditation 禪寺 (to concentrate on meditational exercises), doctrine 講寺 (to study the scriptures to penetrate their meaning), and teaching 教寺 (to go out among the people to preach and conduct rites, especially funerary rites).[13] In the same year, on 10 August, Queen Ma, Emperor Hongwu's wife, passed away.

Several materials regard the death year of Queen Ma as the founding time of the temple. *The Founding Story of the Temple*, a wooden inscription preserved in the temple, marks the founding of the temple as a memorial to the Queen in the 16th year of Hongwu (1383). This inscription records that the prince asked Marquis Yongping 永平侯 to propose to the emperor to erect a new temple in April of the 16th year of Hongwu (1383). It was constructed under the supervision of General Yuan Hong 袁弘 (see Appendix A.1). The *Tulu* also takes the death of Queen Ma as significant evidence that the temple was erected by Prince Gong thereafter. However, since the temple was appointed as Senggang Si of Taiyuan, its political role had to have been affirmed before April of the 15th year of Hongwu. In other words, Queen Ma's death may not be closely related to the temple's erection, whether physically or institutionally.

The 24th year of Hongwu (1391) is another important date in the history of Buddhism during the Ming Dynasty. There are two pairs of iron lions in front of the Lingxing Gate of Wenmiao and the Great Compassion Hall complex, which are dated Xinwei Year of Hongwu 洪武辛未 (1391) (Figure 3). This complies with the completion of the temple in *The Founding Story of the Temple*. In June of that year, one of the most heavy-handed policies on Buddhist consolidation during the Ming Dynasty, the *Declaration of Buddhist List* 申明佛教榜冊, was issued, and its provisions had to be carried out within 100 days. In the following month, another imperial edict was released to "forbid the monastics from having the reside with the lay citizens. A temple with over 30 monks was to house the monks, while a temple with fewer than 20 monks was to be merged with another temple"[14]. It is this thorough national rectification movement that the "consolidation of Buddhism" 清理佛教事 in the wooden inscription *The Founding Story of the Temple* refers to. After this movement was conducted in Taiyuan, Prince Gong entitled the temple Chongshan Chansi 崇善禪寺, appointing it the central Buddhist institution in Taiyuan.

Figure 3. Iron lions in front of the Lingxing Gate of Wenmiao inscribed "made by the princely establishment of Jin in Xinwei Year of Hongwu" 洪武辛未晉府造 (photo by author, 18 January 2020).

Chongshansi became the place for regulating the local Buddhist affairs and staging national Buddhist ceremonies once it was erected. According to the inscription written by Prince Jian in 1563: "after its erection, large ceremonies were held in Chongshansi on every New Year's Day, the Winter Solstice, the emperor's birthday, and the reception of the emperor's envoys for nearly 200 years. It is a place for the Jin Principality to be mourned and should never be abandoned" (see Appendix A.3). It is clear that Chongshansi assumed the responsibility of holding a number of national celebrations. It even acted as the representative of the Jin Principality to receive envoys from the Imperial court. It should be stated that the temple was erected under the guidance of institutional reorganization during the early Ming Dynasty, which accounts for its duty in governing the local Buddhist affairs and conducting state Buddhist rituals within the prince's territory.

Chongshansi suffered rises and declines during the Ming Dynasty. After Prince Gong's death, many monasteries under his support soon fell apart. With the reduction in the princes' military authority starting from the Yongle period (Zhang 1982), Chongshansi also lost its prominent role. It was not until the Chenghua 成化 period (1465–1487) that the temple recovered its strength in local religious institutions with the return of the Jin Principality (see Lü 2020a).

The Founding Story of the Temple also documents land donation to the temple from Prince Ding of Jin 晉定王, the eldest son of Prince Gong (Lü 2020b). In September of the 12th year of Yongle (1414), he granted the temple 9 Qing (ca. 57 ha) of land in memory of his father. He also declared lasting financial support from the family (see Appendix A.1). The *Tulu* incorrectly dated the making of the wooden inscription as the 12th year of Yongle (1414). Considering "Ding" 定 as his posthumous title 謚號, the wooden inscription, therefore, could only have been made after his death in the 10th year of Xuande (1435). Moreover, "September of the 12th year of Yongle" is a date that points to the change of power within the Jin Principality.

After Prince Gong died suddenly in 1398, his heir apparent Jixi 濟熺 inherited his title. After Emperor Yongle took over the empire by force (1402), Jixi was very often framed for revolt by his younger brother Jihuang 濟熿. In September of the 12th year of Yongle (1414), Jihuang was entitled the Prince of Jin, whereas, in November, Jixi was deprived of royal identity and put under house arrest with his son Meigui 美圭. It was not until nine years later, in 1423, that Emperor Yongle released them and granted Meigui the title of Commandery Prince Pingyang 平陽王. They were forced to leave Taiyuan for Pingyang 平陽 (now the city of Linfen 臨汾). In 1427, Jihuang was deprived of his princely title for participating in revolt, and the position of the Prince of Jin was suspended for eight years. It was not until 1435 that Meigui was entitled the Prince of Jin and returned to Taiyuan. Jixi died before the emperor's messenger arrived and received the posthumous title of *Ding*.[15]

Some scholars consider the internal conflicts within the Jin Principality as the result of the centralized autocratic rule and the reduction in the rights of feudal princes in the Yongle period and thereafter (e.g., Sato 1999, pp. 62–76; Zhang 2006; Meng and Zhang 2017). For the Jin Principality, this turmoil lasted more than 20 years. *The Founding Story of the Temple* does not mention anything about Jixi's loss of the position of the imperial prince, but particularly marks the date he donated the land to the temple, the same September that he was deprived of the position by his brother. If the donation date was true, is there a possibility that Jixi transferred his assets voluntarily or involuntarily? Alternatively, was this a deliberate move by Jixi's descendants to rewrite the sorrowful moment of the past? Whatever the truth is, the donation of property from the prince that lost his power to the temple confirms its delicate position between the imperial court and the Jin Principality. It further verifies that the relationship between the Jin Principality and the imperial court determined the rise or fall of Chongshansi.

3. The Ritual Path between Chongshansi and the Princely Palace

Chongshansi is closely related to Jin Principality not only by ritual jurisdiction but also by its geographical relationship with the Princely Palace of Jin 晉王宮. Among the materials gathered in the *Tulu*, the *Plan of Chongshansi* deserves close attention. It is a hanging scroll painting[16] that depicts the whole temple in its heyday (Figure 4). The *Plan* was already photographed in detail in the 1940s (see Li 2003). The renowned architectural historian Liu Dunzhen included the plan together with a diagram of the plan and a restored bird view[17] in *Zhongguo gudai jianzhushi* (the history of ancient Chinese architecture) published in the 1980s (Liu 1984, pp. 13, 372–73). Nonetheless, a deep survey of the *Plan* has been lacking until now. Together with the *Plan of Chongshansi* and other historical materials, the aim was to clarify the ritual order existing among the temple, the Princely Palace, and Taiyuan during the early Ming Dynasty.

Figure 4. The *Plan of Chongshansi* (photo by author, 25 November 2020).

The *Plan* is noted for marking the heyday of Chongshansi after its full recovery in the Chenghua period. Although the *Plan* itself is undated[18], by comparing buildings depicted in the *Plan* and the epigraphical text, an approximate period can be deduced. Two Qielan Halls 伽藍殿 on the east and west sides between the Heavenly Kings Hall 天王殿 and the Vajrapani Hall 金剛殿 are depicted in the painting, which complies with *The Inscription of the Restoration of Chongshansi* (1480) that "build the Qielan Halls facing each other" 增蓋伽藍神祠左右相向 (see Appendix A.2). However, the painting lacks any pavilion as mentioned in *The Inscription of the Restoration of Chongshansi* written by Kong Tianyin (1563) (see (Zhang et al. 2007, pp. 393–94)), that "six pavilions are added for the bell, drum, and tablet" 增置鐘鼓碑亭六座. In this case, the *Plan* was established at a point between these two restorations of the temple (1472–1563). It was Prince Zhuang who carried out this restoration. He was also renowned for his good artistic taste due to a large number of collections of rare editions (Clunas 2013).

Given the tradition that Chinese paintings often tend to represent architecture in an abstract and formulaic manner, it is necessary to examine at first if the carefully painted *Plan* is a faithful depiction of the actual temple. The precision of the *Plan* can be determined from a comparison between the existing early Ming Great Compassion Hall (Zhou and Wen 2021) and its portrait in the *Plan* (Figure 5). Fronted by a pair of iron lions, the Great Compassion Hall is depicted as a double-eave seven-bay hall crowned with a gable-and-hip roof and enclosed by thick walls with three frontal openings, which is exactly what it looks like today. Because all other buildings in the *Plan* are painted as hip-roofed structures for a better painterly effect, the Great Compassion Hall is considered to be represented faithfully in the *Plan*, along with other buildings of the temple. Therefore, we can safely rely on the *Plan* to carry out a restoration of Chongshansi to the cityscape.

Figure 5. The Great Compassion Hall in the *Plan of Chongshansi* (photo by author, 25 November 2020).

One notable feature of the temple shown in the *Plan* is its central axis. Unlike ordinary urban monasteries in China whose southern gates align with the central axis directly open to the city street, the southern gate of Chongshansi was merely symbolic. The axial route was terminated in the south by Pailiang Gate 排梁門 and its screen wall 照壁, while the route turned either to the east or to the west side gate that was the actual opening to the outside. The transition of the axis from north–south to east–west reveals the temple's relation to the expansion of the ancient town of Taiyuan and the location of the Princely Palace of Jin.

During the hasty construction of the principalities during the early Ming Dynasty, the city of Taiyuan changed substantially. The location and the spatial layout of Chongshansi are intricately connected to the construction of the city and the Princely Palace. In April of the third year of Hongwu (1370), Emperor Hongwu granted his sons titles and land. The third son, Zhu Gang, the Prince of Jin in Taiyuan, started to construct his palace there in July. Among the elderly princes, the second son, Zhu Shuang 朱樉, inherited the former administrative office of Shaanxi Province as the Princely Palace of Qin 秦王宮, while the fourth son, Zhu Di, renovated the former palace from Yuan Dynasty as the Princely Palace of Yan 燕王宮. Only the Princely Palace of Jin was newly built.[19] Since the old Taiyuan city was too small to assume its role as a crucial strategic position to the north of the empire, the construction of the Princely Palace of Jin was combined with the task of the expansion of the city. In February of the fourth year of Hongwu (1371), Cao Xing 曹興, the chief princely officer of Jin Principality, proposed to Emperor Hongwu the construction of a new city and palace.[20] The construction of the city started in the eighth year of Hongwu (1375) and generally finished in the following year. During the construction, the original city was extended on three sides toward the east, south, and north. The perimeter of the city wall was enlarged to 24 li with eight gates[21]. The Princely Palace of Jin, occupying vast land in the east of the city, was almost completed at the same time. Similar to the construction of other principalities, the expanded Taiyuan and the Princely Palace enjoyed a huge scale and luxurious decorations, which aroused the deep concern of some far-sighted officers[22]. The design of the new city, characterized by the centrality of the huge palace, generated an entirely new ritual spatial system in the eastern region. This new urban ritual center, separating itself from the Song Dynasty administrative zone, marketing zone, education zone, public zone, and other functional areas, extended its central axis to the southern city gate Cheng'en 承恩門 to form a new ritual axis of the New South Gate Street 新南門街 (Zang 1983; Ma et al. 2013, pp. 32–33) (Figure 6).

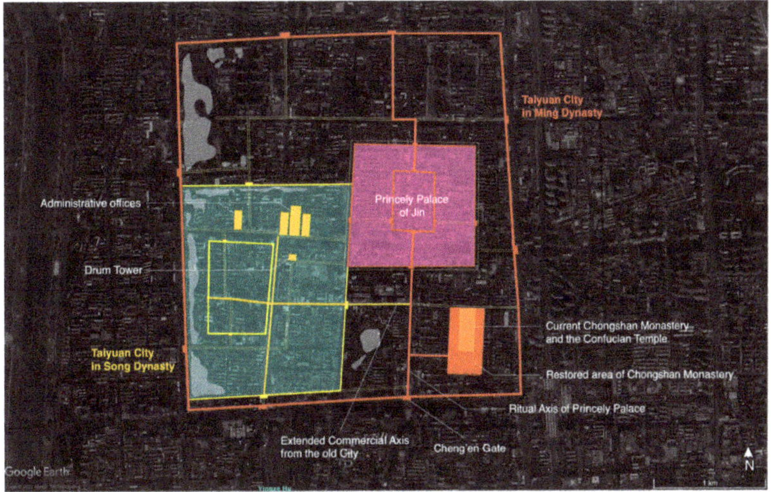

Figure 6. The functional distribution of Taiyuan City in the Ming Dynasty (drawing by author, after (Ma et al. 2013, p. 43), Figures 2.1–4, based on Google Maps 2022).

It is the east–west axis before the Pailiang Gate depicted in the *Plan* that led Chongshansi to the New South Gate Street. Ming princes were obliged to represent the emperor's authority in the regions assigned to them and pray for the empire and their principalities via Buddhist rituals (Luo 2013). In the existing documents, descriptions are lacking for how the Princes of Jin performed rituals in Chongshansi. However, the Great Shuilu Assembly 水陸法會 held by Emperor Hongwu in the capital Nanjing may provide a reliable reference. This ceremony is usually held after war to redeem lost souls from hell by chanting sutras and making offerings. From the first year to the fifth year of the Hongwu period (1368–1372), Emperor Hongwu convoked several great ceremonies in Jiangshansi in Nanjing. Among them, the one held in January of the fifth year was of the biggest scale and the highest rank. The details of the ceremony were documented by Song Lian 宋濂 in *Jiangshan Guangjian Fohui Ji* 蔣山寺廣薦佛會記: "the emperor came to the Fengtian Front Hall 奉天前殿, the main audience hall of the imperial palace, accompanied by the officers. Here, the prayer, written and stamped by the emperor, was sealed and given to the Chief Officer Tao Kai 陶凱 from the Ministry of Rites. Tao walked out of the palace through the central Wu Gate 午門 and took the prayer into the dragon carriage, via which it was sent to the Jiangshansi with guards and an orchestra. It was greeted by the monks of the Great Buddha's Hall." (see Du 2013; He 2013, pp. 354–56). In the early Hongwu period, Taiyuan was a place full of military conflicts and bloody battles where Prince Gong himself led the army to fight with the Mongol troops. It was reasonable for him to hold such great ceremonies there, as Emperor Hongwu did in Jiangshansi. Accordingly, we could speculate a ritual program that began with the prince's procession from the Nanhua Gate 南華門, the south gate of this palace, toward the south, before turning east, going through the west side gate, stopping before Pailiang Gate, and then entering the temple along the central axis.

In *Chengchi tu* 城池圖 (map of the city) in the 1682 gazetteer of Yangqu county, the street directly connecting Chongshansi and the New South Gate Street is clearly depicted. In *Jiexiang tu* 街巷圖 (map of streets and lanes) of the 1843 gazetteer of Yangqu county, the temple is found facing west, intuitively revealing the west as the key position and reflecting the description of political space in the traditional Chinese map (Figure 7).

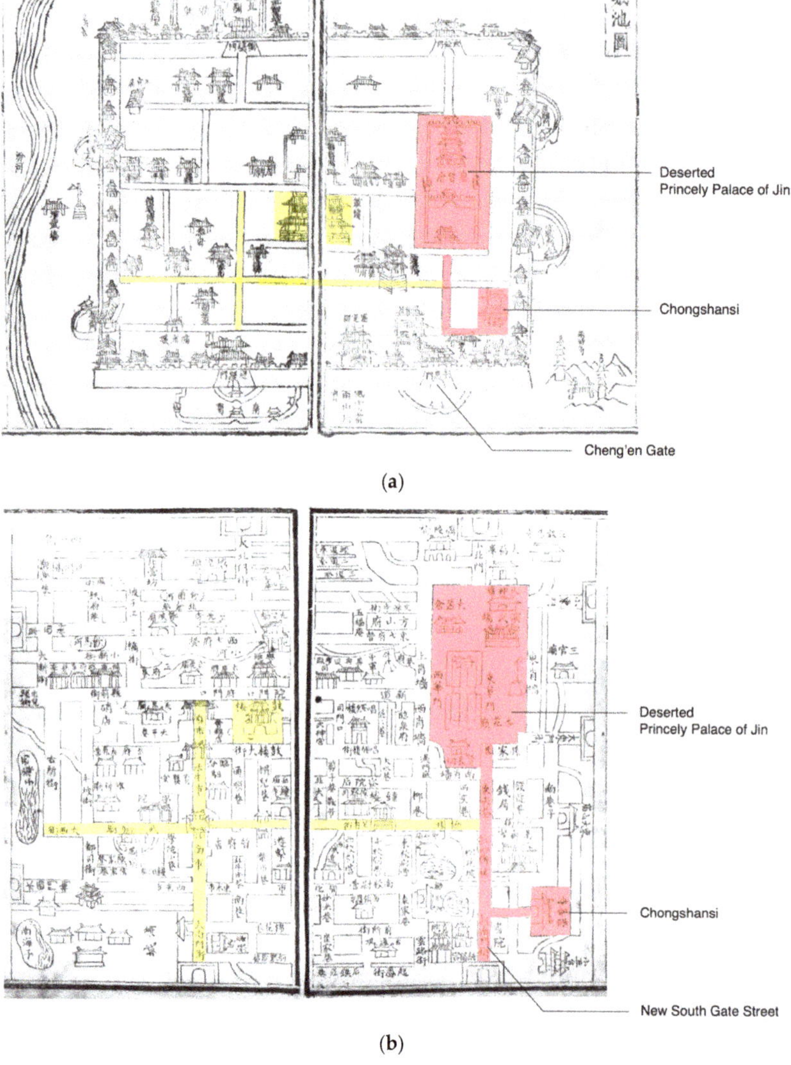

Figure 7. Chongshansi in city maps published during the Qing Dynasty: (**a**) *Chengchi tu* 城池圖 in Kangxi *Yangqu xianzhi* (1682); (**b**) *Jiexiang tu* 街巷圖 in Daoguang *Yangqu xianzhi* (1843).

4. Chongshansi under the Urban Spatial Reproduction

As a royal temple, the close relationship between Chongshansi and the Princes of Jin could be first seen in the location of the temple. However, this physical trace is hardly evident due to the extinction of the Jin Principality, as well as the rapid urban redevelopment of Taiyuan in recent years. Chongshansi underwent a long period of decline from the supreme temple of Taiyuan to the current small-scale temple with only one main hall along its axis.

The current Chongshansi comprises three parallel courtyards. The seven-bay Great Compassion Hall is situated in the middle courtyard. It was built at the founding stage during the Ming Dynasty. The western courtyard, entirely rebuilt after 1992, is the monastic dormitory. The eastern courtyard was built even later after the reclamation of the land in 2005, containing the abbot's living quarter 方丈, the Dharma Hall 法堂, and the canteen. Apart from the Great Compassion Hall, the remainder of the current temple has nothing to do with the Ming Dynasty Chongshansi.

To trace the Ming origin of the temple, the first step is to know its exact scale and boundaries. Fortunately, *The Inscription of the Restoration of Chongshan Temple* written by Prince Zhuang in 1480 is still standing in front of the Great Compassion Hall. The inscription gives a precise measurement of the temple as 344 *bu* 步 long from south to north and 176 *bu* wide from west to east. According to the Ming standard, 1 *bu* 步 is equal to 0.5 *chi* 尺, and 1 *chi* in terms of land measurement is approximately 32.64 cm (see Wu et al. 2005). Accordingly, the temple measures 561 m long from south to north and 287 m wide from west to east[23]. Compared to the current urban blocks, the Ming temple was twice the size of the existing temple from Dilianggong Street 狄梁公街 in the west to Wenmiao Lane 文廟巷 in the east. Furthermore, as the northern boundary of the temple could not go beyond Shangma Street 上馬街 constructed at the beginning of the Ming Dynasty, the southern boundary was likely located around Houjia Lane 侯家巷, very close to the southern city wall (now East Wuyi Boulevard 五一東街) (Figure 8).

Figure 8. Estimated scope of Chongshansi in Taiyuan (drawing by author, based on Google Maps, 2022).

The inscription also documents various important buildings in the temple during Prince Gong's period: "the Main Buddha Hall 正佛殿 is 9-bay wide and around 7-zhang (ca. 22 m) high surrounded by marble balustrades. The roof is covered with dragon- and fish-shaped tiles. There is a 104-bay cloister circling the courtyard of the Main Buddha Hall. Behind the Main Buddha Hall is the 7-bay-wide Great Compassion Hall, whose east and west verandahs were used to worship the 18 disciples of the Buddha. The front gate house of the temple is 3-bay wide[24], where the statues of the Vajrapani stand. The second gate hall is 5-bay wide, where the statues of the Four Heavenly Kings align. The temple boasts all kinds of magnificent religious buildings, including the Scripture Library 经阁, the Dharma Hall 法堂, the abbot's chamber 方丈, the monks' dormitory, the kitchen, the meditation hall 禪室, the well pavilion 井亭, and the revolving sutra cabinet 藏輪" (see Appendix A.2).

The information provided by the inscription can be used to compare with the *Plan*. The *Plan* depicts the scope of the grand temple and its various buildings. On the central axis, there are six halls marking out different parts of the plan: the Main Hall complex surrounded by a cloister in the middle, the Great Compassion Hall complex in the north, 16 small-scale courtyards in the east and the west, the affiliated courtyards such as warehouses in the south, and several gardens of different sizes. On the basis of the *Plan*, further physical evidence can also be traced to the in-situ investigation. All buildings in the Main Hall complex, including the Hall of Heavenly Kings, the Main Hall, the Vairocana Hall 毘盧殿, the Eastern and Western *Tuan* Halls 东西團殿, and the cloister in the painting were found lifted upon a base much higher than the ground. The leveling difference can be detected in Wenmiao to the south, separated from the temple only by an alley. Wenmiao, built in 1882, was constructed on the leftovers of the Main Hall complex of the temple burnt down by a devastating fire in 1864. The current foundation level of Chongsheng Shrine 崇聖祠 to the north of the temple is 2.6 m higher than that of the alley in front of Chongshansi, while the ground level of Lingxing Gate 欞星門 of the temple is also higher than the forefront plaza. Moreover, the two octagonal pavilions in front of Lingxing Gate match the two well pavilions behind Vajrapani Hall described in the *Plan*. Although the pavilions were partially renovated, the form of the bracket sets and the evident incline of the pillars 柱身側腳 show typical features of official Ming style. The plot to the east side of the temple is now a residential community constructed in the 1990s, where six aligned square stones were excavated adjacent to Wenmiao Lane. The top of the stones is cut flat. The side length is 600–686 mm long and the spacing between two stones is 4.4–4.5 m. They were most likely the pillar bases of the cloister to the Main Hall. There are some stone structures included in the west wall of the community close to the Wenmiao. They may have been pillar bases and stone strips at the periphery of the foundation. The top level is 0.9 m lower than the ground level of the Dacheng Hall 大成殿, while it is 1.01 m higher than the pillar base of the cloister. Accordingly, they might be the remains of *Tuan* Hall in the east (Figure 9).

The scope of the Main Hall complex can, thus, be located according to the *Plan*, the inscription, and evidence found on the site. Wenmiao Lane and Dilianggong Street to the east and west are exactly the east and west paved lanes 甬路 alongside the Main Hall complex in the *Plan*. Vairocana Hall to the north of the Main Hall complies with the current Chongsheng Shrine, while the Heavenly Kings Hall was situated at Lingxing Gate.

In mid-Ming, the city of Taiyuan started to decay. The walls and the gate towers were damaged, and the demographics during the Wanli 萬曆 period (1573–1620) fell to one-quarter of those during the Hongwu period; after the Ming Dynasty, and only the market zone was still prosperous in Taiyuan (Wang 2004). The crucial political role of Chongshansi changed greatly due to the fall of the Jin Principality. Although the temple maintained its position as the prefectural Buddhist registry Senggang Si of Taiyuan, its cultural significance was greatly lost. During the Ming–Qing upheaval, the temple provided a meeting place for the Ming loyalties to plan rebellions[25]. In April of the third year of Shunzhi 順治 (1646), the Princely Palace of Jin was burnt down. It was expropriated as a troop camp in the 10th year of Yongzheng 雍正 (1732)[26]. By that time, the ritual order of the temple no longer existed. The decline of the temple is clearly seen in the Qing Dynasty literary works. After a visit around 1727, the poet Wei Yuanshu depicted the temple as "with empty corridors and rotten wall paintings, the wind roars like ghosts without fear"[27]. In the 35th year of Qianlong 乾隆 (1770), Dilianggong's Shrine 狄梁公祠 was moved to the empty plot to the west of the temple[28], which was the origin of the name of Dilianggong Street. In other words, at that time, many of the courtyards in the west of the temple had been abandoned. In the third year of Tongzhi 同治 (1864), the Main Hall complex was destroyed in a fire, and the temple was converted into the educational institution of Chongxiu Academy 崇修書院 the following year.[29] In 1881, Wenmiao was rebuilt on the ruined site of the Main Hall complex.[30] At that time, only the Great Compassion Hall survived but was isolated from the surrounding urban environment.

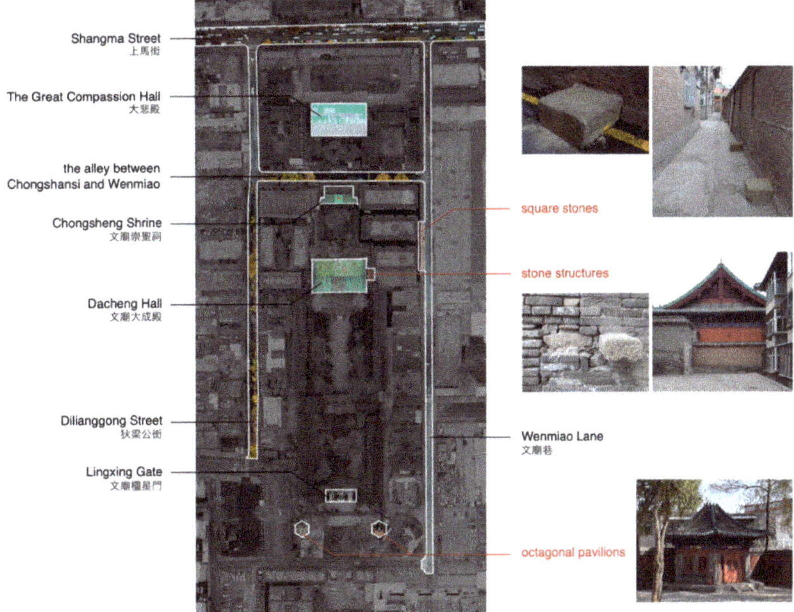

Figure 9. Traces of the Main Hall Complex left in the Chongshansi Community (aerial photo by author, 21 November 2020; drawing and photos by author, 23 March 2021).

The southeastern corner in the ancient town of Taiyuan also changed substantially after the Qing Dynasty. The land of the temple was gradually encroached upon. In the 13th year of Guangxu 光緒 (1887), the new Manchurian City 新滿城 was erected there. Its west wall extended to the east boundary of the temple (Zhu and Han 2006). The area to the west of the temple gradually became an education district starting from the setup of Shanxi Academy. Many schools found themselves a place in the area between Wenmiao and the New South Gate Street (Taiyuan Shi Jiaoyu Weiyuanhui 1990, p. 14; Jia 2015, pp. 96–111).

The street connecting the New South Gate Street and the temple no longer works in contemporary Taiyuan city. However, there is an L-shaped street called Xinsi Lane 新寺巷 between Shangguan Lane 上官巷 and Houjia Lane 侯家巷. According to the gazetteer of Taiyuan prefecture (1783), the local people at that time preferred to call the temple "the new temple" 新寺[31]. The lane, therefore, was named after the temple (Hao 1956, p. 75). As early as 1919, Xinsi Lane can be seen in the *Shanxi shengcheng xiangtu* (detailed map of Shanxi capital city). To the east is the vast land occupied by Shanxi Academy. The academy bought over 200 acres (ca. 1.3 km^2) of empty land around Hou Family Lane and moved out some residents to build the new campus. The school moved in by the fall of 1904 (Wang 2006, pp. 150–52). Before the founding of the academy, large sums of land in the south of the temple had been abandoned for a long time. With the expansion of the academy, the street connecting the temple directly to the New South Gate Street was interrupted. By that time, the ritual order that the temple helped to forge in the city came to a stop (Figure 10).

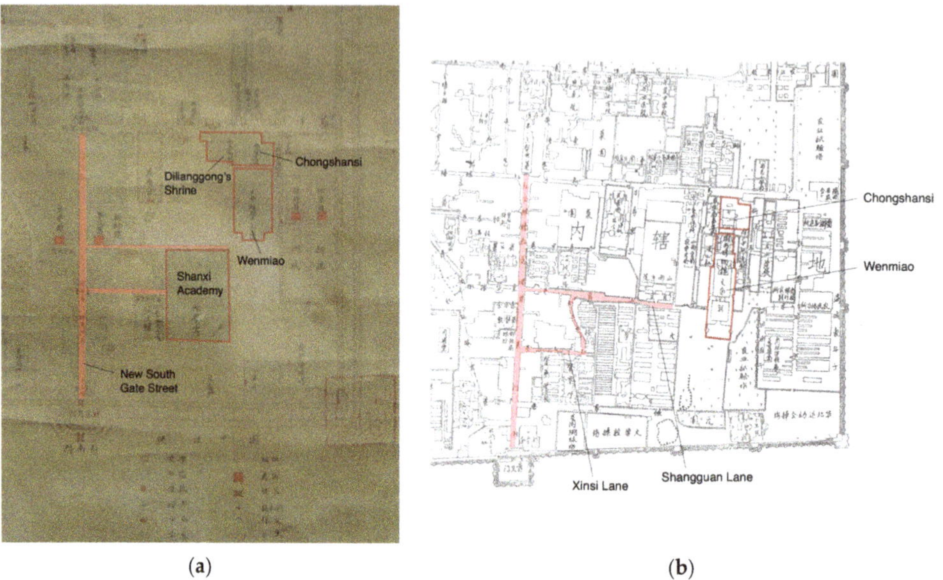

Figure 10. Chongshansi, Wenmiao, and Xinsi Lane in city maps: (**a**) *Shanxi shengcheng quantui* in 1904 (drawing by author based on Zheng 2004, p. 26); (**b**) *Shanxi shengcheng xiangtu* in 1919 (drawing by author based on Taiyuan Shi Nancheng Qu Renmen Zhengfu 1987, Appendix 15).

5. Conclusions

It is indicated that the founding of Chongshansi was deeply connected to the nexus of power in controlling the local government by the central government of the Ming Dynasty. On the one hand, the temple was founded by the Prince of Jin governing Shanxi assigned by the emperor. On the other, under the national Buddhist consolidation, Chongshansi worked from the very beginning as the central Buddhist institution in Taiyuan. Chongshansi enjoyed glory upon its erection under the patronage of Prince Gong of Jin, as early as the Hongwu period. Prince Gong was regarded as a reliable guard of the imperial boundary by his father, while the Jin Principality owned a powerful military force and had a strong political impact. The erection of Chongshansi was an opportunity for Prince Gong to implement the emperor's will by governing the state in his region and to show loyalty to the emperor in Nanjing, while it was also a showcase of royal power to his political rivals in constructing a grand building. In considering this, it is implied that the memorial to Queen Ma should be regarded as one of the functions of Chongshansi rather than the founding reason, let alone its core function. Therefore, the founding of Chongshansi as an institution was earlier than the death of Queen Ma in the 15th year of Hongwu, which also helps to rethink several theories upon the founding time of Chongshansi.

On the other hand, the location of Chongshansi reveals an intimate rapport between the Princely Palace of Jin and the Taiyuan city in the Ming Dynasty. Although the historical documents reveal a former temple at the site of Chongshansi, it was its proper position to the palace that made it the Buddhist registry for the new Dynasty. Prince Gong recovered the strength of the city of Taiyuan ever since its fall after the Song Dynasty. The new city wall included the temple originally located in the outskirts of the inner city, while the newly built Princely Palace of Jin resumed the ritual order of Taiyuan. The central axis of the palace extended to the south into the city, forming the ritual route of the city directed toward the south gate. From the *Plan of Chongshansi*, we can see the north–south ritual axis transferred into an east–west one. By directly connecting to the ritual axis of the city, Chongshansi played a significant role in the ruling system of power.

The collapse of the Ming Dynasty led to the extinction of the palace in the city and the falling apart of the ritual space. Chongshansi became distant from the traditional urban area after the Song Dynasty. Furthermore, it was also marginalized in the new power system and was gradually encroached upon. After the fire during the late Qing Dynasty, most of the temple was reduced to rubble, hindering future renovation. Eventually, it was replaced by the emerging city functions and even lost its connections to the city's main road during the ongoing urban development.

The city of Taiyuan is continuing to embrace renovation and redevelopment. In November 2021, on the site of the former South Gate demolished in 1950, a newly built city gate rose at the cost of removing Wuyi Square 五一廣場, taking with it the collective memory lasting over half a century for the citizens. Its name is not Cheng'en Gate 承恩門 (accepting royal awards) but Shouyi Gate 首義門 (the first place of revolution in Taiyuan) in memory of the modern revolution. Under the impact of this renovated ritual order, the urban regeneration in the southeastern corner of the ancient city is continuously pushing forward. Shangma Street to the north of the temple was widened in 2019, the diminishing cultural relics on the city map are being rebuilt, and Xinsi Lane already has its sign. Chongshansi is facing another round of a construction power system; how it will be manifested in the new system depends upon how its historical value is interpreted.

Funding: This research was funded by Study on the Institution of Buddhist Precept Platform in East Asia, grant number 15AZJ002.

Conflicts of Interest: The author declares no conflict of interest.

Appendix A. Excerpts of Chinese Texts

Appendix A.1. The Founding Story of the Temple 建寺緣由

晉恭王殿下為母后孝慈昭憲至仁文德承天順聖高皇后馬，於洪武十五年八月初十日升霞，無由補報罔極之恩，洪武十六年四月內，令差永平候奏准建立新寺一所，令右護衛指揮使袁弘監修，啓蓋完備。至洪武二十四年清理佛教事，恭王賜額崇善禪寺，就撥施地土一十九頃，永遠與寺里梵修供佛香燈。敬此。永樂十二年九月內晉定王施撥地土九頃，為報先恭王罔極之恩，與崇善寺永遠供佛香燈。敬此。

The mother of Prince Gong of Jin, Queen Ma, died on the 10th of August in the 15th year of Hongwu (1382). In memory of her, the Prince asked Marquis Yongping to propose to the emperor to erect a new temple in April of the 16th year of Hongwu (1383). It was constructed under the supervision of General Yuan Hong. In the 24th year of Hongwu (1391), upon the Buddhist consolidation, the Prince granted the temple its official name Chongshan Chansi on a plaque. He also granted it 19 Qing (ca. 121 ha) land as an eternal offering to the temple. In September of the 12th year of Yongle (1414), the Prince Ding of Jin granted the temple 9 Qing (ca. 57ha) land in memory of Prince Gong, also as an eternal offering from the family.

Appendix A.2. The Inscription of the Restoration of Chongshansi 重修崇善寺碑記, Written by Prince Zhuang of Jin 晉莊王 in 1480, Punctuated and Edited by the Author

佛家者流，謂佛氏西方聖人也，漢明帝時，始入中夏，歷代久遠，蔓延滋甚。當是時，尊禮崇信，釋教大興於世，顧其所尚，夫有自来矣。共惟我太祖高皇帝，奄有四海，百福攸主，佛氏之教以故不遺。太原當西北二邊，山河險固。特勅封先曾祖晉恭王，以仁智英武之資，世守茲邦。不三載威名赫震，羌胡臣伏，邊陲輯寧，軍民熙。誠不負屏藩之托矣。時孝慈昭憲至仁文德承天順聖高皇后崩，先曾祖欲建寺，萃僧修齋禮懺，以報劬勞罔極之恩，及晨昏祝延聖壽。乃詢諸耆耄，僉謂城東隅舊有白馬古，遺基石幢存焉。一聞斯地，遂令闢除榛棘。廣南北長三百四十四步，東西濶一百七十六步。鳩工計材，營建正佛殿九間，週以白石欄楯，螭首承霤，海魚護瓴，合殿穿廊一百零四楹，高十餘仞。後立千手千眼大悲殿七間，東西迴廊有十八羅漢。前門三塑護法金剛，重門五列四大天王，經閣、法堂、方丈、僧舍、廚房、禪室、井亭、藏輪，無不具美矣。成化壬辰夏四月一日，予謁寺炷香，發心捐貲。又增蓋伽藍神祠，左右相向。仍命主持淨金等

眾翻修損壞。金碧丹堊，煥然一新，見者嘆羨不釋於口，屹然山右禪林中第一叢
林也。建在洪武十四年，距今百載矣，惜乎先曾祖恭王薨逝，碑文未勒。迨夫祖
定王、父憲王相繼有此心，未遑暇及。今不舉興，將何以昭後世，遂命匠礱石，
親述建寺始末，鐫諸貞石，庸垂不朽。其釋教之事，世之高明特達者辯論有在，
奚俟予贅自彰矣。

大明成化十六年次庚子秋九月吉日 晉王立

Buddhists believe Buddha is a saint coming from the west. Buddhism was introduced
into China during Emperor Ming's period in the Han Dynasty and was widely spread
thereafter in China for a long time. At that time, people obeyed traditional ethical codes
with deep faith. It is because of this that Buddhism flourished due to its advocating
goodness.

My ancestor, Emperor Taizu, had a huge empire. He was blessed by heaven and
attached great importance to the spread of Buddhism. Taiyuan occupies the western and
northern borders of the empire. In this regard, the emperor appointed my great-grandfather
Prince Gong to guard here. In less than three years, Prince Gong had conquered the enemy.
The battles in the frontier fortress were settled down, while the soldiers and civilians lived
and worked in peace and contentment. He lived up to the emperor's entrustment. After
Queen Ma's death, Prince Gong hoped to build a new temple for commemorating Queen
Ma and praying for the emperor's longevity. After consulting the local elders, Prince Gong
was told that there used to be an ancient temple called Baimasi in the east of the city. It
could be reused, although it was in a state of ruins. Prince Gong then decided to build up a
new temple based on the former Baimasi with an area of 344 steps (ca. 561 m) from north
to south by 176 steps (ca. 287 m) from east to west.

After the temple was built, the Main Buddha Hall is 9-bay wide and around 7-zhang
(ca. 22 m) high surrounded by marble balustrades. The roof is covered with dragon- and
fish-shaped tiles. There is a 104-bay cloister circling the courtyard of the Main Buddha Hall.
Behind the Main Buddha Hall is the 7-bay-wide Great Compassion Hall, whose east and
west verandahs were used to worship the 18 disciples of the Buddha. The front gate house
of the temple is 3-bay wide, where the statues of the Vajrapani stand. The second gate hall
is 5-bay wide, where the statues of the Four Heavenly Princes align. The temple boasts
all kinds of magnificent religious buildings, including the Scripture Library, the Dharma
Hall, the abbot's chamber, the monks' dormitory, the kitchen, the meditation hall, the well
pavilion, and the revolving sutra cabinet.

On the 1st of April, the summer of the 8th year of Chenghua (1472), I visited the temple
and donated money to build the Qielan Halls facing each other. I also ordered the abbot
Jing Jin and other monks to renovate the temple. After the renovation, people praised it and
called it the highest-ranking temple in Shanxi. This temple was founded in the 14th year of
Hongwu (1381) and has enjoyed a history of nearly a hundred years. Unfortunately, my
great-grandfather, Prince Gong, died suddenly and had no chance to write an inscription.
My grandfather, Prince Ding, and my father, Prince Xian, both wanted to erect a monument,
but they were deprived of effort to do it. If I could not document all these histories, they
would not be passed down to posterity. Therefore, I asked a stonemason to make a stone
inscription with my description of the process of building the temple on it, in the hope that
the future generation will remember it and commemorate it. As for Buddhism, there are
lots of much more brilliant people than I already discussing it. So, I would rather spare my
efforts.

Written by the Prince of Jin on an auspicious day of September, in the autumn of the
16th year of Chenghua (1480) in the Ming Dynasty.

Appendix A.3. The Inscription of the Restoration of Chongshansi 重修崇善寺碑記, *Written by
Prince Jian of Jin* 晉簡王 *in 1563, Punctuated and Edited by the Author*

洪武辛酉，我始祖恭王為報聖祖高皇后罔極之恩，奏建寺曰崇善。恪以焚修香
火，祝延聖壽。嗣後凡正旦、冬至、萬壽聖節，率於此習儀，及齎節命使，暫以
駐驛，近二百載。誠一國仰瞻不可廢者。成化弘治間，我高祖莊王、先祖安王蓋

118

嘗增修，兩有碑文，睿製詳悉。今又六十星霜矣。凡塑繪殿廊、莊嚴之具，歷久漸敝。嘉靖乙卯，予命長史馮繼祖、薄世佑、劉顯道、欽賜飛魚服承奉袁定、陳勝、張等詣寺炷香，覩其頹圮。召僧官總督住持備啓爰命，計材集工，筮吉葺理。省諭一下各宗室及內外官庶，亦皆退邁向風，施祿捐金，眾樂欣助，是以上下協和，財用充牣。俾典寶邢欽等分□□□，承奉黃定、溫洋、常保、單賢督率其事，竭力就工，恪誠趨務。典仗桑福等則躬其勤惰而賞罰之。始於丙辰首夏□於庚申孟冬。凡厥殿廊門廡廳堂齋厨及中繪塑，與夫鐘鼓碑樓二百餘楹，悉皆輪奐一新，金碧輝煌，巍峨雄麗，駭心忧目。觀之者自當肅然起敬矣。工既畢，復啓請銘諸石。於惟我國家崇儒重道，化成天下，而文武釋道醫卜亦罔偏廢。然自都下以至藩省郡邑，莫不有寺觀以為焚修祝延之所。此我祖上所建所遺，予當所敬所愛，思不衰者。庸書以垂後。

嘉靖四十二年在昭陽大淵獻莙吉日 晉王立

In the 14th year of Hongwu (1480), my ancestor, Prince Gong, asked the emperor for permission to build Chongshansi in memory of Queen Ma, where he also made offerings and prayed for the emperor's longevity. After its erection, large ceremonies were held in Chongshansi on every New Year's Day, the Winter Solstice, the emperor's birthday, and the reception of the emperor's envoys for nearly 200 years. It is a place for the Jin Principality to be mourned and should never be abandoned. During the reign of Chenghua and Hongzhi (1465–1505), both Prince Zhuang and Prince An ordered additions in the temple and left documenting inscriptions. Sixty years have passed ever since, and the statues, murals, and buildings are dilapidated.

In the 34th year of Jiajing (1555), I ordered Feng Jizu, Bo Shiyou, and Liu Xiandao, the Zhangshi officials of the Jin Principality, and Yuan Ding, Chen Sheng, and Zhang Tang, the Chengfeng officials of the Jin Principality, to make offerings to the Chongshansi. They witnessed the dilapidation of the temple. Therefore, the monks were activated to make a budget and prepare for restoration. All the members of the imperial clan were called and informed of the restoration project. The construction cost was guaranteed by their generous donation. The Dianbao officials Xing Qin, etc. . . . The Chengfeng officials, Huang Ding, Wen Yang, Chang Bao, and Shan Xian were conscientiously involved in managing the project. The Dianzhang official, Sang Fu, etc., supervised the project. They gave rewards or punishments according to the craftsmen's performance. The project started in April of the 35th year of Jiajing (1556) and lasted until October of the 39th year of Jiajing (1560). After the renovation, the temple took on a new look. Buildings such as halls and corridors had a new life, while murals and statues, as well as more than 200 bells and drum pavilions, were almost reborn. All the visitors were surprised at the first sight when they saw the renovated temple and showed their respect afterwards.

After the renovation, I asked a stonemason to make an inscription for documenting the process. Our empire advocates ethical codes for educating its people, while the civil society, the military force, Buddhism, Taoism, medicine, and divination achieve balanced development. From the capital to the provincial capital and the county seat, there are always Buddhist and Taoist institutions for prayers. Chongshansi was founded by my ancestors. I should pay respect to it and put it under good protection away from declining. I hereby document what has happened for future generations.

Written by Prince of Jin on an auspicious day of May, in the 42nd year of Jaijing (1563).

Notes

[1] As early as the 1900s, the German architect Ernst Börschmann included two photos of the temple in his book *Baukunst und Landschaft in China*, depicting the iron lion in front of the temple gate and the Manjusri statue (it is noted as the thousand-armed Kuanyin in the photo, but in fact, it is the thousand-armed Manjusri) in the Great Compassion Hall (Boerschmann 1923, pp. 81, 254). Later, in 1940, the temple became world-renowned following the discovery of the Qisha edition of Tripiṭaka of the Song Dynasty 磧砂藏 (see Sakai 1940; Yoshii 1942).

[2] The founder of the Chongshansi, Prince Gong of Jin, bears several negative records in the official history. However, according to *Taizu Huangdi Qinlu other related literature studies, the records may* 太祖皇帝欽錄 and other related literature studies, the records may have been falsified after Ming Chengzu Zhu Di took the throne by force, as a way to stigmatize his former rival.

[3] Prince Zhuang wrote in 1480: "my great-grandfather, Prince Gong 恭王, died suddenly and had no chance to write an inscription. My grandfather, Prince Ding 定王, and my father, Prince Xian 憲王, both wanted to erect a monument, but they were deprived of effort to do it." (See Appendix A.2).

[4] Yongle 永樂 *Taiyuan fuzhi*, p. 94.

[5] "今掘地而獲石碑，後云延壽寺也" (see (Zhang et al. 2007, pp. 393–94)).

[6] Chenghua成化 *Shanxi tongzhi*, p. 530.

[7] Qinlu Ji, in *Jinling fancha zhi*, p. 49.

[8] *Ming Taizu Shilu*, juan 86, p. 1537.

[9] See Appendix A.2.

[10] See Appendix A.3.

[11] Qinlu Ji, in *Jinling fancha zhi*, pp. 50–52.

[12] *Ming Taizu Shilu*, juan 144, pp. 2262–63.

[13] Qinlu Ji, in *Jinling fancha zhi*, p. 53.

[14] Qinlu Ji, in *Jinling fancha zhi*, pp. 60–63.

[15] See *Ming Taizong Shilu*, *Ming Xuanzong Shilu*, and *Jin Ding Wang kuangzhi*.

[16] The width of the upper side is 865 mm and the lower 880 mm, while the length of the left side is 1407 mm and the length of the right side is 1410 mm.

[17] Drawn by Fu Xinian 傅熹年.

[18] In volume 4 of *Zhongguo gudai jianzhushi* 中國古代建築史 *(第四卷) 元明建築*, Zhang Shiqing 張十慶 refers to the *Plan* made in the 18th year of Chenghua (1482) without evident reference (see Zhang 2001).

[19] *Ming Taizu Shilu*, juan 54, pp. 1060–61.

[20] *Ming Taizu Shilu*, juan 61.

[21] The comprehensive gazetteer of Shanxi (1475) documents the perimeter of the new city wall as "44 li". However, according to current measurements and other historical materials, it should be "24 li" (See Chenghua *Shanxi tongzhi*, p. 238).

[22] Ye Boju葉伯巨reminded Emperor Hongwu that he had "given too much land to the imperial princes", 秦晉燕齊梁楚蜀諸國，無不連邑數十，城郭宮室，亞於天子之都，優之以甲兵之盛, which might become a threat to the empire in the future (see Huang 1961).

[23] Marsha Weidner used the modern chi with 550 yards and 275 yards, respectively, but should have used the chi for land measurement during the Ming Dynasty.

[24] The description of the three-bay-wide front gate does not comply with the *Plan*, which might be a misunderstanding of the inscription. It needs to be further studied.

[25] Chongshansi appeared several times in Fu Shan's articles (see Fu 2016, pp. 16–19; Xie and Ke 2007).

[26] Yongzheng *Shanxi tongzhi*, juan 48, p. 6.

[27] Wei, Yuanshu 魏元樞 *Yu wo zhouxuan ji 與我周旋集*, (published in 1793): "虛廊半落，鬼怪聞嘆喵" (see Yuan 1994, p. 774).

[28] The Chief Officer of Shanxi, Zhu Gui 朱珪 wrote in the *Tang zeng sikong liangguogong diwenhuigong bei 唐贈司空梁國公狄文患公碑* (1770) about the moving of Dilianggong's Shrine to the empty plot west to the temple with old wooden structures, "於城內崇善寺之西隙地移其舊材而建公祠焉".

[29] *Jinzheng jiyao*, juan 23, p. 60.

[30] *Jinzheng jiyao*, juan 37, p. 6.

[31] Qianlong *Taiyuan fuzhi*, p. 1219.

References

Primary Sources

Ming Taizu Shilu 明太祖實錄. *Taipei: Institute of History and Philology, Academia Sinica*, 1962.
Ming Taizong Shilu 明太宗實錄. *Taipei: Institute of History and Philology, Academia Sinica*, 1962.
Ming Xuanzong Shilu 明宣宗實錄. *Taipei: Institute of History and Philology, Academia Sinica*, 1962.
Taizu Huangdi Qinlu 太祖皇帝欽錄. *In Ming Taizu yu Fengyang* 明太祖與鳳陽. *Hefei: Huangshan Shushe*, 2009: 689–712.
Chenghua 成化 Shanxi tongzhi 山西通志 (Comprehensive Gazetteer of Shanxi). 1475. Compiled by Hu Mi 胡謐.
Yongzheng 雍正 Shanxi tongzhi 山西通志 (Comprehensive Gazetteer of Shanxi). 1734. Compiled by Chu Dawen 儲大文.
Guangxu 光緒 Shanxi tongzhi 山西通志 (Comprehensive Gazetteer of Shanxi). 1892. Compiled by Wang Xuan 王軒.
Yongle 永樂 Taiyuan fuzhi太原府志 (Gazetteer of Taiyuan prefecture). 1410. In *Taiyuan fuzhi jiquan* 太原府志集全. Edited by An Jie 安捷. Taiyuan: Shanxi People's Press, 2005: 5–150.

Qianlong 乾隆 Taiyuan fuzhi 太原府志 (Gazetteer of Taiyuan prefecture). 1783. In *Taiyuan fuzhi jiquan* 太原府志集全. Edited by An Jie 安捷. Taiyuan: Shanxi People's Press, 2005: 603–1402.

Kangxi 康熙 Yangqu xianzhi 陽曲縣志 (Gazetteer of Yangqu county). 1682. Compiled by Dai Mengxiong 戴夢熊.

Daoguang 道光 Yangqu xianzhi 陽曲縣志 (Gazetteer of Yangqu county). 1843. Compiled by Li Peiqian 李培謙.

Jinzheng jiyao 晉政輯要. *1887*. Compiled by An Yi 安頤.

Jinling fancha zhi 金陵梵刹志 (*Gazetteer of the Buddhist monasteries of Nanjing*). *1607*. Compiled by Ge Yinliang 葛寅亮. Punctuated and edited by He Xiaorong 何孝榮. Tianjin: Tianjin People's Press, 2007.

Jin Ding Wang kuangzhi 晉定王壙志 (*The inscription of Prince Ding of Jin*). *1435*. In *Sanjin Shike Daquan: Linfen Shi Yaodu Qu Juan* 三晉石刻大全: 臨汾市堯都區卷. Edited by Wang Tianran 王天然 and Wang Jinbao 王金保. Taiyuan: Sanjin Publishing House, 2011: 65–66.

Zhu, Gui 朱珪. 1770. *Tang zeng sikong liangguogong diwenhuigong bei* 唐贈司空梁國公狄文惠公碑. In *Zhizuzhai wenji* 2 知足齋文集. Edited by Wang, Yunwu. Beijing: The Commercial Press, 1936: 21–22.

Secondary Sources

Boerschmann, Ernst. 1923. *Baukunst und Landschaft in China: Eine Reise Durch Zwölf Provinzen*. Orbis terrarum. Berlin: E. Wasmuth a.-g.

Brook, Timothy. 2005. *The Chinese State in Ming Society*, 1st ed. Asia's transformations. New York: RoutledgeCurzon.

Chen, Xuelin 陳學霖. 2003. Guanyu 'Ming Taizu huangdi qinlu' de shiliao 關於《明太祖皇帝欽錄》的史料. *Jinan Shixue* 217–29.

Clunas, Craig. 2013. *Screen of Kings: Royal Art and Power in Ming China*. London: Reaktion Books Ltd.

Du, Changchun 杜常順. 2013. *Mingchao Gongting yu Fojiao Guanxi Yanjiu* 明朝宮廷與佛教關係研究. Beijing: China Social Science Press.

Fu, Shan 傅山. 2016. *Fushan Quanshu* 傅山全書 第2冊. Taiyuan: Shanxi People's Press.

Hao, Shuhou 郝樹侯. 1956. *Taiyuan Shihua* 太原史話. Taiyuan: Shanxi People's Press.

He, Xiaorong 何孝榮. 2013. *Mingdai Nanjing Siyuan Yanjiu* 明代南京寺院研究. Beijing: Forbidden City Press.

He, Xiaorong 何孝榮, and Mingyang Li 李明陽. 2018. Lun Mingchu de fojiao siyuan guibing yundong 論明初的佛教寺院歸並運動. *Nankai Journal (Philosophy, Literature and Social)* 2018: 100–13.

Huang, Zhangjian 黃彰健. 1961. Lun Huangmingzuxun banxing niandai binglun Ming chu fengjian zhuwang zhidu 論皇明祖訓錄頒行年代並論明初封建諸王制度. *Lishi Yuyan Yanjiusuo Jikan* 史語言研究所集刊 32: 119–37.

Jia, Fuqiang 賈富強. 2015. 明清時期太原城市空間演變研究 A Study on Urban Space Evolution of Taiyuan during Ming and Qing Dynasties. Master's dissertation, Northwest Normal University, Lanzhou, China.

Li, Jining 李際寧. 2003. Yibu teshu de tulu: Jieshao Shanxi sheng Taiyuan Chongshansi sheying ji 一部特殊的「圖錄」——介紹山西省太原崇善寺攝影集. *Wen Lü Liu Shang*, April 15.

Liu, Dunzhen 劉敦楨. 1984. *Zhongguo Gudai Jianzhushi 2nd Edition* 中國古代建築史 第2版. Beijing: China Architecture & Building Press.

Lü, Shuang 呂雙. 2020a. Ming dai shanxi jinfan zhengzhi quanshi yanbian yu difang zongjiao de fazhan 明代山西晉藩政治權勢演變與地方宗教網絡的發展. *Folklore Studies* 79–88. [CrossRef]

Lü, Shuang 呂雙. 2020b. Mingdai wangfu xianghuoyuan de tudi youmian wenti: Yi shanxi taiyuan chongshansi weili 明代王府香火院的土地優免問題——以山西太原崇善寺為例. *Journal of Sun Yat-sen University (Social Science Edition)* 60: 19–27. [CrossRef]

Luo, Ying 羅瑩. 2013. Ming dai Qin fan de fodao xinyang tanxi 明代秦藩的佛道信仰探析. *Journal of Southwest Agriculture University (social sciences edition)* 11: 54–60.

Ma, Junhua 馬駿華, Yang Shen 沈旸, Lei Gao 高磊, and Xiaodi Zhou 周小棣. 2013. *Reconstruction of the Historical Impression of Datong, Taiyuan* 雄藩巨鎮 非賢莫居: 太原/大同的城市史意向再造. Nanjing: Southeast University Press.

Meng, Ziliang 孟梓良, and Huanjun Zhang 張煥君. 2017. Shuangchong Shiyuxia Ming Qianqi Jinwang Diwei Xiajiang Yuanyin Fenxi 雙重視閾下明前期晉王地位下降原因探析. *Journal of Yuncheng University* 35: 16–19. [CrossRef]

Ming Qing Shanxi beike ziliao xuan (xu 1) 明清山西碑刻資料選. 2007. Zhang, Zhengming, David Faure, and Yonghong Wang, eds. Taiyuan: Shanxi Guji Chubanshe.

Sakai, Shiro 酒井紫郎. 1940. Sou Sekisahan Daizoukyo ni tsuite 宋磧砂版大藏經について. *Pitaka* 9: 23–26.

Sato, fumitoshi 佐藤文俊. 1999. *Mingdai Ofu no Kenkyu* 明代王府の研究. Tokyo: Kenbun Shuppan.

Taiyuan Shi Jiaoyu Weiyuanhui 太原市教育委員會. 1990. *Taiyuan Jiaoyu Dashiji 1840–1985* 太原教育大事記 1840–1985. Taiyuan: Shanxi People's Press.

Taiyuan Shi Nancheng Qu Renmen Zhengfu 太原市南城區人民政府. 1987. *Taiyuan Shi Nancheng Qu Dimingzhi* 太原南城區地名志. Taiyuan: Taiyuan Shi Nancheng Qu Renmin Zhengfu.

Wang, Lijin 王李金. 2006. From Shanxi Academy to Shanxi University (1902–1937). Ph.D. dissertation, Shanxi University, Taiyuan, China.

Wang, Richard G. 2012. *The Ming Prince and Daoism: Institutional Patronage of an Elite*. Oxford scholarship online. New York and Oxford: Oxford University Press.

Wang, Shejiao 王社教. 2004. Ming Qing shiqi taiyuan chengshi de fazhan 明清時期太原城市的發展. *Journal of Shaanxi Normal University (Philosophy and Social Sciences Edition)* 2004: 27–31.

Weidner, Marsha. 2001. Imperial Engagements with Buddhist Art and Architecture: Ming Variations on an Old Theme. In *Cultural Intersections in Later Chinese Buddhism*. Edited by Marsha Weidner. Honolulu: University of Hawaii Press, pp. 117–44.

Wu, Hui, Xiurui Xia, and Xuehui Zhang. 2005. 明代的度量衡 Ming dai de duliangheng. In *Zhongguo Shangye Tongshi* 3. Beijing: Zhongguo Caizheng Jingji Chuban She, p. 902.

Xie, Xingyao 謝興堯, and Yuchun Ke 柯愈春. 2007. Qing ruguan hou Fushan de huodong yu jiaoyou 清入關後傅山的活動與交遊. In *Fushan Jinian Wenji*. Taiyuan: Shanxi People's Press, pp. 88–97.

Yang, Yongkang 楊永康. 2015. Ming chu Jinwang Zhu Gang shiji bianzheng 明初晉王朱棡事跡辨正——兼及《太祖皇帝欽錄》的史料價. *Journal of Historiography* 2015: 7–16.

Yang, Yongkang 楊永康. 2021. 'Qing shamo zhe Yanwang ye' shuo zhiyi 「清沙漠者，燕王也」質疑——蜀獻王朱椿《與晉府書》所見晉王朱棡北事跡. *Nankai Journal (Philosophy, Literature and Social Science Edition)* 2021: 178–83.

Yoshii, Hohjun 吉井芳純. 1942. Taigen Suzenji Hakken no Sekisahan Zoukyo ni tsuite 太原崇善寺發見の磧砂版藏經に就いて. *Mikkyo Kenkyu* 1942: 81–92. [CrossRef]

Yuan, Xingyun 袁行雲. 1994. *Qingren Shiji Xulu* 清人詩集敘錄 第1冊. Beijing: Wenhua Yishu Chubanshe.

Zang, Xiaoshan 臧筱珊. 1983. Song, Ming, Qing dai Taiyuan cheng de xingcheng he buju 宋、明、清代太原城的形成和佈局. *City Planning Review* 1983: 17–21.

Zhang, Guoning 張國寧. 2006. Ming Jinfan jiqi zongshi zakao 明晉藩及其宗室雜考. In *Jinyang Wenhua Yanjiu Vol. 1*. Edited by Runde Huo. Taiyuan: Shanxi People's Press, pp. 97–127.

Zhang, Jizhong 紀仲, and Ji An 安笈. 1987. *Taiyuan Chongshan si Wenwu Tulu* 太原崇善寺文物圖錄. Taiyuan: Shanxi People's Press.

Zhang, Shiqing 張十慶. 2001. Handi fojiao jianzhu 漢地佛教建築. In *Zhongguo Gudai Jianzhushi Vol.4 Yuan Ming Jianzhu*. Beijing: China Architecture & Building Press, pp. 300–31.

Zhang, Yishan 張奕善. 1982. Duoguo hou de Ming Chengzu yu zhu fanwang guanxi kao 奪國後的明成祖與諸藩王關係考. *Guoli Taiwan Daxue Wenshizhe Xuebao* 31: 1–130.

Zheng, Xihuang. 2004. 中國古代地圖集 *Zhongguo Gudai Ditu Ji*. Xi'an: Xi'an Ditu Chubanshe.

Zhou, Xiaolin 周嘯林, and Jing Wen 温靜. 2021. Formatting and Personalisation: The Architecture of the Great Compassion Hall of Chongshan Temple in the Context of Institution Reorganisation in Early Ming Dynasty. *Heritage Architecture* 22: 59–69. [CrossRef]

Zhu, Yongjie 朱永傑, and Guanghui Han 韓光輝. 2006. Taiyuan mancheng shikong jiegou yanjiu 太原滿城時空結構研究. *Manzu Yanjiu* 2006: 61–70.

Article

Building and Rebuilding Buddhist Monasteries in Tang China: The Reconstruction of the Kaiyuan Monastery in Sizhou

Anna Sokolova

Department of Languages and Cultures, Ghent University, Blandijnberg 2, 9000 Ghent, Belgium; Anna.Sokolova@UGent.be

Abstract: This article explores regional Buddhist monasteries in Tang Dynasty (618–907 CE) China, including their arrangement, functions, and sources for their study. Specifically, as a case study, it considers the reconstruction of the Kaiyuan monastery 開元寺 in Sizhou 泗州 (present-day Jiangsu Province) with reference to the works of three prominent state officials and scholars: Bai Juyi 白居易 (772–846), Li Ao 李翱 (772–841), and Han Yu 韓愈 (768–824). The writings of these literati allow us to trace the various phases of the monastery's reconstruction, fundraising activities, and the network of individuals who participated in the project. We learn that the rebuilt multi-compound complex not only provided living areas for masses of pilgrims, traders, and workers but also functioned as a barrier that protected the populations of Sizhou and neighboring prefectures from flooding. Moreover, when viewed from a broader perspective, the renovation of the Kaiyuan monastery demonstrates that Buddhist construction projects played a pivotal role in the social and economic development of Tang China's major metropolises as well as its regions.

Keywords: regional monasteries; Tang Buddhism; stelae inscriptions; Kaiyuan monastery; Sizhou; ordination platforms

Citation: Sokolova, Anna. 2021. Building and Rebuilding Buddhist Monasteries in Tang China: The Reconstruction of the Kaiyuan Monastery in Sizhou. *Religions* 12: 253. https://doi.org/10.3390/rel12040253

Academic Editors: Shuishan Yu and Aibin Yan

Received: 1 March 2021
Accepted: 1 April 2021
Published: 5 April 2021

Publisher's Note: MDPI stays neutral with regard to jurisdictional claims in published maps and institutional affiliations.

1. Introduction

China's Buddhist community gained extraordinary power and imperial patronage during the Tang Dynasty, which facilitated an unprecedented spread of monasticism throughout the empire[1]. Historical records indicate that there were 4600 state monasteries and approximately 40,000 smaller, private institutions by 845[2]. Tang rulers appreciated that this network of Buddhist monasteries had the potential to sustain the imperial state's power, protect the legitimacy of the ruling clan, boost the economy, and maintain central control over the regions. Consequently, the ruling elite lavishly sponsored new building projects as well as the reconstruction of monasteries that had fallen into disrepair. Moreover, members of the imperial family, officials, and eunuchs all donated their private mansions to the state for conversion into monasteries (Forte 1983). Often enormous in scale, state monasteries could easily accommodate dozens of resident monastics, and the senior monks were increasingly recognized as key members of the upper echelons of Tang society.

To date, scholars have tended to focus on the histories, networks, patronage, and architecture of court-sponsored monasteries that were located in the capitals of successive Chinese dynasties: Pingcheng 平城, Luoyang 洛陽, Yecheng 鄴城, and Chang'an 長安 (e.g., Forte 1992; Xiao 2003; Gong 2006; Zhang 2008; He 2013b; Chen 2015). In contrast, China's regional monasteries have received relatively little attention[3]. This article takes a close

---

1   The literature on monasticism during this period is vast. See, among many other sources: (Goossaert 2000; Heirman and Stephan 2007; Xie and Bai 1990).

2   These are the official figures of monastic institutions that were dismantled and destroyed during the Huichang 會昌 persecution of Buddhism (840–46). See (Weinstein 1987, p. 134).

3   There are a handful of exceptions to this rule, including Evelyne Mesnil's excellent study on the Dashengci monastery 大聖慈寺 in Chengdu (Mesnil 2006).

---

look at the Kaiyuan monastery 開元寺 in Sizhou 泗州 (present-day Jiangsu Province), an important regional institution that was dedicated to the cult of the Buddhist monk Sengqie 僧伽 (628–710) and housed a Buddha tooth relic. After becoming a thriving pilgrimage site in the middle of the eighth century, the monastery was destroyed by fire towards the end of that century, which led to demands for its reconstruction among the local citizenry and foreign pilgrims alike. These calls were eventually answered at the start of the next century with the launch of a large-scale restoration project that was chronicled by three state officials, Bai Juyi 白居易 (772–846), Li Ao 李翱 (772–841), and Han Yu 韓愈 (768–824), who documented the work on behalf of the monastery's clergy. These scholars' detailed accounts demonstrate that the rebuilt Kaiyuan monastery soon became a grand, multi-functional institution, an important ordination center, and one of the major drivers of southern China's religious, social, and economic development during the first half of the ninth century. Moreover, they highlight the significance of China's regional monasteries as sites of considerable influence and power within a broad network of state monasteries that stretched across the Tang Empire.

2. Buddhist Monasteries in Tang China

2.1. Functions

Tang Buddhist monasteries were typically established by the ruling elite to secure supernatural protection for the dynasty through the good offices of communities of authoritative monks. Therefore, many of the religious rites they performed were formulated to protect the emperor and his ancestors from harm, to maintain the prosperity and stability of the imperial state, and to obtain posthumous peace for loyal soldiers who had died in battle (Zürcher 2014, p. 98). In addition, emperors and princes sometimes established state monasteries for more specific purposes, such as to commemorate or safeguard a beloved parent or spouse (Forte 1983, p. 686). A number of scholars have demonstrated that state monasteries were not merely centers of religious activity but also functioned as social, political, and economic institutions (e.g., Yang 1950; Twitchett 1956; Gernet 1995). James Robson provides a useful summary: "Chinese Buddhist monasteries, in addition to being places for traditional Buddhist contemplative practices, also served as granaries, mills, treasuries, orphanages, pawnbrokers, land stewards, auction houses, and sites of marketplaces and community festivals" (Robson 2009, p. 44). Moreover, as they played a crucial role in the translation of Buddhist scriptures, they became important centers of general scholasticism and education. Erik Zürcher has demonstrated that some monasteries started to combine their religious activities with secular education as early as the fourth century (Zürcher 1989), but the phenomenon reached its height during the Tang Dynasty, when many monastic institutions served as educational retreats where scholar-officials could take sabbaticals from their official duties to study a variety of secular as well as religious subjects (Yan 1992, pp. 271–316). In sum, Tang Buddhist monasteries' myriad functions within many areas of religious, social, and political life were closely linked with their architectural development.

2.2. Architecture

Scholarship on medieval Chinese Buddhist architecture demonstrates that the layout of state monasteries was strongly influenced by the urban architecture of Tang China's capital, Chang'an, especially the *Mingtang* 明堂, a central hall of imperial palace where all important ritual ceremonies and secular events were held, and the palace-city (Ledderose 1980, pp. 238–48; Wang 2000; He 2013a, pp. 5–6). A compound (*yuan* 院)—that is, a rectangular courtyard enclosed by rammed-earth walls or a portico—was the basic unit of every typical medieval Chinese Buddhist monastery. As places that had to accommodate both devout religious observance and the practicalities of daily life, the compounds of large monasteries included a number of separate buildings, each of which served a particular function. He (2013b, pp. 61–63) provides detailed descriptions of the main and auxiliary buildings within a typical monastic compound. The main buildings consisted of a middle gate (*zhongmen* 中門), an enclosing corridor (*lang* 廊), a roofed corridor (*langwu* 廊廡), a continuous rammed-earth wall (*hangtuqiang* 夯土牆), a pagoda (*ta* 塔), a Buddha Hall (*fodian* 佛殿), and a lecture hall (*jiangtang* 講堂). Among the auxiliary buildings, there were quarters (*sengfang* 僧房), meditation rooms (*chanshi* 禪室), *sūtra* halls (*jingtang* 經堂), and a bell tower (*zhonglou* 鐘樓).

In his diagram of an ideal monastic compound (Figure 1)[4], the eminent *vinaya* reformer Daoxuan 道宣 (596–667) also includes oil and flour depositories, kitchens, a bathing house, and toilets[5]. Moreover, he insists that monastic ordinations within a monastery must be conducted on an ordination platform (*jietan* 戒壇)—a five-tier structure that would serve as a divine location for the Buddha's presence (McRae 2005, pp. 72, 90–93). Many of these structures were established within the main compounds of China's monasteries in the decades after Daoxuan's death (McRae 2005, p. 88). In the early eighth century, government officials realized that fortunes could be made by authorizing ordinations of those who were willing to pay for the privilege in order to secure exemption from taxes and corvée labor (Ch'en 1964, p. 242). Hence, regional governors adopted the lucrative practice of purchasing ordination certificates and then selling them for personal gain (Barrett 2005, pp. 101–22; Gernet 1995, pp. 48–62; McRae 2005, pp. 69–93; Weinstein 1987, pp. 92, 109), which accelerated the establishment of ordination platforms within monastic complexes throughout China.

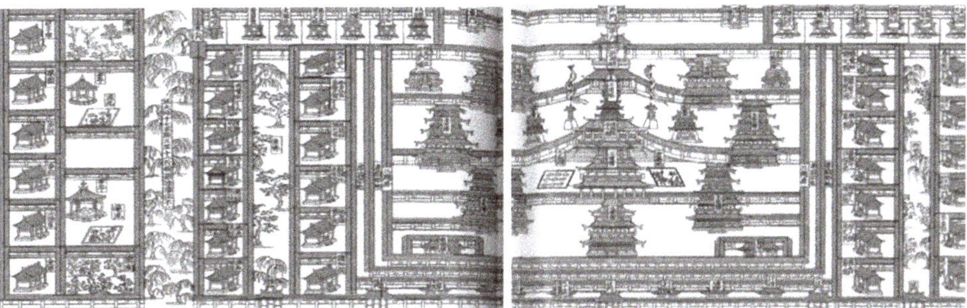

Figure 1. Diagram of Daoxuan's ideal monastery, based on his vision of the Jetavana monastery. Source: (Teiser 2006, pp. 138–39), adapted from Daoxuan's original drawing (T no. 1892.45: 811b10-813b29).

The "multi-compounds" and "multi-halls" layout came to dominate Chinese Buddhist monastic design between the fifth and seventh centuries. This development was closely linked to the evolution of city planning at that time. The typical city was laid out in a grid pattern, with all of the buildings—aside from the Imperial Palace and the most important government buildings—arranged around courtyards, the courtyards grouped into wards (*fang* 坊), and each ward forming a single block (see Figure 2). Lothar Ledderose stresses:

> Each ward contained one or several courtyards, depending on their size and function. Inside were public agencies, monasteries, ancestor temples, and countless larger and smaller residences . . . The similarities in the layout of the courtyards made it easy to exchange functions—for instance, to convert a private residence into a monastery or a monastery into a government office.
>
> (Ledderose 2000, p. 115)

---

4   See Daoxuan's Zhong Tianzhu Sheweiguo Qihuansi tu jing 中天竺舍衛國祇洹寺圖經 (Diagram and Sūtra on the Jetavana Temple of Vaiśālī in Central India), which includes a sketch of his vision of the ideal monastery, the Jetavana monastery in India, where the Buddha lived and preached (Ho 1995). See (Teiser 2006, pp. 140–41) for descriptions of the individual buildings within the complex.

5   On Daoxuan's inclusion of bath and toilet houses on his diagram, see (Heirman and Torck 2012, pp. 37–40).

As monasteries' power increased and their range of social functions steadily grew, their monastic complexes started to expand, too. Moreover, the urban grid pattern meant that new compounds could be inserted into a cityscape with relative ease. For instance, archaeological excavations and historical documents clearly show that a Buddhist monastery was embedded in Luoyang's grid in the early sixth century (He 2013b, p. 201). By the following century, the multi-compound layout had surpassed all others as the preferred form of monastic design in China. The great imperial monasteries might boast a dozen or more compounds, and thousands of buildings, within their confines. For instance, in Chang'an's famous Daci'en monastery 大慈恩寺:

> There were multi-story buildings, halls towering high, and densely built houses. A total of ten or more compounds with one thousand eight hundred and ninety-seven houses altogether.
>
> 重樓複殿，雲閣洞房。凡十餘院，總一千八百九十七間。[6]

A number of drawings from Dunhuang 敦煌 murals indicate that Tang monasteries typically consisted of several compounds separated from one another by walls or porticoes, with each major compound usually having its own hall (see Figures 3 and 4).

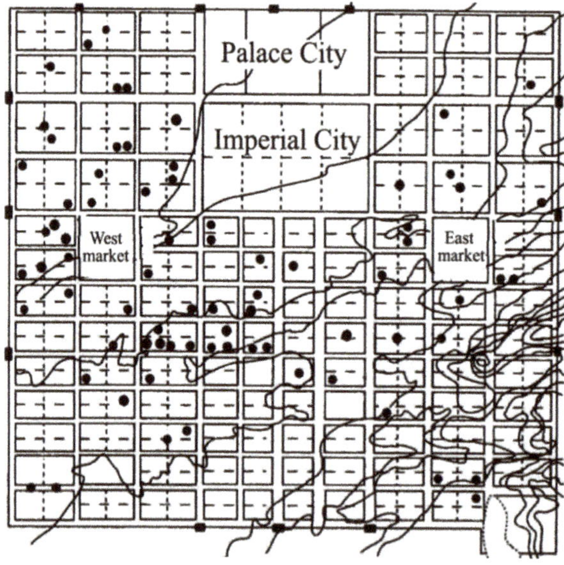

Figure 2. Map of Chang'an, early Tang period, illustrating the distribution of Buddhist monasteries and nunneries in every ward. Source: Adapted from (He 2013b, p. 199).

---

[6] Huili 慧立 (615–c. 677) and Yancong 彥悰 (fl. 688), Da Tang da Ciensi sanzang fashi zhuan 大唐大慈恩寺三藏法師傳 (Biography of the "Master of the Three Canons," Dharma Master [Xuanzang] of Great Cien Monastery [under] the Great Tang), T no. 2053.50: 258a16–17 (He 2013b, p. 71).

Figure 3. Monastic complex in a line drawing of Dunhuang mural, north wall of Cave 231, mid-Tang period. Source: Adapted from (Xiao 2003, p. 70).

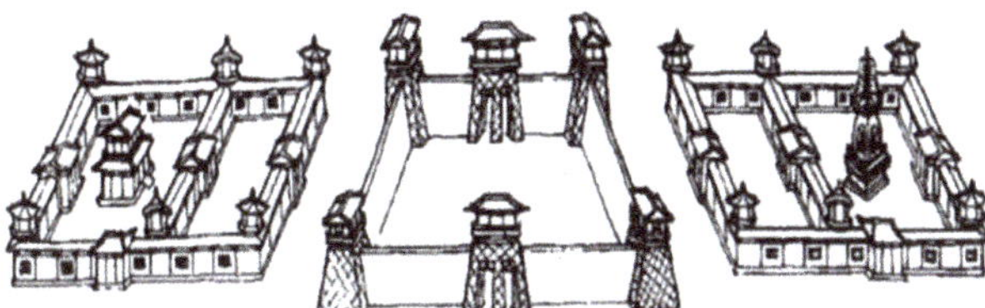

Figure 4. Monastic complex in a line drawing of Dunhaung mural, west ceiling of Cave 85, late Tang period. Source: Adapted from (Xiao 2003, p. 70).

These images are considered to be accurate representations of Tang monastic complexes in the major metropolises of Chang'an and Luoyang as well as Dunhuang itself (Xiao 2003, pp. 35–81). As a result, they are crucial sources of information on the layout of medieval state monasteries, especially in the absence of any major archaeological discoveries. Locally produced sources, such as carved stelae (*bei* 碑) and gazetteers (*difangzhi* 地方誌)[7], are similarly important, as some of them include images of Song Dynasty (960–1279) or later regional monasteries and temples that bear a remarkable resemblance to the institutions depicted in the Dunhuang murals. For instance, a Song stele etching of the Daoist Zhongyue Temple 中嶽 in Dengfeng 登封, near Mount Song 嵩山, Henan Province (Figure 5), and a sketch of the Fangguang monastery 方廣寺 on Mount Hengyue 衡嶽, Hunan Province, in a Ming (1368–1644) gazetteer (Figure 6) both depict an enclosed, rammed-earth monastic compound. Both of these religious institutions were also active throughout the Tang era. Finally, we can supplement this visual evidence with information gleaned from written records that Tang state officials composed on behalf of local monastic communities. Although they lack images, these texts provide comprehensive, eyewitness accounts of regional Buddhism that enable us to trace the development of the monasteries' functions and architectural patterns throughout the empire.

---

[7] Local gazetteers are major sources of information on local monasteries from the Song Dynasty onwards. The term "local gazetteers" was often used collectively to refer to various kinds of geographical texts. These works played a crucial role in reinforcing the links between China's central government and the provinces. Moreover, they provided vital information on strategic locations and military matters because they included comprehensive reports and maps of the whole empire. As a result, they were produced in vast quantities in China's provinces. On the historical development of local gazetteers, see, among others, (Hargett 1996).

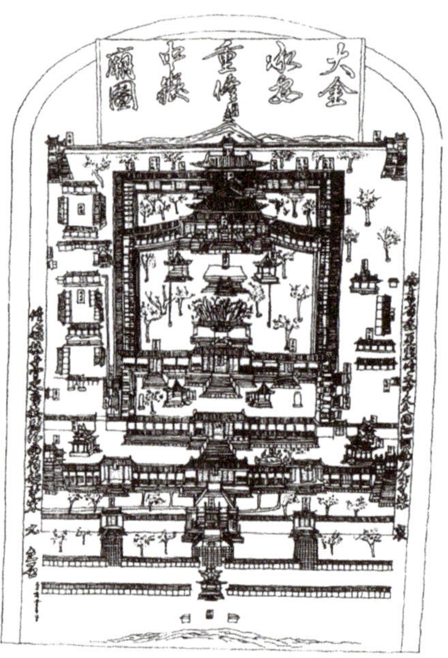

Figure 5. "Da Jin Chengan zhongxiu Zhongyue miao tu bei" 大金承安重修中嶽廟圖碑 ("Stele with a Depiction of the Newly Rebuilt Zhongyue Temple during the Cheng'an [Era] of the Great Jin [Dynasty]"), dated 1200. Source: (Xiao 2003, p. 74).

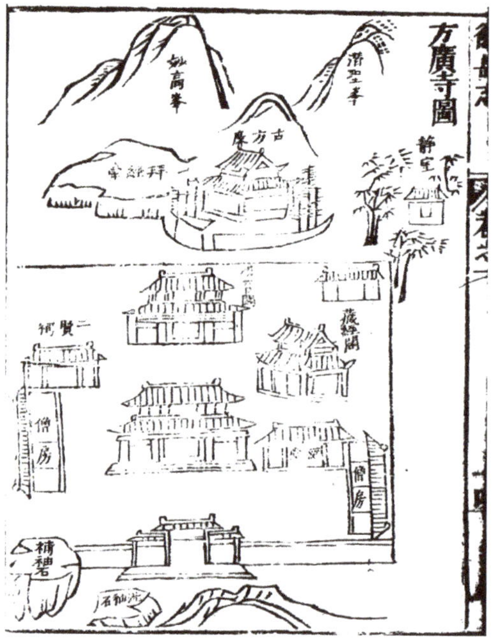

Figure 6. *Hengyue zhi* 衡嶽志 (*Gazetteer of [Mount] Hengyue*), compiled by Peng Zan 彭簪 (?–?), dated 1528. Source: *Siku quanshu cunmu congshu* 四庫全書存目叢書, vol. 229, p. 266.

2.3. Textual Sources of Information on Regional Monasteries

Texts composed by state officials during the Tang Dynasty—both prose and poetry—are unusually rich sources of information on the growth of regional monasticism in medieval China. In particular, scholar-officials' records (*ji* 記) and inscriptions (*beiming* 碑銘) contain a wealth of data on local monastic institutions in the Tang period[8].

The long-standing system of temporary administrative and military appointments as well as the collapse of central government in the wake of a rebellion launched by General An Lushan 安祿山 (703–757) in 755 contributed to an unprecedented dispersal and circulation of the elite to all corners of the empire, especially the southern regions. This mass migration of state officials to the provinces and their subsequent engagement with regional Buddhist communities in the middle of the eighth century led to the emergence of two new literary genres—records and inscriptions—as the relocated administrators supplemented their incomes with commissions from local monks and monasteries[9]. The earliest anthologies of Tang prose—*Wenyuan yinghua* 文苑英華 (*Blossoms from the Garden of Literature*)[10] and *Tang Wencui* 唐文萃 (*The Finest Prose of the Tang Dynasty*)[11]—contain numerous records[12] and inscriptions[13], with specifically "Buddhist" works categorized as "*shishi* 釋氏" in the *Wenyuan yinghua* and "*shi* 釋" or "*fotu* 佛圖" in the *Tang Wencui*. These texts are then subdivided into subjects such as "inscriptions for *sūtra* collections" (*jingzang bei* 經藏碑), "inscriptions for monasteries" (*sibei* 寺碑), "inscriptions for *śarīra stūpas*" (*sheli ta bei* 舍利塔碑), "inscriptions for Buddhist statues" (*xiang bei* 像碑), "inscriptions for Buddhist masters" (*dade bei* 大德碑, *heshang bei* 和尚碑, or *shi bei* 師碑), and so on. In itself, the fact that the anthologies' compilers deemed it necessary to distinguish "Buddhist" texts from the rest signals the scale of the literati's engagement with provincial Buddhism as well as the extent of Buddhist building projects in the regions. In addition, many of these scholar-officials exchanged poems (*shi* 詩) and letters (*shu* 書) with Buddhist monks, particularly in the Jiangnan region 江南[14] (Mazanec 2017). This wealth of correspondence constitutes another valuable source of information on eighth- and ninth-century Buddhism. Taken together, these texts allow us to reconstruct the histories of numerous regional monasteries, visualize their layouts, identify their leaders, and locate them within a broader picture of Buddhism's unprecedented expansion during the Tang era.

3. The Kaiyuan Monastery in Sizhou, Jiangsu Province

3.1. Textual Sources

The Kaiyuan monastery was established in 696 under the name Puguangwangsi 普光王寺 by an imperial decree of Emperor Zhongzong 中宗 (656–710), as recorded by

---

8   For a comprehensive study of monastic records and stelae inscriptions composed by literati during the Tang and Song dynasties, see (Halperin 1997, 2006).

9   For further information on the mid-Tang literati's development of literary genres while in exile from the capital, see, for instance: (McMullen 1988; Tackett 2014; Shields 2015).

10   Wenyuan yinghua was compiled by a team of scholars led by Li Fang 李昉 (925–996) after 980 but not published until 1201–1204. For details of the strategies used in the selection of texts for the Wenyuan yinghua as well as the anthology's compilation and transmission, see: (Owen 2007, pp. 259–326; Ling 2005).

11   Tang Wencui was the work of a single compiler, Yao Xuan 姚鉉 (968–1020), who completed it in 1011. His son presented the manuscript to the emperor in 1020, but it was not published until 1039. See (Shields 2017, pp. 306–35) for recent research into this anthology.

12   The Wenyuan yinghua includes five scrolls of specifically Buddhist ji (juan 817–821). The Tang Wencui boasts a total of nine Buddhist ji on a single scroll (juan 76).

13   The Wenyuan yinghua contains no fewer than ninteen scrolls of monastic stelae inscriptions (juan 850–868). Five of the Tang Wencui's total of fifteen scrolls of inscriptions cover Buddhist topics (juan 61–65). It is striking that the sixth-century literary anthology Wen Xuan 文選 (A Selection of Refined Literature), compiled by Xiao Tong 蕭統 (501–531), an important precursor to the Wenyuan yinghua and the Tang Wencui, contains no texts that could be described as ji. Moreover, it contains just five stelae inscriptions, only one of which—the "Toutuosi beiwen" 頭陀寺碑文 ("Stele Inscription for the Toutuo Monastery"), composed by Wang Jin 王巾 (?–505)—was written for a Buddhist monastery. This points to an unprecedented proliferation of both of these literary genres in the Tang era. For more details, see (Sokolova 2021, pp. 40–43).

14   Jiangnan (literally, "South of the River") refers to the area south of the Yangtze River that stretches from Suzhou and Hangzhou in the east to Nanchang and Jiujiang in the west. This region provided a safe haven for thousands of intellectuals in the wake of An Lushan's rebellion.

the scholar-official Li Yong 李邕 (678–747) on a commemorative stele inscription[15]. The designation was changed to Kaiyuan—the era name of the first phase (713–741) of Emperor Xuanzong's 玄宗 (r. 712–756) reign—a few decades later. The monastery's founder was a renowned Central Asian monk and thaumaturge named Sengqie 僧伽 (628–710) who lived in Sizhou in the early eighth century. Following his death, his body was preserved in lacquer and housed in a *stūpa* within the monastery as a relic[16]. In his seminal study on the Kaiyuan ordination scandal[17], Timothy Barrett suggests that Sengqie's cult continued to grow until, by the start of the ninth century, masses of pilgrims were visiting the monastery to pay homage to its founder (Barrett 2005, pp. 105–6). In addition, a contemporaneous travelogue written by the Japanese pilgrim–monk Ennin 圓仁 (793–864), the *Nittō guhōjunrei kōki* 入唐求法巡禮行記 (*Record of a Pilgrimage to Tang China in Search of the Law*), attests that the Kaiyuan monastery housed a Buddha tooth relic that similarly inspired mass worship and generated lucrative donations from the faithful[18]. Moreover, as Sizhou was a strategically important city on several international trade routes, the monastery came to the attention of hordes of merchants, which served to boost both its coffers and its reputation. After the original Kaiyuan complex burned to the ground towards the end of the eighth century, the local government, in conjunction with the monastery's clergy, decided to reconstruct the monastery on a much grander scale and thereby laid the foundations for it to become one of southern China's foremost centers of religious, social, and economic life.

The earliest surviving text that recounts the reconstruction of the Kaiyuan monastery is a bell inscription composed by Li Ao 李翱 (772–841) in 799[19] and entitled "Sizhou Kaiyuansi zhongming bingxu" 泗州開元寺鍾銘（並序）("Bell Inscription, with Preface, of the Kaiyuan Monastery in Sizhou")[20]. The previous year, Li Ao had passed the imperial examination prior to traveling around southern China in search of an administrative post in local government—a common strategy among new graduates at the time (Barrett 1992, p. 70). When the young scholar arrived in Sizhou, a monk from the newly rebuilt monastery, Chengguan 澄觀 (?–?), asked if he would be interested in writing the bell inscription[21]. Li Ao's response is preserved in a letter entitled "Da Sizhou Kaiyuansi seng shu" 答泗州開元寺僧書 ("A Letter to Answer the Master of the Kaiyuan Monastery in Sizhou")[22]. The following year, 800, a friend of Li Ao, the famous Han Yu 韓愈 (768–824)[23], recorded a meeting with Chengguan in Luoyang in a poem entitled "Song seng Chengguan" 送僧澄觀 ("Seeing off Master Chengguan")[24]. Li Ao's inscription and Han Yu's poem contain a wealth of information on the "first phase" of the reconstruction of the Kaiyuan monastery during the late eighth and early ninth centuries.

Our main source of information on the "second phase" of the reconstruction project (806–827) is a stele inscription entitled "Da Tang Sizhou Saiyuansi lintan lüde Xu, Si, Hao, san zhou sengzheng Mingyuan dashi tabeiming bingxu" 大唐泗洲開元寺臨壇律德徐泗濠三州僧正明遠大師塔碑銘（並序）("Stele Inscription, with Preface, for the Superintendent of Three Prefectures, Xu, Si, [and] Hao, Vinaya Master Mingyuan, Preceptor of the Kaiyuan Monastery in Sizhou of the Great Tang [Dynasty]")[25] that the eminent scholar-official, Bai

---

15 See Li Yong's "Da Tang Sizhou Linhuai xian Puguangwangsi bei" 大唐泗州臨淮縣普光王寺碑 ("Stele Inscription for Puguangwang Monastery in Linhuai County in Sizhou of the Great Tang [Dynasty]"), Quan Tang wen 全唐文 263, pp. 2672–73.

16 Juean 覺岸 (1286–1355), Shishi jigu lüe 釋氏稽古略 (An Outline of Historical Researches into the Śākya Family Lineage), T no. 2037.49: 817c24–25.

17 This scandal is discussed later in this paper.

18 Nittō guhōjunrei kōki 4.137.

19 For a comprehensive study on Li Ao, including his ties with Buddhism, see (Barrett 1992).

20 Wenyuan yinghua 789, p. 4981; Quan Tang wen 637, p. 6427. The prominent Tang literatus Liang Su 梁肅 (753–793) composed an earlier inscription for the Kaiyuan monastery, as is documented by Cui Gong 崔恭 (?–?) (Tang Wencui 92, pp. 381–82; Wenyuan yinghua 789, p. 49881), but this was probably lost in the fire that destroyed the monastery itself.

21 (Tang Wencui 85, pp. 291–292; Wenyuan yinghua 688, p. 4269).

22 (Wenyuan yinghua 688, p. 4269; Quan Tang wen 637, pp. 6423–24).

23 For a study on Han Yu, see (Hartman 1986).

24 See Quan Tang shi 全唐詩 342, p. 3831.

25 Bai Juyi's inscription for Mingyuan is missing from both the Tang Wencui and the Wenyuan yinghua. A version of the text is included in the Quan Tang wen 678, pp. 6935–6936, but I follow the version contained in the Bai Juyi jijianjiao 白居易集箋校 (ed. Zhu 1988, pp. 3729–30).

Juyi 白居易 (772–846)[26], composed to commemorate Kaiyuan's abbot, Mingyuan 明遠 (765–834). Bai Juyi almost certainly witnessed the final stages of this phase of the great reconstruction project at first hand as he was Prefect of Suzhou (*Suzhou cishi* 蘇州刺史) in Jiangsu between 825 and 827. However, he composed the inscription several years later, in 834, during his retirement in Longmen 龍門, near Luoyang, in response to a request from two of the recently deceased abbot's disciples, Liang 亮 and Liangsu 亮素. This text, which was based on "a biographical sketch of the master" that these two monks sent to Bai Juyi (今按弟子僧, 僧亮, 亮素行狀)[27], sheds considerable light on the layout, functions, and architecture of the new monastic complex.

3.2. The First Phase of the Reconstruction: Master Chengguan

The reconstruction of the Kaiyuan monastery began under the supervision of the local monastic who commissioned Li Ao to write his bell inscription, Chengguan. This monk should not be confused with another Chengguan (738–839), who was an esteemed patriarch of the Huayan Buddhist school in the same period. Timothy Barrett identifies Chengguan of the Kaiyuan monastery as a *vinaya* master who was probably a disciple of Jianzhen 鑑真 (688–763), a famous missionary to Japan (Barrett 1992, p. 79)[28]. Although this Chengguan is less renowned than his illustrious namesake, it seems that he still enjoyed considerable authority in his own region. In his bell inscription, Li Ao writes that "Chengguan, along with several fellow monks, oversaw the reconstruction of the monastery as well as the casting of a bell for the monastery in the fifth year of the Zhenyuan era (799)" (僧澄觀與其徒僧若干, 複舊室居, 作大鍾。貞元十五年，厥功成)[29]. The following year, in the preface to the poem he sent to Chengguan, Han Yu alludes to the monk's establishment of Sengqie's *stūpa* (澄觀建僧伽塔於泗州)[30]. Given that Chengguan oversaw a major reconstruction project as well as the casting and delivery of a new bell, it is safe to assume that he was Kaiyuan's abbot at this time[31]. It also seems that both he and the monastery enjoyed the patronage of Wang Zhixing 王智興 (758–836)[32], an officer in Wuning Circuit 武寧 (Jiangsu), who probably viewed the reconstruction enterprise as a means to maximize donations from devotees of Sengqie's cult (Barrett 2005). Chengguan fled to Luoyang after Wang Zhixing seized control of this circuit. However, he returned to the region near the end of his life and participated in a rather less ambitious building project—the construction of a well parapet in Li Yang 溧陽 (Jiangsu) in 811[33]. Thereafter, there are no further traces of him.

3.3. Further Reconstruction and Expansion under Mingyuan

3.3.1. Mingyuan's Early Career: Expansion of the Kaiyuan Monastery and Its Function as a Flood Barrier

Mingyuan succeeded Chengguan as abbot of the Kaiyuan monastery in 806. Bai Juyi's inscription, which is our main source of biographical information on this master, indicates that he was a native of Cuo 鄼 District in Qiao 譙 County (present-day Anhui Province) and that his secular clan name was Bao 暴. However, in 772, at the age of seven, he renounced the secular world under the guidance of Chan Master Pei (*Pei Chanshi* 霈禪師) in his local monastery. Interestingly, Chengguan—the aforementioned Huayan patriarch, not Mingyuan's predecessor as abbot of Kaiyuan—had done the same under a Chan monk

---

26   There has been extensive research into Bai Juyi's life and work, as well as his involvement with Buddhism. For the best studies, see: (Chan 1991; Feifel 1961; Xie 1997; Waley 1951; and Seo 1993).

27   Bai Juyi jijianjiao, p. 3729.

28   Jianzhen was originally from Guangling 廣陵 in Jiangsu. For his biography, see Zanning 贊 (919–1001), Song gaoseng zhuan 宋高僧傳 (Biographies of Eminent Monks [Compiled] under the Song Dynasty), T no. 50.2061: 797a24–c11.

29   Wenyuan yinghua 789, p. 4981. Quan Tang wen 637, p. 6427.

30   Quang Tang shi 342, p. 3831.

31   Typically, only abbots had sufficient authority to commission bells. See (Burdorf 2019, p. 325).

32   For Wang Zhixing's biography, see Jiu Tangshu 舊唐書 156, pp. 4138–41.

33   Han Chong 韓崇, Baotiezhai jinshi wen bawei 寶鐵齋金石文跋尾 (Colophons on Inscriptions on Bronze and Stone from the Baotiezhai [House]), cited in (Barrett 1992, p. 80).

of the same name in the Baolin monastery 寶林寺 on Mount Yingtian 應天 (in Zhejiang)[34] at the age of eleven in 749[35]. Imre Hamar identifies Chengguan's Master Pei as Hongpei 洪霈 (?–?), who also mentored one of Daoxuan's disciples, Xuanyan 玄儼 (Hamar 2002, pp. 32–33). Hence, if Mingyuan renounced under the same master some twenty-three years later, then he may have belonged to an extensive monastic network that also included the patriarch Chengguan[36]. Either way, in his inscription, Bai Juyi characterizes Mingyuan as one of southern China's foremost Buddhist authorities.

In 784, at the age of nineteen, Mingyuan received full ordination from an otherwise unknown *vinaya* master named Lingmu 靈穆 (?–?) in Sizhou. He engaged in extensive study of the *Dharmaguptaka vinaya*[37] and the *Abhidharmakośa-bhāṣya*[38] as part of his monastic training before "ascending the lecturer's seat, then assuming the leadership of the ordination platform" (乃升講座，乃登戒壇)[39]. Thereafter, as Bai Juyi reports in his inscription, Mingyuan was appointed abbot of Kaiyuan, then Great Monastic Rectifier (*sengzheng* 僧正)[40], which gave him the authority he needed to initiate and supervise the construction of new lecture halls and monastic compounds:

> In the first year of the Yuanhe era (806), he was asked by the multitude to become the abbot of this monastery, the next year he was appointed by the government [authorities] as Great Monastic Rectifier in the region, overseeing the twelve divisions. Two hundred steps north from the Kaiyuan monastery, [he] built seven lecture halls [and] six monastic compounds.

> 元和元年，眾請充當寺上座，明年官補本州僧正，統十二部。開元寺北地二百步，作講堂七間，僧院六所。[41]

There is evidence that Mingyuan was an active participant in another large building project during his tenure as abbot of Kaiyuan: the founding of the Lingju monastery 靈居寺 in Yangzhou 揚州 in 813[42]. However, his home monastery remained the main focus of his construction efforts. For instance, under the patronage of the local prefect Su Yu 蘇遇 (?–?), he supervised the building of a major new monastic ward with the express intention of mitigating the flooding that blighted the area each year. Bai Juyi's inscription includes a vivid account of this scheme:

> There are heavy rains in the low land between the Huai and Si [prefectures]; they cause yearly floods. The Master [Mingyuan] planned with Su Yu, a commandery governor, [and] other [officials] to establish a monastic ward in the wasteland to the west of the Shahu [area] in order to prevent water flow. [They] constructed two hundred [buildings, including] gates, corridors, halls, kitchens, [and] stables; [they] planted ten thousand pine, cedar, willow, [and] cypress trees. Since then, the monks and the laity have been in no danger of flooding.

> 又淮,泗間地卑多雨潦，歲有水害，師與郡守蘇遇等謀於沙湖西隙地創避水僧坊，建門廊廳堂廚廄二百間，植松杉楠樫檜一萬本，由是僧與民無墊溺患。[43]

---

[34] Song gaoseng zhuan, T no. 2061.50: 737a6.

[35] See the biography of Chengguan in Song gaoseng zhuan, T no. 2061.50: 737a5–6.

[36] For a discussion of this network, see (Hamar 2002, pp. 31–42).

[37] Daoxuan's Sifen lü 四分律, Dharmaguptaka vinaya (Vinaya in Four Parts; T no. 1428.567), is frequently cited in stelae inscriptions and the biographies in Song gaoseng zhuan as a text that monks were required to study prior to ordination and to qualify as vinaya masters.

[38] Xuanzang 玄奘 (600?–664) translated Vasubandhu's fifth-century text Jushe lun 俱捨論 (Treasury of the Abhidharma; full title Apidamo jushe lun 阿毘達磨舍論; T no. 29.1558), one of the most important classical Indian works on abhidharma, into Chinese in 651.

[39] Bai Juyi jijianjiao, p. 3728.

[40] The Great Monastic Rectifier was a monastic supervisor whose principal responsibility was to maintain the moral standards of his fellow monks and nuns. He was recruited from within the saṃgha and appointed by the emperor. See (Forte 2003) for further details.

[41] Bai Juyi jijianjiao, p. 3928.

[42] Shu Sunju 叔孫矩 (?–?) "Da Tang Yangzhou Liuhexian Lingjusi bei" 大唐揚州六合縣靈居寺碑 ("Stele of the Lingju Monastery in the Lingju County in Yangzhou of the Great Tang [Dynasty]"), Quan Tang wen 745, p. 7714.

[43] Bai juyi jianjiao, p. 3728.

This passage illustrates that the Kaiyuan monastery was actually an extensive complex with multiple wards. The new flood-barrier ward would have consisted of a group of main buildings around a central courtyard as well as numerous auxiliary structures. This sort of multi-compound architectural design allowed for almost infinite rearrangement and expansion. In the case of the Kaiyuan monastery—which, as we have seen, attracted an unusually large number of religious and secular visitors on account of its relics and its location on major trade routes—it may be assumed that the new ward served a dual function. In addition to protecting the local population from flooding, these buildings would have greatly increased the monastery's capacity to accommodate the crowds of pilgrims and traders who made their way to Sizhou each year.

3.3.2. The Ordination Platform and the New Complex

After Mingyuan's construction of the new flood-barrier ward, the whole monastic complex was once again destroyed by fire, as Bai Juyi reports:

> Soon after that, the monastery burned in a fire. For a few years, the monastery was in a state of disrepair, the statues were destroyed, [and] the monks scattered.

旋屬災焚本寺，寺殱像滅僧潰者數年。[44]

However, Wang Zhixing seized power in Wuning Circuit 武寧 in 822[45], whereupon he gave Mingyuan religious authority over the whole region. The two men then wrote to the imperial court to request permission to establish an ordination platform in the Kaiyuan monastery. Approval was granted and the platform was erected right next to Sengqie's pagoda in 824[46]. This location was no accident, as the intention was to maximize donations from pilgrims to Sengqie's shrine. That said, Wang Zhixiang and Mingyuan's next fundraising scheme was even more lucrative, not to mention scandalous. They decided to sell ordination certificates to anyone who was willing to pay the requisite fee, regardless of the applicant's devotion to Buddhism, study of the scriptures, or monastic training. Indeed, they did not even bother to hold an ordination ceremony: each applicant simply handed over the money and received his certificate. Demand was high, because ordained monastics were exempt from taxation: in total, some 600,000 men bought one of the false certificates[47]. Each (genuine) monk who participated in the scam was then rewarded with 2000 in cash from the proceeds[48]. Although Wang Zhixing was reported to the emperor for selling ordination certificates for personal gain[49], Bai Juyi suggests that the money was actually used to rebuild the Kaiyuan complex. He writes:

> The Master [Mingyuan] and Wang [Zhixing], a military commander in Xuzhou, were destined [to meet]. United in their intentions, [they] joined forces to rebuild the monastic compound. Thus, the master was invited to accept the position of Rectifier of Monks of the three prefectures [and] a petition was presented [to the emperor] with a request to establish an ordination platform without delay. The profits from the donations enabled [rebuilding] on a larger scale, the Palace Attendant [Wang Zhixing] also assisted [by donating] household goods amounting to ten thousand [in cash], [and the reconstruction] was completed.

---

[44] Bai juyi jianjiao, p. 3728.

[45] Jiu Tangshu 156, pp. 4139–40.

[46] A commentary by Hu Sanxing 胡三省 (1230–1302) (Zizhi tongjian 資治通鑑 234, p. 7840) reads: "There is a stūpa of the Great Sage (Sengqie) in Sizhou, it is venerated by the people, therefore Wang Zhixing requested permission to establish an ordination platform right next to it" (泗州有大聖塔，人敬事之，故王智興請於此置戒壇). According to the Shishi jigu lüe, T no. 2037.49: 835c14–15, Wang Zhixing established the platform in the twelfth month of the fourth year of the Changqing 長慶四年 era (i.e., 824) in honor of the emperor's birthday.

[47] See Li Deyu 李德裕 (787–859), "Wang Zhixing du sengni zhuang" 王智興度僧尼狀 ("Memorial on the Ordination of Buddhist Monks and Nuns by Wang Zhixing"), Quan Tang wen 706, pp. 7242–43. For English translations, see (Weinstein 1987, pp. 60–61; Barrett 2005, pp. 103–4).

[48] See Cefu yuangui 府元龜 689, p. 7940.

[49] See Cefu yuangui 689, p. 7940.

師與徐州節度使王侍中有緣，遂合願力，再造寺宇，乃請師三郡僧正，奏乞連
置戒壇，因其施利，廓其規度，侍中又以家財萬計助而成之。[50]

As Bai Juyi explains in this passage, the monks' and Wang Zhixing's fundraising efforts—whether legitimate or not—meant that the monastic complex could be rebuilt on a much grander scale. Moreover, he reports that the work proceeded at an unprecedented pace, as it "began in the spring of the fifth year of Changqing era (825), and it was completed in the autumn of the first year of Taihe era (827)" (長慶五年春作，太和元年秋成)[51]. When it was finished, the complex boasted more than two thousand individual structures, including main and auxiliary buildings, which meant that it now exceeded some of the great monasteries of Chang'an and Luoyang in size. Bai Juyi describes what must have been a truly impressive sight:

> From tower halls, residential halls, corridors, kitchens, [and] granaries to houses for monks, servants[52], workers, [and] livestock. There were a total of two thousand and several hundred buildings. Inside [these buildings], there were ample statues [and] utensils . . . Star-shaped decorations adorned the buildings; [they seemed to have] emerged from beneath the earth, or descended from heaven. Donations arrived every single day; the sound of bells and chanting never ceased. The four *varga*[53] know [where to find] refuge, [an] uncountable [number of] people [have] converted [to Buddhism].

> 自殿閣堂亭廊庖藏，洎僧徒臧獲傭保馬牛之舍，凡二千若干百十間，其中像設之儀，器用之具，一無闕者 . . . 輪奐莊嚴，星環棋布，如自地踊，若從天降。供施無虛日，鍾梵有常聲，四眾知歸，萬人改觀。[54]

From this description, it is evident that the new complex could accommodate hundreds of pilgrims and merchants as well as many local workers who were essentially the monastery's employees. Thus, by the late 820s, Kaiyuan had not only reestablished its reputation as a major pilgrimage site but also broadened its range of social activities. Moreover, it had become an important ordination center for the vast region that lay between the Jiang 江 and Huai 淮 rivers. Bai Juyi's record of Mingyuan's career attests that the abbot observed eight ordination ceremonies and gave fifteen lectures on *vinaya* to a total of thirty thousand monks and nuns[55]. All told, he "carried out the Teaching in the Jianghuai [region] for forty years" (江, 淮行化者四十年)[56].

The Kaiyuan monastery remained the undisputed focal point of the Sengqie cult and continued to house the Buddha tooth relic long into the Song Dynasty. Indeed, a number of Song emperors initiated further building and renovation projects at the monastery[57]. However, it was Mingyuan's two building programs that enabled Kaiyuan to establish itself as a major center of Sizhou's religious, social, and economic life for centuries to come.

4. Conclusions

In conclusion, this paper has discussed the reconstruction and expansion of the Kaiyuan monastery in Sizhou during the late eighth and early ninth centuries. These grand building projects illustrate the prevailing tendencies in Buddhist architectural design as well as the Tang imperial elite's monastic construction and renovation strategies. The Kaiyuan monastery became an important hub within the network of Buddhist institutions

---

[50] Bai Juyi jijianjiao, p. 3928.

[51] Bai Juyi jijianjiao, p. 3928.

[52] The precise meanings of zang 臧 and huo 獲 are uncertain here, although both were used as abusive terms for slaves. Pu (2016) has demonstrated that individual monks and monasteries acquired slaves during the Tang Dynasty, and the Chang'an's slave-market was the largest in the world at the time. It seems highly likely that Sizhou, which was an important center of trans-Asian trade during the Tang era, would have had a similar market.

[53] The four groups of every monastic community: monks, nuns, male devotees, and female devotees.

[54] Bai Juyi jijianjiao, pp. 3928–29.

[55] Bai Juyi jijianjiao, p. 3929.

[56] Bai Juyi jijianjiao, p. 3929.

[57] On the history of the Kaiyuan monastery during the Song Dynasty, see (Zhang 2013).

that spread across medieval China's landscape in a wave of mass building. Benefiting from its strategically important location in the city of Sizhou, during the eighth century the monastery gained renown as the final resting place of Sengqie and the home of a Buddha tooth relic. As a result, when it burned down at the end of that century, the Tang court, the local government authorities, the monks themselves, and countless devotees launched an extensive reconstruction campaign. This was overseen by two successive abbots—Chengguan and Mingyuan—under the auspices of the local authorities and especially Wang Zhixing, a powerful provincial general. Over the course of nearly thirty years, their combined efforts secured Kaiyuan's status as a key pilgrimage site, ordination center, and economic force in southern China.

Descriptions of the Kaiyuan monastic compound in contemporary sources, especially Bai Juyi's stele inscription for Mingyuan, reveal that it conformed to the multi-compound design that typified Chinese state monasteries in the Tang period. New lecture halls and monastic compounds were erected alongside the main compound during the "first phase" of reconstruction. Thereafter, an entirely new ward of two hundred buildings was constructed on low-lying wasteland to act as a flood barrier and provide accommodation for the monastery's numerous pilgrims. This enlargement of the monastic complex demonstrates the inherent flexibility of the typical grid-based design, so it is certainly feasible that similar expansion projects were proceeding elsewhere in regional China around the same time.

Although Kaiyuan suffered another devastating fire halfway through the "second phase" of reconstruction, by the end of the 820s, the monastery boasted more than two thousand buildings and performed a wide variety of social as well as religious functions. This grand architectural project was funded by unprecedented donations from the local population, pilgrims to Sengqie's shrine and the Buddha tooth relic, and passing merchants, along with the highly lucrative sale of ordinations, which were performed on the platform that Wang Zhixing and Mingyuan established in 824.

The biographies of Kaiyuan's two abbots, Chengguan and Mingyuan, help to illuminate regional monks' contributions to monastic and other construction projects in the Tang period. Although our information on Chengguan is rather limited, it seems that he was a well-traveled and well-connected monk who studied under the cosmopolitan missionary Jianzhen, migrated to Luoyang, and exchanged letters and poems with state officials. Mingyuan, in contrast, spent his whole life in his home region yet still managed to attain the esteemed positions of abbot of the Kaiyuan monastery and Great Monastic Rectifier. In part, this was because he enjoyed the support of Wang Zhixing, the region's civil governor, and shared the latter's ambitious economic and political goals. Thus, in the course of his fifty-year career, he became not only a major monastic authority in southern China but also a prominent member of the local elite.

Finally, in addition to shedding light on the histories of local monastic institutions and allowing us to locate their leaders within broader intellectual networks, the information contained within inscriptions, records, poems, and letters obliges us to reassess non-canonical literature's ability to supplement and enrich our knowledge of monasticism in medieval China. These sources reveal that monasteries assumed a range of religious, social, and political functions and adopted an almost infinitely flexible form of architectural design that enabled them to become powerful forces for urbanization, economic growth, and development throughout the regions of the Tang Empire.

Funding: This research was funded by the Dissertation Fellowship from The Robert H. N. Ho Family Foundation Program in Buddhist Studies administered by the American Council of Learned Societies.

Institutional Review Board Statement: Not applicable.

Informed Consent Statement: Not applicable.

Data Availability Statement: Not applicable.

Conflicts of Interest: The author declares no conflict of interest.

References

Primary Sources

Bai Juyi jijianjiao 白居易集箋校 (*Annotated and Collated Edition of Bai Juyi's Collected Works*). Edited by Zhu Jincheng 朱金城. Shanghai: Shanghai guji chubanshe, 1988.

Cefu yuangui 册府元龜 (*The Imperial Book Treasury as Grand Tortoise*). Compiled by Wang Qingruo 王欽若. Nanjing: Fenghuang chubanshe, 2006.

Jiu Tangshu 舊唐書 (*Old History of the Tang*). Compiled by Liu Xu 劉昫. Beijing: Zhonghua shuju, 1975.

Nittō guhōjunrei kōki 入唐求法巡禮行記 (*Record of a Pilgrimage to Tang China in Search of the Law*). Composed by Ennin 圓仁. Guilin: Guangxi shifan daxue chubanshe, 2007.

Quan Tang shi 全唐詩 (*Complete Poems of the Tang*). Compiled by Peng Dingqiu 彭定求.Beijing: Zhonghua shuju, 1960.

Quan Tang wen 全唐文 (*Complete Prose of the Tang*). Compiled by Dong Hao 董浩. Beijing: Zhonghua shuju, 1983.

Siku quanshu cunmu congshu 四庫全書存目叢書 (*Collectanea of Books that Were Noted in the Catalogue of the Complete Books in the Four Treasuries*). Jinan: Qi Lu shushe chubanshe, 1996.

Taishō shinshū daizōkyō 大正新修大藏經 (*New Edition of the Buddhist Canon in the Taishō Era*). Edited by Takakusu Junjirō 高楠順次郎 and Watanabe Kaikyoku 渡邊海旭. Tōkyō: Taishō Issaikyō Kankōkai, 1924–1932.

Tang Wencui 唐文萃 (*The Finest Prose of the Tang Dynasty*). Compiled by Yao Xuan 姚. Changchun: Jilin renmin chubanshe, 1998.

Wen Xuan 文選 (*A Selection of Refined Literature*). Compiled by Xiao Tong 蕭統. Shanghai: Shanghai guji chubanshe, 1986.

Wenyuan yinghua 文苑英華 (*Blossoms from the Garden of Literature*). Compiled by Li Fang 李昉. Beijing: Zhonghua shuju, 1966.

Zizhi tongjian 資治通鑑 (*Comprehensive Mirror for the Aid of Government*). Compiled by Sima Guang 司馬光. Beijing: Guji chubanshe, 1956.

Secondary Sources

Barrett, Timothy H. 1992. *Li Ao: Buddhist, Taoist, or Neo-Confucian?* Oxford: Oxford University Press.

Barrett, Timothy H. 2005. Buddhist Precepts in a Lawless World: Some Comments on the Linhuai Ordination Scandal. In *Going Forth: Visions of Buddhist Vinaya*. Edited by William Bodiford. Honolulu: University of Hawai'i Press, pp. 101–22.

Burdorf, Suzanne. 2019. Alerting the Masses: An Inquiry into Buddhist Communication through Bells in the Song Dynasty (960–1279): China from the Perspective of Material Culture and Sound. *Monumenta Serica–Journal of Oriental Studies* 67: 319–61. [CrossRef]

Ch'en, Kenneth Kuan Sheng. 1964. *Buddhism in China: A Historical Survey*. Princeton: Princeton University Press.

Chan, Chiu Ming. 1991. Between the World and the Self-orientations in Po Chü-i's (772–846) Life and Writings. Ph.D. dissertation, University of Wisconsin, Madison, WI, USA.

Chen, Jinhua. 2015. The Multiple Roles of the Twin Chanding Monasteries in Sui–Tang Chang'an. *Studies in Chinese Religions* 1: 344–56. [CrossRef]

Feifel, Eugène. 1961. *Po Chü-i as a Censor: His Memorials Presented to Emperor Hsien-tsung during the Years 808–10*. The Hague: Mouton.

Forte, Antonino. 1983. Daiji 大寺 (The Masters of Great Virtue). In *Hōbōgirin—Dictionnaire encyclopédique du bouddhisme d'après les sources chinoises et japonaises*. Paris and Tōkyō: L'Académie des Inscription et Belles-Lettres, Institute de France, vol. VIII, pp. 682–704.

Forte, Antonino. 1992. Chinese State Monasteries in the Seventh and Eighth Centuries. In *Echō ō Go-Tenchikukyū den kenkyū* 慧超往五天竺國伝研究. Edited by Kuwayama Shōshin 桑山正進. Kyoto: Jinbun kagaku kenkyūjo, pp. 213–58.

Forte, Antonino. 2003. Daisōjō 大僧正 (The Grand Monastic Rectifier). In *Hōbōgirin—Dictionnaire encyclopédique du bouddhisme d'après les sources chinoises et japonaises*. Paris-Tōkyō: L'Académie des Inscription et Belles-Lettres, Institute de France, vol. VIII, pp. 1043–70.

Gernet, Jacques. 1995. *Buddhism in Chinese Society: An Economic History from the Fifth to the Tenth Centuries*. New York: Columbia University Press.

Gong, Guoqiang. 龔國強. 2006. *Sui Tang Chang'an cheng fosi yangjiu* 隋唐長安城佛寺研究. Beijing: Wenwu chubanshe.

Goossaert, Vincent. 2000. *Dans Les Temples de la Chine: Histoire des cultes, vie des communautés*. Paris: Albin Michel.

Halperin, Mark. 1997. Pieties and Responsibilities: Buddhism and the Chinese Literati: 780–1280. Ph.D. dissertation, University of California-Berkeley, Berkeley, CA, USA.

Halperin, Mark. 2006. *Out of the Cloister: Buddhism in Elite Lay Discourse in Sung China, 960–1279*. Cambridge: Harvard University Asia Center.

Hamar, Imre. 2002. *A Religious Leader in the Tang: Chengguan's Biography*. Tokyo: International Institute for Buddhist Studies, International College for Advanced Buddhist Studies.

Hargett, James M. 1996. Song Dynasty Local Gazetteers and Their Place in the History of Difangzhi Writing. *Harvard Journal of Asiatic Studies* 56: 405–42. [CrossRef]

Hartman, Charles. 1986. *Han Yü and the T'ang Search for Unity*. Princeton: Princeton University Press.

He, Liqun. 2013a. Cong Beiwuzhuang foxiang maicangken lun Yecheng zhaoxiang de fazhan jieduan yu Yecheng moshi 從北吳莊佛像埋藏坑論鄴城造像的發展階段與"鄴城模式". *Kaogu* 考古 5: 76–87.

He, Liqun. 何利群. 2013b. Buddhist State Monasteries in Early Medieval China and their Impact on East Asia. Ph.D. dissertation, Heidelberg University, Heidelberg, Germany.

Heirman, Ann, and Peter Bumbacher Stephan, eds. 2007. *The Spread of Buddhism*. Leiden: Brill.

Heirman, Ann, and Mathieu Torck. 2012. *A Pure Mind in a Clean Body: Bodily Care in the Buddhist Monasteries of Ancient India and China*. Gent: Academia Press, Ginkgo.

Ho, Puay-peng. 1995. The Ideal Monastery: Daoxuan's Description of the Central Indian Jetavana Vihāra. *East Asian History* 10: 1–18.

Ledderose, Lothar. 1980. Chinese Prototypes of the Pagoda. In *The Stūpa: Its Religious, Historical and Architectural Significance*. Edited by Anna Libera Dallapiccola and Stephanie Zingel-Ave Lallemant. Wiesbaden: Franz Steiner Verlag, pp. 238–48.

Ledderose, Lothar. 2000. *Ten Thousand Things: Module and Mass Production in Chinese Art*. Princeton: Princeton University Press.

Ling, Chaodong. 凌朝棟. 2005. *Wenyuan yinghua yanjiu* 文苑英華研究. Shanghai: Shanghai guji chubanshe.

Mazanec, Thomas J. 2017. The Invention of Chinese Buddhist Poetry: Poet–Monks in Late Medieval China (*c.* 760–960 CE). Ph.D. dissertation, Princeton University, Princetoni, IL, USA.

McMullen, David. 1988. *State and Scholars in T'ang China*. Cambridge: Cambridge University Press.

McRae, John R. 2005. Daoxuan's Vision of Jetavana: The Ordination Platform Movement in Medieval Chinese Buddhism. In *Going Forth: Visions of Buddhist Vinaya*. Edited by William Bodiford. Honolulu: University of Hawai'i Press, pp. 68–100.

Mesnil, Evelyne. 2006. Didactic Paintings between Power and Devotion. In *Essays on East Asian Religion and Culture*. Edited by Christian Wittern and Shi Lishan. Kyoto: Festschrift Committee for Nishiwaki Tsuneo, pp. 97–148.

Owen, Stephen. 2007. The Manuscript Legacy of the Tang: The Case of Literature. *Harvard Journal of Asiatic Studies* 67: 259–326.

Pu, Chengzhong. 2016. Slaves (*nubi* 奴婢) in Daoxuan's Vinaya Writings. *Studies in Chinese Religions* 2: 18–44. [CrossRef]

Robson, James. 2009. Introduction: Neither too Far, nor too Near: The Historical and Cultural Contexts of Buddhist Monasteries in Medieval China and Japan. In *Buddhist Monasticism in East Asia: Places of Practice*. Edited by James Benn, Lori Meeks and James Robson. London: Routledge, pp. 1–17.

Seo, Tatsuhiko. 妹尾達彦. 1993. Haku Kyoi to Choan, Rakuyō 白居易長安, 洛陽. In *Haku Kyoi kenkyū kōza* 白居易研究講座. *Vol. I: Haku Kyoi bungaku to jinsei* 白居易文人生. Edited by Ōta Tsugio 太田次男. Tokyo: Benseisha, pp. 270–96.

Shields, Anna M. 2015. *One who Knows Me: Friendship and Literary Culture in Mid-Tang China*. Cambridge: Harvard University Press.

Shields, Anna M. 2017. The "Supplementary" Historian? Li Zhao's *Guo shi bu* as Mid-Tang Political and Social Critique. *T'oung Pao* 103: 413–18. [CrossRef]

Sokolova, Anna. 2021. State, Bureaucracy, and the Formation of Regional Monastic Communities in Tang Buddhism. Ph.D. dissertation, Ghent University, Ghent, Belgium.

Tackett, Nicolas. 2014. *The Destruction of the Medieval Chinese Aristocracy*. Cambridge: Harvard University Asia Center.

Teiser, Stephen F. 2006. *Reinventing the Wheel: Paintings of Rebirth in Medieval Buddhist Temples*. Seattle: University of Washington Press.

Twitchett, Denis C. 1956. Monastic Estates in T'ang China. *Asia Major* 5: 123–46.

Waley, Arthur. 1951. *The Life and Times of Po Chü-i*. London: George Allen and Unwin.

Wang, Yintian. 王銀田. 2000. Bei Wei Pingcheng mingtang yizhi yanjiu 北魏平城明堂遺址研究. *Zhongguoshi yanjiu* 中國史研究 1: 37–44.

Weinstein, Stanley. 1987. *Buddhism under the T'ang*. Cambridge: Cambridge University Press.

Xiao, Mo. 蕭默. 2003. *Dunhuang jianzhu yanjiu* 敦煌建築研究. Beijing: Jixie Gongye chubanshe.

Xie, Siwei. 謝思煒. 1997. *Bai Juyi ji zonglun* 白居易集綜論. Beijing: Zhongguo shehui kexue chubanshe.

Xie, Chongguang. 謝重光, and Wengu Bai. 白文固. 1990. *Zhongguo sengguan zhidu shi* 中國僧官制度史. Qinghai: Qinghai Renmin chubanshe.

Yan, Gengwang. 嚴耕望. 1992. Tangren xiye shanlin siyuan zhi fengshang 唐人習業山林寺院之風尚. In *Yan Gengwang shixue lunwen ji* 嚴耕望史學論文集. Taibei: Lianjing, pp. 271–316.

Yang, Lien-sheng. 1950. Buddhist Monasteries and Four Money-Raising Institutions in Chinese History. *Harvard Journal of Asiatic Studies* 13: 174–91. [CrossRef]

Zhang, Yuhuan. 張馭寰. 2008. *Zhongguo Fojiao siyuan jianzhu jiangzuo* 中國佛教寺院建築講座. Beijing: Dangdai zhongguo chubanshe.

Zhang, Dewei. 2013. An Unforgettable Enterprise by Forgotten Figures: Making the Zhaocheng Canon 趙城藏 in North China under the Jurchen Jin Regime. *ZINBUN: Memoirs of the Research Institute for Humanistic Studies (Kyoto University)* 44: 1–38.

Zürcher, Erik. 1989. Buddhism and Education in T'ang Times. In *Neo-Confucian Education: The Formative Stage*. Edited by W. Theodore de Bary and John W. Chaffee. Berkeley: University of California Press, pp. 19–56.

Zürcher, Erik. 2014. Buddhism in a Pre-modern Bureaucratic Empire: The Chinese Experience. In *Buddhism in China: Collected Papers of Erik Zürcher*. Edited by Jonathan A. Silk. Leiden: Brill, pp. 89–103.

religions

MDPI

Article

The Translation of Buddhism in the Funeral Architecture of Medieval China

Shuishan Yu

School of Architecture, Northeastern University, Boston, MA 02115, USA; sh.yu@northeastern.edu

Abstract: This article explores the Buddhist ritual and architectural conventions that were incorporated into the Chinese funeral architecture during the medieval period from the 3rd to the 13th centuries. A careful observation of some key types of sacred architectural forms from ancient East Asia, for instance, pagoda, *lingtai*, and *hunping*, reviews fundamental similarities in their form and structure. Applying translation theory rather than the influence and Sinicization model to analyze the impact of Buddhism on Chinese funeral architecture, this article offers a comparative study of the historical contexts from which certain architectural types and imageries were produced. It argues that there was an intertwined mutual translation of formal and ritual conventions between Buddhist and Chinese funeral architecture, which had played a significant role in the formations of both architectural traditions in Medieval China.

Keywords: Buddhist architecture; funeral architecture; Chinese architecture; pagoda; *lingtai*; *xiangtang*; *mubiao*; *hunping*; *mingqi*; *mingtang*

Citation: Yu, Shuishan. 2021. The Translation of Buddhism in the Funeral Architecture of Medieval China. *Religions* 12: 690. https://doi.org/10.3390/rel12090690

Academic Editor: Todd Lewis

Received: 19 July 2021
Accepted: 23 August 2021
Published: 27 August 2021

Publisher's Note: MDPI stays neutral with regard to jurisdictional claims in published maps and institutional affiliations.

1. Introduction

Since its legendary introduction to the imperial court during the Eastern Han dynasty (25–221 CE),[1] Buddhism had a profound impact on every aspect of the Chinese civilization in the following millennium of Medieval China.[2] In the physical world, cities and landscapes were transformed by the spread of Buddhist practices and the introduction of new architectural types; in the spiritual world, Buddhism changed the way people perceived human life in terms of both its synchronic relationships with heaven (gods) and earth (societies), and its diachronic relationships with the past (ancestors) and the future (descendants). In the field of architectural history, much scholarship has been dedicated to the translation of Indian and Central Asian Buddhist architecture into Chinese contexts, especially the creation and development of the pagoda, the adoption and evolvement of traditional Chinese courtyard conventions in the construction of monasteries,[3] and their influence on the development of traditional Chinese architecture. With great contributions to our knowledge on Chinese architecture, such approaches, however, presume an interpretative model of influence vs. Sinicization that treats Buddhist and Chinese architecture as two separate systems. Buddhism, as would be argued here, not only influenced but also directly participated in the very formation of Chinese architectural conventions.

This article explores the Buddhist ritual and architectural conventions that were incorporated into the Chinese funeral architecture during the period from the 3rd to the 13th centuries. A careful observation of some key types of sacred architecture from ancient East Asia, both before and after the Buddhist introduction, reviews fundamental similarities in their forms and structures. For instance, the early pagodas of West China from the 3rd to 10th centuries share great resemblances with both the sacrificial halls, or *xiangtang* 享堂, in the royal tombs of the Zhongshan kingdom during the Warring States period (476–221 BCE) of the Eastern Zhou dynasty (771–256 BCE), and the tomb mound, or *lingtai* 陵臺, in the royal mausoleums of the Western Xia kingdom (1032–1227), in their concentric plans and the combination of an earth core with the multilayered wooden verandas. Similarly, the highly architectural elements in the crowns of the funeral jars, or *hunping* 魂瓶, in the areas of Jiangsu and Zhejiang provinces from the 4th century resemble both the Indian stupa and the Han *mingtang* 明堂. What is the nature of such resemblances? Rather than using the influence and Sinicization model, I prefer to use the concept of "translation" for the discussion on the interactions among different cultures. Translation is not simply a substitution of words between the original and targeted languages, but a reconstruction of meanings in new contexts based on the understanding of a text in a different cultural and linguistic milieu. Borrowed from the linguistic practice, the translation model in the analysis of cultural exchanges forsakes the static view on a given culture as singular and predefined, thus allowing us to treat both Buddhist and Chinese architecture as multiple and evolving traditions.[4] Through the study of the historical contexts from which certain architectural types and imageries were produced and a comparative analysis of their forms and functions, intertwined mutual translations of formal and ritual conventions can be observed between Buddhist and Chinese funeral architecture, which played a significant role in the formations of both architectural traditions in Medieval China.

2. The Shifting Underground: *Mingqi, Hunping*, Stupa, and the Early Translation of Buddhism in Local Tombs

Before the introduction of Buddhism to China during the Eastern Han dynasty, the Chinese buried their deceased's bodies without cremation. In the Qin and Western Han period (221 BCE–8 CE), the rich and powerful built enormous aboveground tumuli with elaborate underground chambers. The Mausoleum of the First Emperor of Qin (*Qin Shihuangdi*), for instance, had two layers of rectangular enclosure of walls with gates on the cardinal directions, which measured 2165 m north–south and 940 m east–west for the outer layer, and within the inner layer, a pyramidal rammed earth mound (封土 *fengtu*) with a base square of 350 m each side and an extant height of 76 m.[5] Commoners had much humbler tombs, buried with at least several specially made objects or utensils they once had during their lifetimes to accompany their afterlives. Buddhists, on the other hand, cremated the remains of the deceased, especially the venerated ones, to create the relics known as *sharila*, which was believed to be a confirmation of *nirvana*, the ultimate enlightenment. On such occasions, such as in the case of Shakyamuni Buddha, structures such as the stupa would be built to honor such a spiritual achievement.[6] During the early Medieval period of the 3rd to 6th centuries, with the progressive Buddhist conversion and the gradual establishment of Buddhist communities and monasteries, pagodas were built to honor both relics of Buddha and the remains of great Buddhist masters.[7] Burial customs of common Buddhists also started to bear the imprints of the new faith.

The reception of Buddhism was incorporated into the Chinese funeral tradition by the substitution of the previous Confucian and Daoist imageries with Buddhist deities and forms. The images of Buddha appeared in Chinese tombs as early as the late 2nd century. In the cave-tomb from Mahao in Sichuan province, for instance, a seated Buddha carved on the lintel carries the unmistakable Buddhist iconographic marks of *ushnisha* and *abhaya mudra* (see Figure 1).

Among other narrative carvings of Confucian ideology and morality, such as the famous Jing Ke's attempted assassination of the First Emperor of Qin, it occupies a place previously occupied by the Daoist gods of Dong Wanggong 東王公 and Xi Wangmu 西王母. Although the icon is Buddhist, the concept and functionality behind the Buddha image is Daoist, as Professor Wu Hung pointed out that "this holy image is no longer an object of worship on public occasion, but is a symbol of the deceased individual who had hoped to attain immortality after his death".[8] Buddhism was translated into the Chinese architectural context of the 2nd to 3rd centuries through the substitution of traditional Chinese deities with the Buddha. Such an understanding of the foreign god was also confirmed by the contemporary intellectual discourse on Buddhism. In the famous monograph *Lihuolun* 理惑論, Mu Rong, the late 2nd-century Buddhist scholar, frequently cited Confucius and Laozi in the defense of Buddhism and argued that Buddha is an honorary title for the enlightened one, much like the title of *Shen* 神 for the Three *Huang* 三皇, the title *Sheng* 聖 for the Five *Di* 五帝, and the supreme origin of *Dao* and *De* 道德之元祖.[9] Deities and saints of both Confucianism and Daoism were evoked to help for a translation of Buddha into the Chinese religious contexts.

Figure 1. Mahao tomb, front chamber, lintel carved with a Buddha image, Leshan, Sichuan province, later 2nd century CE. (Drawing by author, after Figure 1 in Wu Hung, "Buddhist Elements in Early Chinese Art".)

Grave goods of the early Medieval China experienced a similar Buddhist translation. The previously mentioned specially made objects buried to serve the afterlife of the tomb masters are known as *mingqi* 明器, which literally means the "bright utensils". Since the Han dynasty, a type of *mingqi* made in the form of house model became increasingly popular. Archaeologists discovered such architectural *mingqi* in large quantities, especially from the tombs of the Eastern Han dynasty and after.[10] Some *mingqi* house models feature multi-story towers with painted or sculpted details of wooden structure, such as post-and-lintel frames and the *dougong* 斗拱 brackets. These multi-story tall buildings were either interpreted as the watchtowers for defensive purpose and scenery enjoyment[11] or the reference to the dwelling of immortals.[12] The former fulfills the Confucian filial service of the deceased in the similar way as they were alive, while the latter was deeply imbedded in the Daoist practice aiming for immortality. A house model from Xiangyang in Hubei province, on the other hand, features Buddhist architectural details (see Figure 2).

Figure 2. *Mingqi* from Xiangyang, collection of the Xiangyang municipal Museum, Hubei province, Eastern Han to Three Kingdoms period, 2nd–3rd centuries CE. (Drawing by author, after Figure 1 in Luo Shiping, "Xianren hao louju: Xiangyang xin chutu xianglun taolou yu Zhongguo futuci leizheng".)

This house shares with other Eastern Han and Three Kingdoms period *mingqi* model in the combination of a main tower with a walled and gated courtyard. Both the tower and the gate have wooden frames supporting tiled roofs. The main tower is a two-story structure. The first story has a rectangular plan and an overhanging two-slope roof, supporting a wooden platform on brackets, above which the second story with a square plan and a four-slope roof sits. The unique feature is a pole with seven umbrellas, rising from the center of the four-slope roof, which clearly resembles the vertical axis and *chatras* in Indian stupa such as the one from the Great Stupa at Sanchi.

The typical roofs of *mingqi* houses from this period often have sculpted birds standing the tops and ends of the ridges. Here, in the Xiangyang model, while the center of the top horizontal ridge supports a Buddhist axis, the ends of both the horizontal and slanting ridges turn into a leaf-like shape. According to Luo Shiping from the Centra Academy of Fine Arts in Beijing, these leaves represent the bodhi tree, under which Shakyamuni Buddha meditated and became enlightened.[13] Thus, the Buddhist concepts of enlightenment and *nirvana* were both translated into the Chinese funeral practice through the incorporation of iconic imageries from the life of Buddha and the features of the stupa into the *mingqi* house. The terminal of the axis in the Xiangyang model is a crescent, which is neither a Chinese nor an Indian architectural feature. Rather, some stupa images from Central Asia bare crescent forms on the railing poles. The formal translation of Buddhist architecture into the Chinese funeral practice was not direct and straightforward. Given the great distance and diverse cultures in between China and the original land of the Buddha, it might had gone through many mediating stages.[14]

Another type of funeral object that yields rich architectural images is called *hunping* 魂瓶, the bottle of the soul. Though made in ceramic with a body shaped like a jar, *hunping* was not made to be used as a container. Small holes often appear in the middle part of the body and the round mouth is often sealed with a highly sculptural complex of additional decorations. The sculptural forms forming the crown of *hunping* are often tall, concentric in plan, and quite architectural, featuring a two to three-story tower-pavilion amid a crowd of animal and human figures of both real and mythical, birds, lions, phoenixes, musicians, dancers, singers, etc. The appearance of *hunping* was in the 3rd century, much later than the architectural *mingqi*, and spread in the following centuries of the early Medieval period mainly in the lower Yangtze provinces of Jiangsu and Zhejiang. Wu Hung argued that an important prototype of the *hunping* jar might be the reliquary used in the Indian world such as the famous Reliquary of Kanishka from the 2nd century.[15] He suggested that *hunping* was primarily used by the northern Chinese exiled to the southern Yangtse area during the catastrophic collapse of the Western Jin dynasty, who had lost the bodies of the deceased and adopted the local tradition to provide the resting places for their souls using such vessels. *Hunping*, thus, became the embodiment of a specific funeral practice amalgamating the Confucian filial rites, the ancient shamanistic soul-calling (*zhaohun* 招魂) of the southern (Chu and Wu) areas, and the Buddhist belief in the separation of the physical remains and the true self.[16]

The tangible Buddhist decorative motifs are human figures on some *hunping* jars that can be identified as Buddha or bodhisattvas through their attributes such as the *ushnisha*, halo, or *mudras*. While many previous discussions focus on the stylistic and iconographic analysis of imageries, this article pays special attention to the architectural elements in the *hunping*. A typical crown structure of the *hunping* has a concentric plan with a pyramidal three-dimensional form. In a 3rd-century Western Jin dynasty *hunping* jar from the Museum of Fine Arts Boston, the peaking center is an urn with a wide circular opening, surrounded by four smaller urns of similar shape and proportion (see Figure 3).

The four small urns form the corners of a square, defining the four cardinal directions. On one direction, four layers of sloping roofs create a façade of a wooden tower, under which a square opening serves as the only entrance hidden behind a mounted figure, who is seemingly emerging from the gate of the tower, defined by two solid pillars. Two additional mounted figures frame on the outer sides of the pillars, seemingly guarding the gate. On both sides of the four-story tower are singers and musicians playing various instruments, divided into four groups by three flocks of birds. All the birds are heading up with widely open wings, creating a strong soaring momentum. The base of the crown complex and the openings of the top central urn are both circular, while the four corner urns define a square set in between the larger lower circle of the base and the smaller upper circle of the central urn (see Figure 4).

Figure 3. *Hunping*, collection of the Museum of Fine Arts Boston, Western Jin dynasty, late 3rd century CE (photo by author).

While the urns and the architectural, anthropomorphic, and zoomorphic figures are typical decorative motifs on *hunping* jars, the entrance tower in the MFA Boston item strongly resembles a later pagoda in East Asia, and the combination of square and circle, as well as the emphasis on both the cardinal directions and the axial centrality, share much in common with the Indian stupa. Like a tower-pavilion style pagoda (*lougeshi-ta* 樓閣式塔), the MFA *hunping* tower has multiple layers of tiled roofs diminishing in size and width toward the top. Attached to the wall of the jar, it especially resembles the wooden edifice added to make a façade for a large Buddhist cave, for instance, the Nine-Story Pagoda (*jiucengta* 九層塔) at the Mogao Caves in Dunhuang, Gansu province (see Figure 5).

Figure 4. *Hunping*, view from above, collection of the Museum of Fine Arts Boston, Western Jin dynasty, late 3rd century CE (photo by author).

The cave behind the pagoda façade was created during the early Tang dynasty (618–907), much later than the Boston *hunping*, and the wooden structure has been rebuilt many times. It was also unlikely that the Dunhuang pagoda copied the images of the Eastern Jin dynasty *hunping*. The resemblance must be a reflection of a common architectural prototype they both shared, such as the context for the generation of meanings in a linguistic translation to which both the original and targeted languages refer.

The grouping of a large central stupa surrounded by four smaller ones on the corners can be traced back to the stupa at the Mahabodhi Temple in Bodhgaya, the place where historical Buddha attained enlightenment. The stupa with a tall *shikara* tower surrounded by four smaller one elevated on a square platform had undergone much rebuilding and repairs, but the original construction was at least since the Gupta dynasty (c. 319–467), and its prototype might even be earlier. The Diamond Throne, *vajrasana*, covered by the Bodhi tree, under which Buddha was believed to have meditated, was said to be established by King Ashoka (r. c. 268 to 232 BCE).[17] It is still a popular style in Theravada stupas of Southeastern Asia. Additionally, some scenes in Dunhuang murals from the Northern dynasties period (386–577) contain pagoda-like structures with similar assemblage. In late imperial China, such an assemblage in the Buddhist relics structure can be found in the Vajrayana school (*Mizong* 密宗) known as *Jingangbaozuo-ta* 金剛寶座塔, the Diamond Throne Pagoda, which became especially popular among the monasteries related to the Tibetan practice since the Yuan dynasty.[18]

A *hunping* jar in the National Museum in Beijing from the same region and period, indeed, has the five urns replaced by five pavilions, with the central one elevated on a tower (see Figure 6). *Hunping* jars with five pavilions arranged in similar concentric plan can also be found in other museum collections, for instance, the one from the Museum of the Six Dynasties in Nanjing.[19] With the *que* gates added to mark the entrance and axiality, the whole complex strongly resembles a *Jingangbaozuo-ta* such as the one from the Ming dynasty Zhenjue Monastery in Beijing or the one from the Qing dynasty Five Pagodas Monastery in Hohhot, Inner Mongolia (see Figure 7).

Figure 5. The Nine-Story Pagoda (*jiucengta*) at the Mogao Caves, Dunhuang, Gansu province (photo by author).

Figure 6. *Hunping*, Celadon Hunping (Soul Jar), Western Jin dynasty (Creative Commons: https://commons.wikimedia.org/wiki/File:Western_Jin_Celadon_Hunping_(Soul_Jar).jpg; accessed on 23 May 2021).

Figure 7. *Jingangbaozuo-ta* from the Five Pagodas Monastery, Hohhot, Inner Mongolia, Qing dynasty, 18th century (photo by author).

3. The Transformed Passage: *Que* Gate, *Mubiao*, and the Ashoka Column

The early incorporation of Buddhist ideas and imageries in Chinese funeral architecture, in the 2nd to 3rd centuries as discussed in the previous section, concerns mostly the afterlife of the deceased, remains on objects that were not meant to be engaged by the livings, and concentrates largely underground. In the following 4th to 6th centuries, the translation of Buddhist architecture into the Chinese funeral context became more psychological, life engaging, ritually related, and manifests mainly along the aboveground passage.[20] While objects discussed in the previous section were mostly from tombs of common well-to-do families, relatively of humble origins, cases sampled in this section can be related to the funeral practice of the imperial rulers.[21] Compared to those of the previous Han dynasty, imperial mausoleums from the Eastern Jin (266–420) and the Southern Dynasties (420–589) were much smaller in terms of both the aboveground mounds and the underground chambers.[22] The sculptural scheme framing the spirit path leading from the entrance to the tomb mound, however, became more refined and elaborate with concentrated Buddhist translation of architectural elements.

The sculptures framing a typical Southern Dynasties spirit path *shendao* 神道 for the imperial mausoleum start with the chimerical creature combining the body of a lion and a pair of wings, a mythical animal known as *qilin* 麒麟 or *pixie* 辟邪. They are followed in a certain distance by a pair of free-standing columns and ended with a pair of stone steles on turtle bases[23] (see Figure 8).

The stone column often bears a rectangular tablet inscribed with the name and titles of the tomb master, thus known as *mubiao* 墓表, the sign of the tomb. The *mubiao* pillars are rich in details of European, Indian, or West Asian origins (see Figure 9).

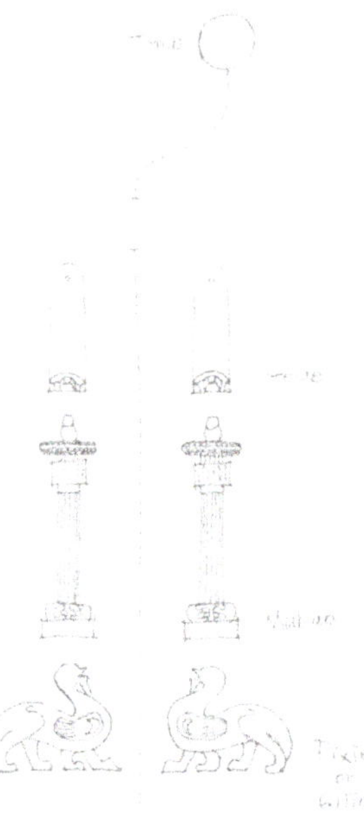

Figure 8. Spirit path in a typical plan of a Southern Dynasties royal tomb (drawing by author, after Figure 5.1 in Wu Hung, *Monumentality in Early Chinese Art and Architecture*, 252).

Figure 9. Prince Xiao Jing (477–523) *mubiao*, Nanjing, Jiangsu province, Liang Dynasty (Creative Commons: https://commons.wikimedia.org/wiki/File:Tomb_of_Xiao_Jing_-_P1200017.JPG; accessed on 29 May 2021).

The Prince Xiao Jing (477–523) *mubiao* from the Liang Dynasty, for instance, has a fluted shaft similar to that of an ionic column in classical Greco-Roman architecture. The earliest fluted column shaft came from the Eastern Han dynasty, such as the *mubiao* of the Cleric Qin of the You Prefecture 東漢幽州書佐秦君墓表 from the modern-day Beijing area[24] (see Figure 10).

Figure 10. Cleric Qin of the You Prefecture *mubiao*, Shijingshan District, Beijing, Eastern Han dynasty, 105 CE (Creative Commons: https://upload.wikimedia.org/wikipedia/commons/f/fe/%E5%B9%BD%E5%B7%9E%E6%9B%B8%E4%BD%90%E7%A7%A6%E5%90%9B%E7%9F%B3%E9%97%95_03.jpg; accessed on 19 June 2021).

The round details at the end of the fluting resemble the egg-and-dart in a classical ionic capital. The twin animal figures framing the neck connecting the fluted shaft below and the inscribed tablet above resemble the incorporation of symmetrical animal figures in the carved capitals from the Persian Persepolis.[25]

The framing of a passageway with free-standing columns can be considered a Buddhist translation of the Chinese *que* 闕 gate.[26] A standard spirit path of an Eastern Han cemetery often starts with a pair of stone *que* towers, followed by mythical animals and stelae and ending at the shrine in front of the pyramidal rammed earth tumulus (see Figure 11).

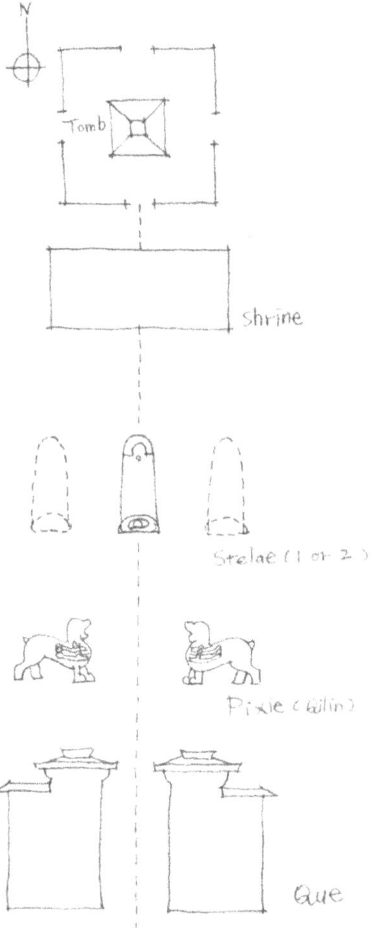

Figure 11. Standard plan of an Eastern Han dynasty tomb (drawing by author, after Figure 4.1 in Wu Hung, *Monumentality in Early Chinese Art and Architecture*, 191).

The *mubiao* of the Cleric Qin of the You Prefecture, for instance, was originally part of a *que* gate complex. Fragments of Eastern Han funeral *que* had been discovered in different areas, especially from the southeastern Sichuan province.[27] The stone *que* used in the funeral complex was already a translation of its counterpart with wooden structure serving in the living quarters. In an actual wooden *que* gate, the two towers framing the passageway are often connected in the middle above the path with roofs and sometimes verandas, such as the ones in the crown complex of the *hunping* jar from the museum in Nanjing. They might also be connected with enclosing walls thus serve the function of an actual gate. In the tombs from the period of Eastern Jin and the Southern Dynasties (266–589), however, the free-standing stone *que* towers preceding the mythical animal sculptures largely disappeared and the *mubiao* columns were added following them.

These *mubiao* pillars bear great similarities with the Ashoka columns in India from the 3rd century BCE.[28] The combination of animal sculptures at the summit, a slender shaft with inscriptions planted on the ground, and a seat decorated with lotus petals in-between is a characteristic feature for the columns ordered to be founded by Ashoka (r. 268–232 BCE), the model king for Buddhist sponsorship and the quintessential example of *chakravarti*, a religious universal ruler. Known as Ayuwang 阿育王 in Chinese, many monasteries in his name were established in his honor in early Medieval China, including the one rebuilt by Emperor Wu of Liang 梁武帝, the cousin of Prince Xiao Jing.[29] The Xiao Jing *mubiao* has a sculpture of a winged chimerical creature standing on the top of a large disk with lotus petal decoration. A comparison of the Xiao Jing *mubiao* (early 6th century) with the Ashoka pillar (3rd century BCE) (see Figure 12), the Persepolis columns (5th–4th centuries BCE), and the Naxos Sphinx pillar (6th century BCE) (see Figure 13) clearly indicates the cognate relationships among them and the possible transmission of the chimerical column from the West to the East. Buddhism served as a very important medium for the formal translation in this monumental construction.

Figure 12. Lauria Nandangarh pillar of Ashoka, India (photo date: 27 June 2018; photo author: M. Vidyut Prakash Maurya; Creative Commons: https://commons.wikimedia.org/wiki/File:Lauria_Nandangarh_pillar_of_Ashoka_side_view.jpg; accessed on 19 June 2021).

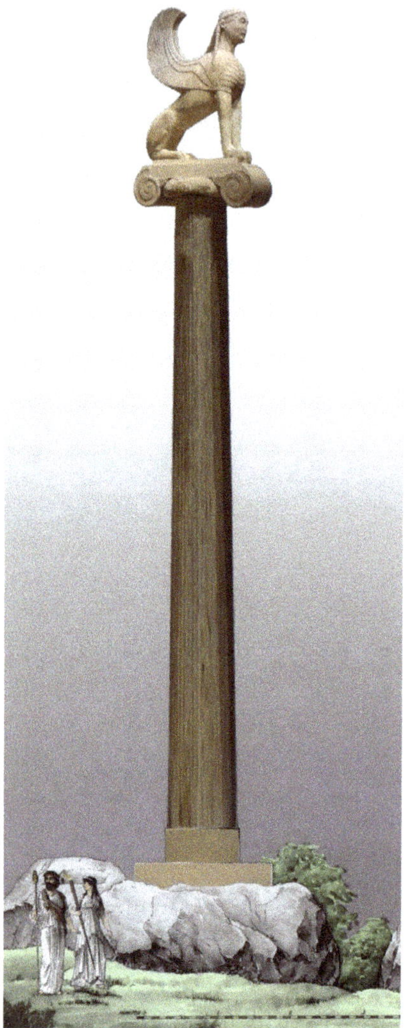

Figure 13. Reconstructed view of the Naxian Sphinx on its 12.5 m Ionic column, erected next to the Temple of Apollo in Delphi, Greece, 560 BCE, collection of the Delphi Archaeological Museum (Creative Commons: https://www.wikiwand.com/en/Sphinx_of_Naxos; accessed on 19 June 2021).

While the major formal features of the Xiao Jing *mubiao* were mainly of foreign origins, the tablets inscribed with the tomb master's name and official titles, a traditional Confucian ritual for honoring the deceased ancestors, also went through the Buddhist translation. With spirit path running from south to north, two *mubiao* pillars used to stand on the east and west sides of the passage. Today, only the one on the west side survived. The inscriptions on its tablet, however, would be unreadable to any literate Chinese on the first sight. The twenty-three Chinese characters were written in mirror-image, or *fanshu* 反書, exactly symmetrical stroke by stroke with those on the tablet of the missing east *mubiao* pillar. In traditional Chinese cosmology, the east, which is associated with the element wood in the circle of Five Elements *wuxing* 五行, symbolizes life, while the west is the direction associated with death and afterlife. The mirror-image of standard legible Chinese characters on the west *mubiao* of the Xiao Jing tomb, thus, was not meant for the living—the visitors who pass through the spirit path to pay homage to the deceased, but for the very tomb master who was buried behind. Wu Hung argued that the mirror-image inscriptions suggest a transparent stone slab and presume a vision from the direction of the tomb, the perspective of the deceased.[30]

The resurrection of the vision of the dead in funeral architecture was the result of a revised understanding about life and death brought by the new faith of Buddhism. Buddhists believe death was not the end of life but a necessary step to transcend life toward enlightenment. Buddhism gained enormous popularity in south China during the Liang dynasty, whose ruling classes include the imperial Xiao clan. Xiao Jing's cousin, Emperor Wu of Liang, was the famous Buddhist emperor who had founded numerous monasteries and personally served in Buddhist temples only to be ransomed by his courtiers to return to throne.[31] Xiao Jing himself also wrote letters in support of Buddhism.[32] In discussion of the mirror-image in the *mubiao* inscriptions, Wu Hung argued, "The important point is that this reading/viewing process forces the mourner to go through a psychological dislocation from this world to the world beyond it . . . The discovery of the mirror relationship between the two inscriptions forges a powerful metaphor for the opposition between life and death . . . More important, to completely fulfill the ritual transformation, the material existence of the gate has to be rejected".[33]

Indeed, the Buddhist translation of funeral architecture in China during this period made more tangible transformation of the physical structure engaging ritual activities aboveground. The *mubiao* pillars were no longer replicas of timber towers in the Chinese palace, as the stone carved *que* gates did in previous funeral complexes. Rather, they derived their forms from foreign Buddhist architecture. In terms of decorative motifs, sacred figures and historic scenes representing Confucian ideologies and morality were replaced by chimerical creatures both guarding the ground and soaring in the sky; architectural details imitating palatial edifices were replaced by the symbolic Buddhist image of the lotus flower. As Wu Hung observed, "a funerary gate no longer related itself to a counter-image in the living world and derived its meaning from this opposition; rather, it directly expressed the idea of transcendence and enlightenment. Most important, the mirror inscriptions on the gate completely alter the relationship between a monument and its audience: instead of presenting readable texts confirming the shared values of filial piety, they 'reverse' the conventional way of writing and challenge the viewer's perception, forcing him to reinterpret a funerary monument and to view it with fresh eyes".[34] The new monumentality after the Buddhist translation blurred the boundaries between life and death, highlighting the active process of the funeral rites rather than simply providing a static familiar living environment for the afterlife. The inscriptions along the spirit path were written for both the eyes of the living and the soul of the deceased, making a spiritual transformation for the stone and earth of the tomb structure. It is a powerful metaphor about transcendence and enlightenment.

4. The Shared Superstructure: *Lingtai, Xiangtang, Mingtang,* and the Pagodas

The most characteristic architectural symbol of Buddhism in East Asian is the pagoda. Translating the Buddhist relics structure into the Chinese timber building context, the *Louge* 樓閣 (tower-pavilion) style[35] wooden pagoda features a multi-story tower, often with an odd number of levels each with a sloping tiled roof diminishing in both heights and width toward the top. The number of bays for each level often remain the same with the intercolumnar distances diminish as the height rises. Completing the submit of such a pagoda is the wooden axis with *chatras*. Originated in China, such wooden pagodas are popular in all East Asian countries, including Korea and Japan, for instance, the 7th-century five-story pagoda from Horyuji in Nara,[36] which is the oldest extant wooden pagoda in the world (see Figure 14). In China, three-dimensional images of early Louge style pagoda can be found in Buddhist grottoes, for instance, the 5th-century Yungang caves.[37] Although the oldest extant wooden pagoda in China, the Wooden Pagoda of Yingxian (see Figure 15) is from the 11th century, some masonry pagodas from the Tang dynasty also mimic wooden frames in decorative details on the walls.[38]

Figure 14. Five-story pagoda, Horyuji, Nara, Japan, 7th–8th centuries (photo by author).

Buddhist relics structures first appeared in China in the western regions along the ancient Silk Road. The region, known as *Xiyu* 西域, the Western Realm, refers to a vast area west of the Hexi Corridor in today's Gansu province, including Xinjiang in modern China and areas in many other Central Asian countries. Buddhism was first introduced to China from India via Central Asia through the ancient Silk Road.[39] In the early 20th century, Buddhist ruins were discovered by Western exploders[40] such as Marc Aurel Stein (1862–1943).[41] Since the 1950s, Chinese archaeologists made systematic surveys of Buddhist sites in Xinjiang, excavating and documenting ancient structures including stupas. These early stupas in Western China, for instance, the Mauri-tim stupa near Kashgar,[42] were often constructed with adobe. Sharing with the Indian stupa in features such as the solid core and a tripartite vertical division, however, they often had a square plan with an emphasis on the continuity from the base to the upper axis, shapes taping gradually toward the top instead of featuring a large hemispherical main body (*anda*) in the middle part. They share more in common with the Central Asian, especially the Gandara region stupa from the Kushan period (1st to 4th centuries).[43] A 3rd-century stupa from the Subash monastery near Kuche has a tall square base, on top of which a cylindrical middle drum supports a domical top similar to the *anda* in Indian stupa[44] (see Figure 16). Small holes appear on different levels in the square base, the cylindrical drum, and the domical top, which suggest there might be additional wooden or ceramic surface materials attached. Compared to the Sanchi stupa, the tomb mound feature was played down and the soaring verticality was highlighted. The square base has an arch opening on one side, clearly defining a frontal façade and giving the concentric plan an axial orientation. The plan and architectural image give equal emphasis on the geometric motifs of square and circle.[45] All these features make the *Xiyu* stupas closer to the ritual architecture of China and further differentiated from the Indian prototype. Similar stupas had also been discovered in other *Xiyu* sites such as Miran and Khotan.[46]

Figure 15. Wooden Pagoda of Yingxian, Ying county, Shanxi province, Liao dynasty, 11th century (photo by author).

Figure 16. Subash monastery stupa, Kuche, Xinjiang (Creative Commons: https://commons. wikimedia.org/wiki/File:Subashi_BLP472_PHOTO1125_16_671.jpg; accessed on 28 June 2021).

In China Proper, the earliest wooden pagodas in the Louge Style were constructed around a rammed earth core. Professor Fu Xinian suggested that the pre-Buddhist *taixie* 臺榭 tradition in Chinese architecture made it natural for the Chinese Buddhists to adopt the forms of the *Xiyu* earth stupa and combine them with the Chinese wooden construction.[47] *Taixie* is a construction method in early Chinese architecture for building multi-level monumental structures combining rammed earth platforms with surrounding wooden structures. Wrapping wooden verandas around the lower levels of platforms and elevating wooden towers on the top, the *taixie* method created an appearance of great scale and height without the actual creation of multiple-story interior space.[48] Widely discovered in archaeological sites of the Zhou (1046-221 BCE), Qin (221-206 BCE), and Han (206 BCE to 220 CE) periods, they gradually disappeared since the Medieval time.[49] Like Luo Shiping and many other scholars, Fu also believes that the high wooden towers were built for the hope to achieve immortality. The Daoist belief that immortals favored tower dwellings had contributed to the creation of the Louge Style pagoda.[50]

The famous nine story pagoda of the Yongning Monastery from the Northern Wei capital Luoyang was constructed with such a *taixie* method. After excavation, archaeologists discovered on the site of the Yongning Monastery pagoda rammed earth platform of 20 by 20 m on top of a rammed earth base of 38.2 by 38.2 m. The upper inner platform should be the earth core for the first floor of the pagoda, which was originally surrounded by a wooden veranda; the wider lower platform should be the base of the pagoda. Based on historical records such as the *Water Classics* (*Shuijingzhu* 水經註), contemporary pagoda images from the Northern Wei Buddhist caves such as those form Yungang and Dunhuang, and the archaeological data, scholars reconstructed the plan, elevation, and section of this famous early wooden pagoda. Fu Xinian's reconstruction is a nine-story pagoda with a square plan and a central pillar within a rammed earth core with wooden reinforcement. The lower seven stories are wooden verandas around earth platforms and the upper two stories are pure wooden structures. Each floor has a nine-bay façade on each side with both a sloping roof and a wooden balcony[51] (see Figure 17).

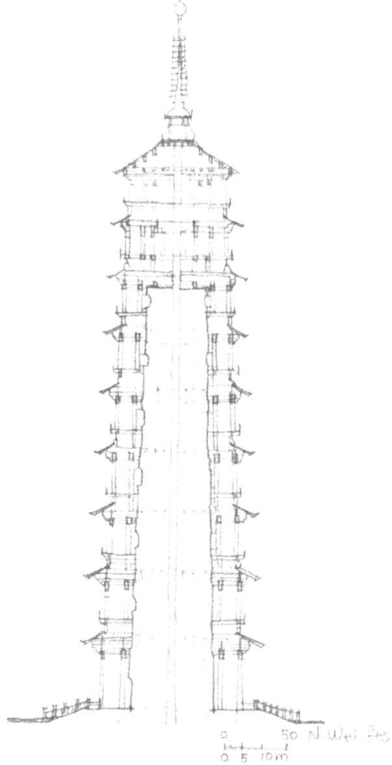

Figure 17. Reconstructed section of the Yongning Monastery pagoda (drawing by author, after Figure 2-7-24 in Fu Xinian, *Zhongguo gudai jianzhushi di 2 juan: Liangjin, Nanbeichao, Sui, Tang, Wudai jianzhu*, 188).

The Yongning Monastery pagoda was built in 516 and burned down in 534. Half a millennium later, rulers of a regional regime in Northwestern China, the Western Xia Kingdom, built the main structures of their mausoleums in a way similar to both the pagoda in Luoyang and the stupas along the Silk Road. In today's Ningxia Hui Autonomous Region of northwest China, at the foot along the eastern slopes of the Helan Mountain, an area of some 50 square kilometers provides the resting places for nine rulers of the Western Xia kingdom from the late 10th to the early 13th centuries. It was not entirely clear about the identity for each of the nine royal tombs, which share the same basic plan and the same north–south orientation. Each mausoleum has two layers of walls with corner towers, defining an inner court and an outer court such as in the living cities. The outer layer has only one gate located at the center of the south wall while the inner layer has gates on all four cardinal directions. The plan has a strong axiality and is highly symmetrical. The north and south gates are located on the main north–south axis and other key elements, such as the corner towers, *quetai* 雀臺 platforms, and stele pavilions, symmetrically frame the central axis on both sides (see Figure 18). The main tomb structures, the sacrificial hall *xiandian* 獻殿 and the tomb mound *lingtai* 陵臺, however, are deliberately off the main axis on the west side.[52] The geometric center for the inner court is the central platform *zhongxintai* 中心臺.[53]

Figure 18. Model of a Western Xia royal tomb in the Mausoleum Museum, Ningxia (photo by author).

Most structures in the Western Xia royal tombs were constructed with rammed earth, with the additions of bricks, paints, and glazed tiles for protection and decoration. The Tomb No. 1, which was believed to be the mausoleum for the founding ruler Taizu (Li Jiqian 李繼遷), for instance, has an octagonal rammed earth platform of nine steps, the sizes of which reduce progressively toward the top. On the base level, each of the eight sides of the octagon measures approximately 12 to 14 m. Horizontal lines of holes appear on each side of every level. The holes measure approximately 20 cm in diameter for each, which must have been the traces left by the wooden rafters supporting a tiled roof. During the archaeological excavations, large number of ceramic tiles, tile ends, animal sculptures for roof decoration, charcoal, and rotten wood were discovered on the upper steps of the rammed earth platform, as well as on the ground near the *lingtai*.[54] Thus, it is highly possible that the steps of the earth platform were surrounded by wooden verandas with tiled roof, transforming the exterior of the earth platform into a wooden pagoda (see Figure 19). Like the Yongning Monastery pagoda, the structure of wooden façades around an earth core in the Western Xia royal tombs is basically a *taixie* method. Seen from the exterior, the *lingtai* mound used to be a multi-story tower, with all levels featuring a sloping tiled roof diminishing in both heights and width toward the top. In a similar way as the system of Louge style Buddhist pagodas, the number of levels for each *lingtai* is always an odd number, either 9, 7, or 5. The *lingtai* for emperor Taizu, for example, is the highest nine story, same as the Yongning Monastery pagoda.

Figure 19. *Lingtai* and the inner walls, Western Xia royal tomb, Ningxia (photo by author).

Unlike the pre-Buddhist Chinese tombs such as the Mausoleum for the First Emperor of Qin, the *lingtai* mound in the Western Xia mausoleum is not built on top of the underground tomb chambers *mushi* 墓室. Due to the damage made by earlier tomb robbers, the tomb chambers for Tomb No. 6 were revealed and archaeologists followed to make excavation and documentation. Tomb chambers are located right in front of the *lingtai*, approximately 18 m to its south. Symmetrically planned with a north–south axis, the central chamber is framed by two smaller side chambers on the east and west sides, and connected with another group of chambers to its south through a corridor. South of the tomb chambers is sloping passageway *mudao* 墓道, linking the floor of the underground chambers with the ground surface, which is approximately 25 m above.[55] (Figure 20) The sacrificial hall *xiandian* is very close to the entrance of the *mudao* passageway. The three central parts from south to north, *xiandian*, the underground *mudao* and *mushi*, and the *lingtai*, all follow the same north–south axis, which is to the west of the main north–south axis established by the walls and gates, parallel but not overlapped. Without covering the coffins of the tomb masters in the underground tomb chambers, the functionality of the *lingtai* mound is purely symbolic. Like the pagoda in Buddhism, *lingtai* in the Western Xia royal tomb is not a structure to protect the physical remains of a reincarnated sentient body, but a timeless monument dedicated to the hope for a future enlightenment.

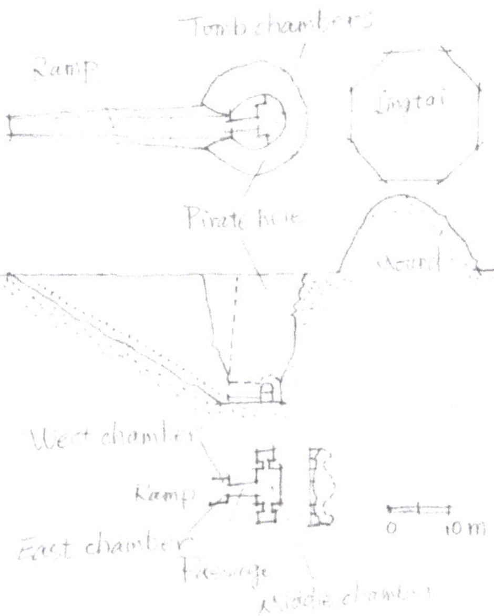

Figure 20. Plan and section of the tomb chambers, Western Xia royal tomb No. 6 (drawing by author, after Figures 5–60 in Guo Daiheng, *Zhongguo gudai jianzhushi di 3 juan: Song, Liao, Jin, Western Xia jianzhu*, 223).

According to *Xixiaji* 西夏紀, the Records of the Western Xia, the rulers of the Western Xia were descendants of the Tuoba clans of the Northern Dynasties, the imperial clan of the Northern Wei empire who built the Yongning Monastery in the 6th century.[56] Like their remote ancestors, the imperial families of the Western Xia were also Buddhists. In 1007, after his mother Wangshi 罔氏 died, King Taizong (Li Deming 李德明) sent envoys to the Song court, asking for permission to repair the ten great temples in Mount Wutai.[57] Mount Wutai was located at the border area among Song, Liao, and Western Xia then under the Song jurisdiction. It is one of the four sacred Mountains in China and revered as the seat of bodhisattva Manjushri, the Bodhisattva of Wisdom. By the Tang dynasty (618–907), Mount Wutai had become one of the most important Buddhist centers in China and an international pilgrimage site, drawing Buddhist pilgrims from such faraway places as Japan and India.[58] The Song emperor responded to Li's request by sending mourning envoys as well as sacrificial items to Mount Wutai, contributing to the religious rituals performed for Wangshi's afterlife.[59] The Repairing or making monetary contribution for the repair of statues and buildings in temples and monasteries is a major merit for Buddhists to build up good *karma* 業 for desirable future reincarnations and the eventual enlightenment. Such religious contributions were part of the funeral service for Buddhists in East Asia. The adoption of the form of Buddhist pagoda for the construction of *lingtai* in Western Xia royal tombs was the result of the same intention for desirable *karmic* retribution in honor of the deceased.

The Yongning Monastery pagoda is not the only prototype for the *lingtai* from the Western Xia royal tombs. There was an even more remote model in Chinese funeral architecture. During the Warring States period (476–221 BCE), the kings of the Zhongshan state also built their mausoleums within two layers of rectangular walls, gated by the *que* towers in the middle of the southern wall. *Zhaoyutu* 兆域圖, the map of the entire mausoleum complex carved in a bronze plate discovered from the very archaeological site, indicates that within the inner court, five multi-story structures were elevated on top of a shared platform (see Figure 21). Each of the five structures features two additional concentric layers of rammed earth platforms surrounded by wooden verandas. The lower layer measures 92 by 110 m, a near square rectangle. The upper layer recedes from the lower platforms. The sacrificial hall, or *xiangtang* 享堂, is built on top of the upper layer rammed earth platform. The pointed roof of the *xiangtang* and the single slope roofs of the two verandas expand in size toward the lower base, forming a pyramidal profile for the tomb structure[60] (see Figure 22) The concentric plan and multilayered verticality of the *xiangtang* shared much in common with the *lingtai* from the Western Xia Royal tombs, which were built more than a thousand years later. Combining a supporting earth core with a wooden frame wrapping, they were both manifestations of the *taixie* structure that had been widely used for the construction of monumentality since the Zhou dynasty.

Figure 21. The inscribed plan with inscriptions on the bronze plate discovered on the site of the Zhongshan Mausoleum, Pingshan county, Hebei province (Creative Commons: https://commons.wikimedia.org/wiki/File:Zhongshan_Zhaoyutu.jpg; accessed on 27 May 2021).

Figure 22. Reconstructed drawing for the Royal Mausoleum of the Zhongshan state (drawing by author, after Figures 3–101 in Liu Xujie, *Zhongguo gudai jianzhushi di 1 juan: Yuanshi shehui, Xia, Shang, Zhou, Qin, Han jianzhu*, 268).

5. Buddhism and the Transformation of Chinese Funeral Architecture

The Yongning Monastery pagoda was chronologically sandwiched in-between the mausoleums for the kings of the Zhongshan state and the Western Xia royal tombs. A common feature for the three types of monumental constructions as represented by these three complexes is a ritual hierarchical order that all Chinese monumentality subscribes to. The ritual hierarchy may be expressed in the size of the walled area, the scale of the central monument, the number of buildings and gates for a complex, or the number of stories for the main tower. The Yongning Monastery pagoda has nine layers, the Great Wild Geese pagoda in Tang capital Changan has seven, and the wooden pagoda in the Ying county from the Liao dynasty has five.[61] Among the Western Xia royal tombs, the *lingtai* mounds for Taizu and Taizong were nine-story tall, and the rest of them were either seven- or five-story tall, depending on the posthumous evaluations of the tomb masters by their decedents. According to the *Book of Rites*, or *Liji* 禮記, the Son of Heaven *Tianzi* 天子 can have seven ancestral temples, a ruler of a local state *Zhuhou* 諸侯 five, a minister *Dafu* 大夫 three, a common aristocrat *Shi* 士 one; for the elevation above the ground of the main ceremonial hall's floor, the Son of Heaven nine feet (*chi* 尺), a local state ruler seven feet, a minister five, and a common aristocrat three.[62] The legendary architecture for such a ritual system is *mingtang* 明堂, the building complex where the Son of Heaven was supposed to make sacrifice to Heaven and the royal ancestors, performed the monthly rituals in different halls to harmonize the world according to the seasons, met homage-paying local rulers and court subjects, and delivered the most important edicts and lectures.[63] The structure of the Han dynasty *mingtang* was also of the *taixie* type, a combination of a stepped rammed earth core with an envelope of wooden frame and sloping tiled roofs.[64] Legitimizing the regime as holding the Mandate of Heaven, the construction of *mingtang* is one of the most significant architectural projects a Confucian dynastic ruler would have embarked on, including the non-Chinese speaking rulers of the Northern Wei and Western Xia.[65]

The translation of Buddhist concepts and forms into East Asian architecture was deeply imbedded in the Confucian context. It was a process operating upon the vocabularies and grammars of various local building traditions, for which the Medieval Chinese funeral architecture bears one of the most tangible fruits. The transformation started from underground, substituting local motifs in the tomb chambers with Buddhist symbols without altering the funeral space and structures. As Buddhist concepts further incorporated into the Chinese thoughts, revised views on life and death brought new elements to the spirit path, the passageway framing active funeral rites. When, finally, the main funeral structures were given Buddhist touches, it was already hard to tell whether the Buddhist became Chinese or the Chinese became Buddhist. Like the translation of Buddhist sutras that had created new vocabularies and brought new meanings to the Chinese language, the translation of Buddhism in the built environments had also participated in the very formation of architectural traditions in China. It blurred the boundaries between Buddhist architecture and other architectural typological divisions, which were largely based on the differentiation of functionality, the "content" of architecture.

In his 1939 article "Iconography and Iconology: An Introduction to the Study of Renaissance Art", Erwin Panofsky discussed the relationship between form and subject matter in European art and argued that on the deepest level, there is no "form as such" but just a continuous series of multilayered "contents". He argued that "in whichever stratum we move, our identifications and interpretations will depend on our subjective equipment, and for this very reason will have to be supplemented and corrected by an insight into historical processes the sum total of which may be called tradition".[66] In the field of architectural history, the content is often understood as the functionality of a building, practical or symbolic. The application of a split between form and content in the discussion of traditional Chinese architecture obscures the profound impact Buddhism made to the Chinese architectural tradition, in which formal consistency often overshadows the differences in content, just like the power of the Chinese characters have done to every textual translation. On the other hand, Buddhist architecture, or *fojiao jianzhu* 佛教建築, and funeral architecture, or *lingmu jianzhu* 陵墓建築, are often categorized as two different types in the discipline of Chinese architectural history. An uncomfortable subtype within these two categories is the pagoda, which is obviously of funeral origin but often strictly discussed under the Buddhist umbrella. The other side of the dilemma, those classified as funeral architecture but primarily constructed and decorated within the Buddhist contexts, is often neglected, remaining a blind spot unable to be recognized as a discomfort at all. The uncovering of such categorical discomfort as imbedded in Buddhist vs. funeral architecture is meaningful for a fuller understanding of Chinese architecture as, like what Panofsky said, the "historical processes" and "the sum total of which may be called tradition". It is an index of the great impact that the introduction of Buddhism had on Chinese architecture, like Xuanzang's translation of the *Prajnaparamita Sutra* 波若心經, which has become inseparably Buddhist and Chinese.

Funding: This research received no external funding.

Conflicts of Interest: The author declares no conflict of interest.

Notes

[1] For a collection of legends and historical records on the introduction of Buddhism during the Eastern Han period, see (Wang 2016, pp. 8–15).

[2] The word "medieval" is derived from the European historical context, and its application to Chinese history has been controversial. I use the word "Medieval China" to refer to the period from the end of the Eastern Han dynasty in the early 3rd century to the end of the Southern Song dynasty (1127–1279) in the late 13th century. Early Medieval China corresponds to the period of the 3rd to 6th centuries, which includes the periods of the Three Kingdoms, Western Jin, Eastern Jin and the Sixteen Kingdoms, and the Northern and Southern Dynasties.

[3] For examples of this type of scholarship, see (Miller 2015).

[4] (Salguero 2014, pp. 4–11).

[5] (Liu 2003); see also (Shi 2014).

[6] For the Buddhist concepts of *nirvana*, relics, and the early development stupa in the Indian subcontinent, see (Huntington 1999, pp. 61–100).

[7] Buddhism might have been introduced to the Han court as early as the 1st century. It seems, however, Buddhism during the Eastern Han period was practiced mainly in the form of worshiping a new deity, and was often conceptualized in the same religious frameworks of the Daoist school of Huangdi and Laozi (*Huang-Lao zhishu*). During the Three Kingdoms period (pp. 220–80), however, records from both the dynastic and the Buddhist indicate the building of tall relics towers with chatras. See, e.g., Fan Ye 范曄 ed., Hou Han shu 後漢書 [History of the Later Han], Southern Song Shaoxi edition, 419; Shi Daoshi 釋道世 ed., Fayuan zhulin 法苑珠林 [Jewel Forest of the Buddhist Garden], Ming Wanli edition, p. 640.

[8] (Wu 1986).

[9] Mou Rong 牟融, *Lihuolun* 理惑論 (https://zh.wikisource.org/wiki/%E7%90%86%E6%83%91%E8%AB%96; accessed on 27 May 2021).

[10] (Liu 2003, pp. 486–97).

[11] (Liu 2003, pp. 494–95).

[12] (Luo 2012).

[13] (Luo 2012, pp. 13–14).

[14] Luo believes that the architectural form and style of the Xiangyang tomb model resembles the earliest Chinese Buddhist temple *futuci* 浮屠祠, which is believed to be a direct forerunner of the pagoda. See (Luo 2012, pp. 10–26).

[15] (Wu 1986).

[16] (Wu 1986).

[17] (Huntington 1999).

[18] (Pan 2001); see also (Wang 2016).

[19] (Steinhardt 2019).

[20] For a comprehensive introduction to Chinese arts and architecture during the age of disunion from the 3rd to 6th centuries, see (Steinhardt 2014).

[21] For a comprehensive introduction to the Chinese imperial tombs, see (Luo 1993)

[22] (Fu 2001).

[23] (Wu 1995).

[24] Donghan Youzhou shuzuo Qinjun mubiao shizhu tanwei 東漢幽州書佐秦君墓表石柱探微 [Detailed study of the Eastern Han dynasty *mubiao* pillar from the Cleric Qin of the You Prefecture], http://www.zggsdjd.com/wap/show.asp?id=29, accessed on 1 June 2021; see also Fu Xinian, *Zhongguo gudai jianzhushi di 2 juan*, p. 128.

25 For a comprehensive introduction to the foreign motifs in Chinese arts and architecture from this period, see (Steinhardt 1998); see also (Steinhardt 2005). For volumes that cover both Indian and Chinese architecture, see (Fergusson 1967).

26 For examples of stone *que* towers from the Han dynasty, see (Chavannes 2020); see also (Sekino 2017).

27 (Wu 1995); see also (Fu 2001).

28 There were free standing columns other than *mubiao* for specific Buddhist funeral purposes in China during this period as well, for example, the stone pillar at Dingxing county from Northern Qi dynasty (550–577), see (Liu 1934).

29 (Wang 2016, pp. 50–126).

30 (Wu 1995, pp. 251–62).

31 See Shi Daoxuan (Tang dynasty), ed., Guang Hongming Ji 廣弘明集, vol. 15 (https://zh.wikisource.org/wiki/%E5%BB%A3%E5%BC%98%E6%98%8E%E9%9B%86/15); vol. 3 (https://zh.wikisource.org/wiki/%E5%BB%A3%E5%BC%98%E6%98%8E%E9%9B%86/04).

32 See Yan Kejun (Qing dynasty) ed. Quan Liang wen 全梁文, vol. 22 (https://zh.wikisource.org/wiki/%E5%85%A8%E6%A2%81%E6%96%87/%E5%8D%B7%E4%BA%8C%E5%8D%81%E4%BA%8C)

33 (Wu 1995, p. 255).

34 (Wu 1995, p. 278).

35 For different types of Buddhist pagodas in China, see (Liang 1984); see also (Liang 2005).

36 (Ito 1940); see also (Mizuno 1974) & (Ito 1942).

37 For a thorough survey of archiecrural images in the stone carvings of the yungang caves, see (Chavannes 2020, pp. 333–422); See also (Steinhardt 2019, p. 86).

38 See, for examples, the Great Wild Goose Pagoda in (Chavannes 2020, p. 1204); see also (Steinhardt 2019, p. 116).

39 For a detailed discussion on the long-distance trade among India, Central Asia, and China the transmission of Buddhist, see (Neelis 2011).

40 For sample lives, itineraries, and activities of Western and Japanese exploers in China during the late 19th to early 20th century, see (Yu 2015a); see also (Yu 2015b) and (Tokiwa 2017).

41 See, for instance, (Stein and Archaeological Survey of India 1968); see also (Stein 2019).

42 (Wriggins 1996, pp. 162–63, 218); see also (Chen 2021).

43 (Stein and Archaeological Survey of India 1968, pp. 221–66).

44 (Fu 2001, pp. 176–77).

45 Chen Xiaolu, "On the Origin and Development of Three-back-shaped Buddhist Temple in Western Region" (https://new.qq.com/omn/20210122/20210122A02WNO00.html; accessed on 27 May 2021).

46 (Stein and Archaeological Survey of India 1968, pp. 332–62).

47 (Fu 2001, p. 177).

48 (Fu 2001, pp. 1–30).

49 *Mingtang* and *Biyong*, the quintessential Confucian architectural type for the legitimatization of the Son of Heaven since the Zhou dynasty, was constructed using the *taxie* method during the Han dynasty. See the conclusion section of this article; see also (Liu 2003, pp. 429–32).

50 (Fu 2001, p. 177).

51 (Fu 2001, pp. 184–88).

52 According to the contemporary Song scholar-official Shen Kuo (1031–1095), the custom of the ruling Tangut people of Xiaxia was to leave the center of the palatial complex for sacrifice to spirits, which might be the reason for such an off-center location. See Shen Kuo, Mengxi bitan 夢溪筆談.

53 (Guo 2003, pp. 217–29).

54 (Guo 2003, p. 221).

55 (Guo 2003, p. 221–222).

56 *Xixiaji* 西夏紀, the Records of the Western Xia, was compiled during the Republican period (1911–1949). The twenty-eight-volume monumental compilation, however, made reference to hundreds of historical documents, including not only the official dynastic histories, for instance, Xin Wudaishi 新五代史 (New History of the Five Dynasties), Songshi 宋史 (History of the Song Dynasty), Liaoshi 遼史 (History of the Liao Dynasty), and Jinshi 金史 (History of the Jin Dynasty), but also local gazetteers, literature, anecdote collections, etc. See Dai Xizhang, Xixiaji (https://zh.wikisource.org/wiki/%E8%A5%BF%E5%A4%8F%E7%B4%80/%E5%8D%B7%E9%A6%96, accessed on 27 May 2021).

57 *Xixiaji*, vol. 4 (https://zh.wikisource.org/wiki/%E8%A5%BF%E5%A4%8F%E7%B4%80/04).

58 For a detailed account on the history, art, and architecture of Mount Wutai as a sacred Buddhist mountain, see (Lin 2014).

59 *Xixiaji*, vol. 4 (https://zh.wikisource.org/wiki/%E8%A5%BF%E5%A4%8F%E7%B4%80/04).

[60] (Liu 2003, pp. 206–69).

[61] For a collection of images of different Buddhist pagodas and other old buildings from China's long imperial past that were still extant in the early 20th century, see (Boerschmann 1982), (Münsterberg 1910), & (Münsterberg 1912).

[62] See Chen, Hao, ed. 1985. Liji jishuo 禮記集説 [Book of Rites with a collection of annotations], No. 10 Liqi 禮器 [The Ritual Utensils]. In Sishu Wujing 四書五經 [Four Books and Five Classics]. Beijing: Zhongguo shudian, pp. 132–41.

[63] See Chen Hao ed., Liji jishuo, No. 6 Yueling 月令 [The Monthly Commands], in Sishu Wujing, pp. 83–100.

[64] See (Liu 2003, pp. 429–32).

[65] For a history of the mingtang design and construction during the Zhou through Ming dynasties, see (Zhang 2007).

[66] (Panofsky 2009, pp. 220–35).

References

Boerschmann, Ernst. 1982. *Old China in Historic Photographs*. New York: Dover Publications, Inc.

Chavannes, Emmanuel Edouard. 2020. *Huabei kaoguji/Mission archeologique dans la Chine Septentriomale*. Translated by Junsheng Yuan. Beijing: Zhongguo Huabao Chubanshe.

Chen, Xiaolu. 2021. On the Origin and Development of Three-back-shaped Buddhist Temple in Western Region. Available online: https://new.qq.com/omn/20210122/20210122A02WNO00.html (accessed on 29 May 2021).

Fergusson, James. 1967. *History of Indian and Eastern Architecture, vol. I & II*. Delhi: Munshiram Manoharlal. First published in 1910.

Fu, Xinian. 2001. *Zhongguo gudai jianzhushi di 2 juan: Liangjin, Nanbeichao, Sui, Tang, Wudai jianzhu [History of ancient Chinese architecture, vol. 2: Two Jins, Southern and Northern Dynasties, Sui, Tang, and Five Dynasties]*. Beijing: Zhongguo Jianzhu Gongye Chubanshe.

Guo, Daiheng, ed. 2003. *Zhongguo gudai jianzhushi di 3 juan: Song, Liao, Jin, Western Xia jianzhu [History of Ancient Chinese Architecture, vol. 3: Song, Liao, Jin and Western Xia Dynasties]*. Beijing: Zhongguo Jianzhu Gongye Chubanshe.

Huntington, Susan L. 1999. *The Art of Ancient India*. New York: Weatherhill Inc.

Ito, Chuta (伊東忠太). 1940. *Horyuji* 法隆寺. Tokyo: Sogensha (創元社).

Ito, Chuta (伊東忠太). 1942. *Nihon kenchiku no kenkyū* 日本建築究. Tokyo: Ryuginsha (龍吟社).

Liang, Sicheng. 2005. *Zhongguo jianzhu shi (History of Chinese Architecture)*. Beijing: Zhongguo Jianzhu Gongye Chubanshe. First published in 1954.

Liang, Sicheng. 1984. *Chinese Architecture: A Pictorial History*. Cambridge: The MIT Press.

Lin, Wei-Cheng. 2014. *Building A Sacred Mountain: The Buddhist Architecture of China's Mount Wutai*. Washington: University of Washington Press.

Liu, Dunzhen. 1934. Dingxingxian Bei Qi shizhu (A stone pillar of Northern Qi in Dingxing county). *Zhongguo Yingzao Xueshe Huikan* 5: 28–66.

Liu, Xujie, ed. 2003. *Zhongguo gudai jianzhushi di 1 juan: Yuanshi shehui, Xia, Shang, Zhou, Qin, Han jianzhu [History of Ancient Chinese Architecture, vol. 1: Primitive Society, Xia, Shang, Zhou, Qin, and Han Dynasties]*. Beijing: Zhongguo Jianzhu Gongye Chubanshe.

Luo, Zhewen. 1993. *China's Imperial Tombs and Mausoleums*. Beijing: Foreign Languages Press.

Luo, Shiping. 2012. Xianren hao louju: Xiangyang xin chutu xianglun taolou yu Zhongguo futuci leizheng [Immortals favor tower dwelling: A comparative study on the newly unearthed ceramic building model with *chattras* from Xiangyang and the *futuci* of China]. *Gugong Bowuyuan Yuankan [Palace Museum Journal]* 4: 10–26.

Miller, Tracy. 2015. Of Palaces and Pagodas: Palatial Symbolism in the Buddhist Architecture of Early Medieval China. *Frontiers of History in China* 2: 222–63.

Mizuno, Seiichi. 1974. *Asuka Buddhist Art: Horyu-ji*. Translated by Richard L. Gage. New York: Weatherhill, Tokyo: Heibonsha.

Münsterberg, Oscar. 1910. *Chinesische Kunstgeschichte, Band 1: Vorbuddhistische Zeit, die hohe Kunst, Malerei und Bildhauerei*. Esslingen a.N.: Paul Neff Verlag (Max Schreiber).

Münsterberg, Oscar. 1912. *Chinesische Kunstgeschichte, Band 2: Die Baukunst, das Kunstgewerbe*. Esslingen a.N.: Paul Neff Verlag (Max Schreiber).

Neelis, Jason. 2011. *Early Buddhist Transmission and Trade Networks: Mobility and Exchange within and beyond the Northwestern Borderlands of South Asia*. Leiden: Brill.

Pan, Guxi, ed. 2001. *Zhongguo gudai jianzhushi di 4 juan: Yuan Ming jianzhu [History of Ancient Chinese Architecture, vol. 4: Yuan and Ming Dynasties]*. Beijing: Zhongguo Jianzhu Gongye Chubanshe.

Panofsky, Erwin. 2009. Iconography and Iconology: An Introduction to the Study of Renaissance Art. In *The Art of Art History*. Edited by Donald Preziosi. Oxford: Oxford University Press, pp. 220–35.

Salguero, C. Pierce. 2014. *Translating Buddhist Medicine in Medieval China*. Philadelphia: University of Pennsylvania Press.

Sekino, Tadashi. 2017. *Zhongguo gudai jianzhu yu yishu [Studies of Ancient Chinese Architecture and Arts]*. Translated by Zhen Hu, and Shanshan Yu. Beijing: Zhongguo huabao chubanshe.

Shi, Jie. 2014. Incorporating All For One: The First Emperor's Tomb Mound. *Early China* 37: 359–91. [CrossRef]

Stein, Aurel Sir. 2019. *Central-Asian Relics of China's Ancient Silk Trade*. London: E.J. Brill.

Stein, Aurel, and Archaeological Survey of India. 1968. *Ruins of Desert Cathay: Personal Narrative of Explorations in Central Asia and Westernmost China*. New York: Benjamin Blom.

Steinhardt, Nancy S. 1998. Early Buddhist Architecture and Its Indian Origins. In *Flowering of a Foreign Faith*. Edited by Janet Baker. New Delhi: New Delhi Publishers, pp. 38–53.

Steinhardt, Nancy S. 2005. Twin Pillars Tomb. *Journal of Inner and East Asian Studies* I: 25–57.

Steinhardt, Nancy Shatzman. 2014. *Chinese Architecture in an Age of Turmoil, 200–600*. Honolulu: U of Hawaii.

Steinhardt, Nancy S. 2019. *Chinese Architecture: A History*. Princeton: Princeton University Press.

Tokiwa, Daijo. 2017. *Zhongguo fojiao shiji [Buddhist Historical Relics of China]*. Translated by Yizhuang Liao. Beijing: Zhongguo Huabao Chubanshe.

Wang, Guixiang. 2016. *Zhongguo hanchuan fojiao jianzhu shi [History of Han Chinese Buddhist Architecture]*. Beijing: Tsinghua University Press.

Wriggins, Sally Hovey. 1996. *Xuanzang: A Buddhist Pilgrim on the Silk Road*. Boulder: Westview Press.

Wu, Hung. 1986. Buddhist Elements in Early Chinese Art (2nd and 3rd Centuries A.D.). *Artibus Asiae* 47: 263–352.

Wu, Hung. 1995. *Monumentality in Early Chinese Art and Architecture*. Stanford: Stanford University Press.

Yu, Shuishan. 2015a. Ito Chuta and the Narrative Structure of Chinese Architectural History. *JA (Journal of Architecture)* 20: 1–35. [CrossRef]

Yu, Shuishan. 2015b. The Formation of the Basic Narrative Structure of Chinese Architectural History as Reflected in Ito Chuta's Scholarship. *Zhongguo jianzhu shilun huikan (The Journal of Chinese Architecture History)* 11: 3–30.

Zhang, Yibing. 2007. *Mingtang zhidu yuanliu kao [The Origin and Historical Development of the Mingtang System]*. Beijing: Renmin chubanshe.

Article

Secular Dimensions of the Aśoka Stūpa from the Changgan Monastery of the Song Dynasty

Yue Dai

Department of Art History and Archaeology, Washington University in St. Louis, St. Louis, MO 63130, USA; yued@wustl.edu

Abstract: In 2008, in the course of excavating the site of the pagoda foundations of the former Nanjing Da Bao'en Monastery 南京大報恩寺, archaeologists discovered Buddhist relics enshrined in nested reliquaries along with some two hundred offering objects. The most impressive finding was a specially designed, richly decorated reliquary stūpa, known as the Seven-Jeweled Aśoka Stūpa 七寶阿育王塔, created in the Song Dynasty (960–1279 CE). This paper begins with the history of the site where a series of famous Buddhist structures had been built since the Wu Kingdom (222–280 CE), and which has long been associated with the cult of King Aśoka and relic worship. It then goes on to examine the form and features of the reliquary stūpas prevalent in the Wuyue period (907–978). Through comparisons between the Aśoka stūpas commissioned by Wuyue King Qian Chu 錢俶 (929–988) and those by laypeople around the same time, I will demonstrate that the Seven-Jeweled Aśoka Stūpa is distinct in its secular features. It is not a Buddhist reliquary that strictly conforms to the conventions of reliquary-making in terms of scale, inscription, and functionality; besides relic worship, it also features a remarkable manifestation of laypeople's beliefs and expectations, sacred or secular. Viewed in its historical context, in which the Song emperors imposed political control over religious affairs and Buddhism became increasingly secular, the stūpa was a product of negotiation between the political authorities and local Buddhist communities in the Song Dynasty.

Keywords: Changgan Monastery; Aśoka Stūpa; Wuyue Kingdom; Nanjing; Buddhist reliquary

Citation: Dai, Yue. 2021. Secular Dimensions of the Aśoka Stūpa from the Changgan Monastery of the Song Dynasty. *Religions* 12: 909. https://doi.org/10.3390/rel12110909

Academic Editors: Shuishan Yu and Aibin Yan

Received: 21 July 2021
Accepted: 5 October 2021
Published: 21 October 2021

Publisher's Note: MDPI stays neutral with regard to jurisdictional claims in published maps and institutional affiliations.

1. Introduction

During the excavation of the site of the former Nanjing Da Bao'en Monastery in 2008, archaeologists discovered the crypt of the True-Body Pagoda 真身塔 of the Changgan Monastery 長干寺built during the Song Dynasty. Through this excavation, the crypt, along with several Buddhist relics and over two hundred offering objects, was revealed to the public after an interval of nearly a thousand years. The crypt's airtight condition had helped to preserve the nested reliquaries and other precious objects therein for centuries, protecting them from natural and manmade destruction. As a result, most objects remained intact, and information on the crypt is well preserved on a stone stele titled "A Record on the Stone Casket Encasing the Relics of the True-Body Pagoda of the Jinling (Nanjing) Changgan Monastery", 金陵長幹寺真身塔藏舍利石函記 (Qi and Gong 2011, 2012; Zeng 2011; Nanjing Municipal Institute of Archaeology NMIA 2015). Among the archaeological findings, the Seven-Jeweled Aśoka Stūpa (stūpa: a dome-shaped structure that contains the relics of Buddha), also known as the Changgan stūpa (now in the collection of Nanjing Museum), has attracted the most attention and discussion.

At present, scholarship on the Changgan crypt is dominated by archaeologists, whose research concentrates on its history, structure, and extant objects. The most comprehensive publication is the excavation report by the Nanjing Municipal Institute of Archaeology (NMIA), covering the site's Buddhist history and functions as a monastery in the Ming Dynasty (1368–1644), as well as the crypt's structure and its nested reliquaries and their contents in great visual and textual detail (NMIA 2015). Several scholars have conducted more specific investigations into the archaic structure of the crypt and the stele inscriptions (Qi and Gong 2011, 2012; Zeng 2011). By contrast, a few scholars, such as Katherine Tsiang and Hattori Atsuko 服部敦子, have contextualized the Changgan stūpa in relation to the history of the construction of Aśoka stūpas in China (Hattori 2011; Tsiang 2017). Aside from this brief analysis, no comprehensive study of this reliquary's form, imagery, content, and inscription has so far been carried out.

In this paper, I intend to illuminate the stūpa's secular dimensions by analyzing its unique features. By "secular", I refer to features of relic worship in Chinese Buddhist history that deviate from canonical rules and meet the worldly needs of devotees, including the political purposes of rulers and the personal interests of laypeople. More specifically, "secular" refers to the distinctive aspects of the Changgan stūpa that do not conform to the strict regulations or conventions of reliquary-making in terms of scale, inscription, and functionality. Traces of public and governmental involvement can also be seen here and there, from the stūpa's exterior ornament to its interior contents. In light of these features, the Changgan stūpa is not merely a Buddhist reliquary, but a complex embodiment of religious devotion, political praise, and the wishes of individuals.

On one hand, a growing number of scholars agree that secular and sacred were not mutually exclusive in the Buddhist practices of medieval China.[1] "Secular" does not mean a complete deviation from canonic concepts, prescriptions, and traditions, employed without restrictions. On the contrary, many Buddhist practices in medieval China integrate religious theories, traditions, and up-to-date interpretations. In particular, laypeople would incorporate their daily experiences into religious activities, drawing on Buddhist language, images, and rites to meet their own needs. For example, Buddhist believers in Dunhuang were engaged in copying scriptures and building caves to attain merit and thereby bring blessings to their ancestors, families, and themselves.[2]

On the other hand, the secular dimension of relic worship was not new in the Song Dynasty, nor could it be used to differentiate between Song and pre-Song Buddhism. Prior to the Song Dynasty, both Emperor Wen of the Sui Dynasty 隋文帝 (541–604) and Wu Zetian 武則天 (624–705), the female emperor of the Zhou Dynasty (690–705), had identified themselves with the *cakravartin*, the wheel-turning king 轉輪聖王, and built Aśoka stūpas to legitimate their rule.[3] Three times during his reign, Emperor Wen sent more than one hundred relics in total to all the prefectures in the country and instructed special envoys, local officials, monks, and ordinary people to participate in the worship and enshrinement of these relics.[4] Judging from the historical records in these cases, large-scale relic worship probably often involved all classes of society, and the Changgan stūpa was no exception.

The Changgan stūpa was the product of collective fundraising and donation collection, as its introductory stele indicates. In addition, the inscriptions that cover the stūpa's surface are not limited to Buddhist expressions: they include blessings for the state and political leaders and votive offerings by individuals. Both aspects will be closely examined in the following sections. This kind of inclusiveness was not common in reliquary-making prior to this. In fact, the majority of Aśoka stūpas previously made in China were commissioned by and dedicated to individuals or their families only. The most representative of this kind were the miniature stūpas favored by the ruling class and ordinary people during the Wuyue period. These stūpas, which enshrined either bodily relics of the Buddha or sutra (Dharma relics), were inscribed with the donor's name and occupation, and the date. Though the Changgan stūpa mimics the traditional form of the Wuyue stūpas in terms of its square shape, tripartite structure, Jātaka tale engravings, and other compositional elements, its unusual size (120 cm) and inscriptions for various subjects, as well as miscellaneous donations, make it different from its Wuyue prototype.

This study benefits from the rich collection of images and detailed textual materials provided by the archaeological report and uses them to study the site and reliquary. To explain these unusual features, I will start by introducing the history of relic worship at this site and the politico-religious tensions characteristic of its tenth-century historical context. On one hand, as many scholars have pointed out, the site of the Changgan Monastery had a long history of relic worship dating back to the late fourth century (Yang 2009; Qi and Gong 2011). This historical background explains the internment of Buddhist relics beneath the True-Body Pagoda of the Changgan Monastery. On the other hand, some scholars demonstrate that the Wuyue kings compared themselves to the Indian King Aśoka (303–232 BCE) by copying his feat of constructing Aśoka stūpas on a large scale, seeking to demonstrate their political legitimacy under a religious guise (Brose 2015; Shen 2019; Lee 2021). This research, however, focuses on the Wuyue kings' actions of building of stūpas, ignoring the later presence of Aśoka stūpas in the early Song Dynasty.

Unlike the Wuyue miniature stūpas, the Changgan stūpa embodies the various wishes of the regional Buddhist community, consisting of local officials, monks, and ordinary people. I articulate this idea through two comparisons. The first compares the Changgan stūpa with the Aśoka stūpas commissioned by Wuyue King Qianchu. Some scholars have conducted a visual analysis of Buddhist reliquaries in East Asia made during the tenth century (Choi 2003; Li 2009; Chen 2011). Their scrutiny of the details facilitates our comparison between the Changgan and Wuyue stūpas, highlighting three major secular features in particular.

The second comparison is between the Changgan stūpa and the Wuyue-style stūpas made by the laity around the tenth century, according to which the unconventional features of the Changgan stūpa can be explained secularly. It is in light of this social context, such as the intensified intervention of politics into religious practice, that the auspicious inscriptions blessing the emperor and his ministers can be largely understood. In addition, some scholars have shown that Chinese believers during the Tang and Song dynasties showed pragmatic and utilitarian tendencies when participating in Buddhist activities (Zheng and Lin 1996; Li 1999; Zhou 2005; Nakamura 2013). This unique characteristic is evident in the laypeople's inscriptions on and inside the Changgan stūpa, engraved on the surfaces of their offering objects or written on their wrapping textiles.

The Changgan stūpa is thus an eclectic product of secular and religious pursuits, created by the central government and the local Buddhist community. For local devotees, the concept of the Aśoka stūpa changed from a pure reliquary to a materialized embodiment of political aspirations and ultimately laymen's wishes.

2. The History of the Site of the Changgan Monastery

The site of the former Nanjing Da Bao'en Monastery is in the southern part of today's Nanjing. It is just outside the ancient city wall, near the south city gate—now known as Zhonghua Gate 中華門—in a place formerly known as Changgan District長干里. The best-known Buddhist complex at this site is the Da Bao'en Monastery, with its world-famous Porcelain Pagoda 琉璃塔 built during the reign of Emperor Yongle of the Ming Dynasty 永樂(1360–1424, r. 1402–1424). This location had been occupied by a series of important Buddhist structures before the monastery was constructed. According to the *Liang shu* (History of the Liang Dynasty 梁書), a small vihara (an early type of small-scale Buddhist monastery) with a stūpa had been on this site since the Wu Kingdom (222–280), but was destroyed shortly after. In the Western Jin period (266–316), monks constructed a monastery with a stūpa on the same site and named it the Changgan Monastery, after the district in which it was located. The most legendary episode that ever occurred on that site was during the reign of Emperor Xiaowu of the Eastern Jin Dynasty 東晉孝武帝(r. 372–396), when the monk Liu Sahe劉薩訶 (active late fourth to early fifth century), sometimes confused with another monk, Huida 惠達 (c. 342–423) in many Buddhist narratives, arrived in Nanjing. Seeing a strange light radiating from the Changgan District, Liu unearthed two true bodily relics of the Buddha—a fingernail and a thread of hair—that had been contained in nested reliquaries beneath the stūpa of the Changgan Monastery. Liu's fruitful excavation was essentially due to his quick identification of the stūpa as one of those commissioned by Indian King Aśoka, built to enshrine the bodily relics of the Buddha (Yao, 1973, p. 791; Shi Huijiao, 1983; Jao 1990).[5] After this event, the stūpa at the Changgan Monastery was regarded as one of the few of King Aśoka's stūpas in China in which relics of the Buddha's true body were enshrined.

From a present-day perspective, this episode of Liu's discovery of the bodily relics comes across as theatrical and thus of questionable reliability. More importantly, as a monk who had traveled throughout the central, eastern, and western regions of China and unearthed the bodily relics of the Buddha, Liu was constantly deified and idolized in a variety of narratives, ranging from officially codified history to popular folklore appearing shortly after his death. In some northwestern areas, such as Dunhuang, a cult of Liu Sahe that treated him as a Buddhist saint emerged alongside the cult of relic worship among local laypeople (Shang 2007). Despite Liu's widespread popularity, however, scholarship has long questioned the credibility of historical accounts regarding Liu and his miraculous deeds. As Wu Hung and Shang Lixin尚麗新 point out, Liu Sahe played the role of a mysterious monk with supernatural powers ever since he appeared in the very early document *Gaoseng zhuan* (Biographies of Eminent Monks 高僧傳). In the historical narratives, the reality and fictionality of this figure are so deeply intertwined that even contemporaries could not overcome this mystification; they simply accepted him as a legendary figure, albeit to different degrees (Wu 1996, pp. 32–37; Shang 2007). Moreover, the later the documents were composed—such as the *Ji shenzhou sanbao gantong lu* (Collected Records of the Miracles of the Three Jewels in Shenzhou (China) 集神州三寶 感通錄), complied during the Tang Dynasty—the more legendary elements appeared in lieu of historical facts (Shang 2007, p. 81). Nevertheless, though Liu's discovery of relics might be exaggerated or even largely manipulated, this episode indeed helped to construct the reputation of the Changgan Monastery and its significance in relic worship. Famous people who accepted this story include Emperor Wu of the Liang Dynasty and senior monks such as Huijiao (慧皎 497–554) and Daoxuan (道宣 596–667) of later generations, as can be seen in the early document *Biographies of Eminent* Monks and the officially codified History of Liang Dynasty.

During the Liang Dynasty (502–557), Emperor Wu 梁武帝 (464–549), a pious Buddhist, unsurprisingly paid a great deal of attention to this place and initiated a project to repair the aged structures and expand the scale of the monastery.[6] He also reverently removed the relics from the stūpa several times in order to worship them personally and then had them re-encased in a grand ceremony (Yao, 1973, pp. 789–93). According to the *Ji shenzhou sanbao gantong lu*, after the Six Dynasties (222–589), the relics once enshrined in the Changgan Monastery were removed and distributed to other places at different times, leading to the gradual decline and desolation of the site (Shi Daoxuan, 1983, pp. 405–6). Nonetheless, the site has been significant to Buddhists since the Eastern Jin. It is full of historic events and symbolic meanings, which have in turn fostered Buddhist beliefs and culture over the years in southeast China.

It was not until the early Song Dynasty, after multiple petitions to the central government, that local monks managed to reconstruct the Changgan Monastery. According to "A Record on the Stone Casket Encasing the Relics of the True-Body Pagoda of the Jinling (Nanjing) Changgan Monastery", engraved on a stone stele in the crypt, the crypt was dedicated to enshrining the uṣṇīṣa (the round protuberance on the head of the Buddha)—one of the most precious relics of the Buddha donated by Western monk Danapala (?–1017), known as Shihu 施護in Chinese.[7] With the permission of Emperor Zhenzong 宋真宗 (968–1022), local monks built a nine-story brick pagoda, the "True-Body Pagoda", a name granted by the court, completing the project in 1011. At the same time, they dug an earthen pit beneath the pagoda as a crypt. The crypt enshrined the relics in nested reliquaries, one of them the Seven-Jeweled Aśoka Stūpa (Figure 1). After the Song Dynasty, no records of the crypt or its true-body relics are found in historical documents or scholarly works. Despite how celebrated they had once been, both clergy and laity gradually forgot their existence.

Figure 1. Anonymous. The Seven-Jeweled Aśoka Stūpa, 1011. Sandalwood and gilt silver, 117 × 45 cm. Courtesy of Zhou Baohua 周保華.

The crypt remained undisturbed, even when a majestic pagoda, commissioned by Emperor Yongle of the Ming Dynasty, was built on the site.[8] With the excavation of 2008, this invaluable construction recaptured the attention of both Buddhists and scholars whose research benefited from the intact status and complete state of the buried objects.

3. The Association between the Aśoka Cult and Relic Worship at the Site

The history of the Changgan Monastery site shows it to be a site famed for Buddhist affairs in Southeast China, a site that had always been associated with the Aśoka cult and relic worship. The ruling class also played an important role in the Buddhist projects at the site through personal direction, financial support, and political endorsement. According to this history, the moment at which the site achieved its symbolic meaning in the history of Chinese Buddhism was in the Eastern Jin period (317–420), when Liu Sahe discovered the relics of the Buddha's fingernail and hair hidden beneath the stūpa of the Changgan Monastery. Since then, historical records of the site have usually gone hand in hand with descriptions of Buddhist activities centered on relic enshrinement and worship. For instance, after the distribution of the relics to other monasteries, ordered successively by Emperor Yang of the Sui Dynasty 隋煬帝 (569–618) and Li Deyu 李德裕 (787–850) of the Tang Dynasty, the monastery lost much of its past glory and was deserted.[9] Ownership of the relics, distinguished by their scarcity and numinous power, had had far-reaching effects on local Buddhists, and evidence suggests it was the bodily relics that determined the fate of the site.

The link between the Changgan stūpa and the eighty-four thousand stūpas commissioned by King Aśoka also dates back to the mythical monk Liu Sahe's excavation of the stūpa's bodily relics (whose housing was indeed a primary function of the Aśoka stūpas). This particular role of the stūpa at the Changgan Monastery has since been accepted and reinforced from time to time. In the "Foji xu lue" (Brief Preface to the Buddhist Anthology 佛集序略) to the *Guang hongming ji* (Expanded Collection on the Propagation and Clarification [of Buddhism] 廣弘明集), author Shen Yue (沈約, 441–513) details nineteen of the eighty-four thousand Aśoka stūpas located in Chinese territory. He frequently uses the words "aged" and "numinous signs" to describe them. Unsurprisingly, the stūpa at the Changgan Monastery is on his list (Shen, 1983, pp. 201–2).

The very title of the "Seven-Jeweled Aśoka Stūpa" offers a clue to the function of the newly excavated stūpa and contextualizes it in the tradition of the Aśoka cult at the

Changgan Monastery. Unlike the Aśoka stūpas commissioned by Wuyue King Qian Chu, this Aśoka stūpa was not the product of a ruler's religio-political aspiration, the assumed primary motivation behind King Aśoka's massive constructions of stūpas[10] (which led rulers of later ages to legitimize and defend their political authority under the guise of devotion to Buddhism).[11] Instead, the Seven-Jeweled Aśoka Stūpa was a product of a public, collective Buddhist project, patronized by the local Buddhist community, which had a strong secular dimension.

4. The Crypt of the Song Changgan Monastery

The crypt of the Song Changgan Monastery lies on the central axis of the former Da Bao'en Monastery, beneath the foundation of its Porcelain Pagoda (Figure 2). The crypt is in the form of a round vertical earthen shaft measuring 6.74 meters deep, making it the deepest relic crypt ever found in China (Figure 3).[12] The shaft features a deep pit without additional components, and so its structure is simpler than the horizontal crypts and underground palaces common at Tang Dynasty sites such as the Famen Monastery. Qi Haining 祁海甯 and Gong Juping 龔巨平 argue that this form was commonly adopted for crypts in the early Buddhist history of China, such as that of the Yongning Monastery 永寧寺 in the Western Wei period (386–534). Gradually, a more developed form—the horizontal brick chamber—replaced the previous plain form and dominated the spatial design of relic deposits from the Six Dynasties onwards. In Qi and Gong's opinion, the Changgan crypt's preference for the classical form suggests a revitalization of the ancient crypt style of the Six Dynasties period (Qi and Gong 2012, pp. 76–81). In other words, the form was selected out of respect for the tradition of the Aśoka cult and relic worship at the site.

Figure 2. Bird's-eye view of the foundation of the Porcelain Pagoda before excavation. From outer to inner parts: foundation slot, outer earth rim, sandwich stone rammed layer, inner earth rim, and crypt. Photo: courtesy of Zhou Baohua. Lines added by the author.

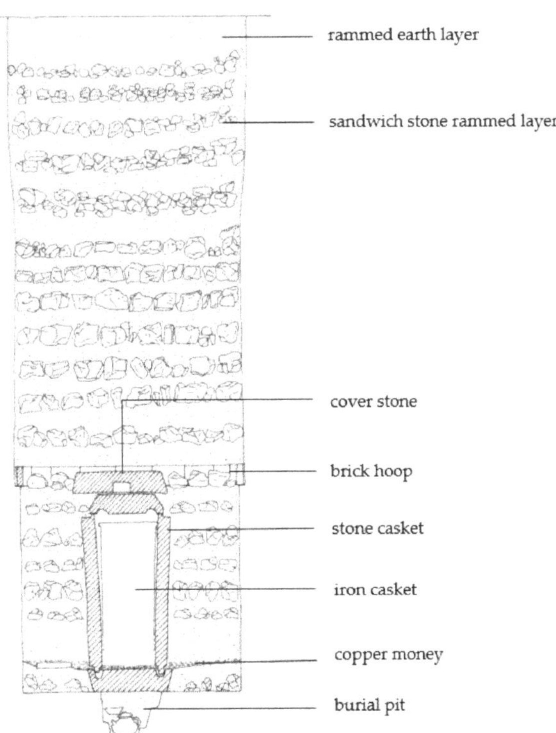

Figure 3. Profile of the crypt structure. Photo: courtesy of Zhou Baohua. Texts added by the author.

At the bottom of the crypt sits a large stone casket, 1.83 meters high, consisting of five stone slabs and a square stone base. The slab facing the northern wall is engraved with a lengthy inscription, "A Record on the Stone Casket Encasing Relics of the True-Body Pagoda of the Jinling Changgan Monastery", which functions as a merit-recording stone stele. Inside the stone casket and only slightly smaller is an iron casket in the form of a sealed cube. The most important object inside this is the Seven-Jeweled Aśoka Stūpa, measuring 117 centimeters high and 45 centimeters wide. The stūpa has a sandalwood frame; its surface is entirely covered by sheets of gilt silver, richly decorated with a variety of jewels. The stūpa is hollow and contains the uṣṇīṣa of the Buddha, numinous-response relics 感應舍利, and the relics of holy monks, as well as some two hundred offering objects. The principal object of worship—the uṣṇīṣa—was put here with other relics in nested reliquaries, made of gold and silver and shaped like coffins. In short, the Seven-Jeweled Aśoka Stūpa is a masterpiece of Song Dynasty reliquary because of its impressive shape, precious material, elaborate ornaments, delicate craftsmanship, and abundant contents.

Since the crypt safely kept its relics until modern times, most of its items remained untouched and in good condition. Hence, the inscription on the stone stele has become a useful textual reference for examining the crypt and its buried objects and restoring them to their proper historical context. The inscription reads, in part:

> ⋯⋯ 大事既周，提河示寂，碎黃金相為設利羅，育王鑄塔以緘藏耶。舍手光而分佈，總有八萬四千所，而我中夏得一十九焉。金陵長幹寺塔，即第二所也 ⋯ 舊基空列於蓁蕪，嵓級孰興於佛事。每觀藏錄，空積感傷.。
>
> 聖宋之有天下，封禪禮周，汾陰祀畢，乃有講律演化大師可政，塔就蒲津，願興墜典。言告中貴，以事聞天，尋奉綸言，賜崇寺、塔。同將仕郎、守滑州助教王文，共為導首。率彼眾緣，於先現光之地，選彼名匠，造建磚塔，高二百尺，八角九層，又造寺宇。□□進呈感應舍利十顆，並佛頂真骨泊諸聖舍利，內用金

棺，周以銀槨，並七寶造成阿育王塔，□以鐵□□函安置。 即以大中祥符四年太
歲辛亥六月癸卯朔十八日庚申，備禮式設闔郭大齋，於皋際，庶□名數，永鎮坤
維。

....

塔主演化大師可政。助緣管勾賜紫善來，小師普倫。導首將仕郎、守滑州助教王
文，妻史氏十四娘，男凝、熙、規、拯 ... 僧正賜紫守邀宣慧大師齊吉，賜紫文
仲，僧仁相，紹之。舍舍利施護、守正、重航 ... ... [13]

... When his business (teaching the laws) was done, the Buddha entered nirvāṇa.
His golden body was broken to make Śarīras (relics), and King Aśoka constructed
stūpas to encase them. The stūpas were broadly distributed and in the total
number of eighty-four thousand, there are nineteen of them in our country—
Zhongxia [China]. The stūpa at Changgan Monastery of Jinling is the second
[Aśoka stūpa of China] ... The old foundation of the stūpa [pagoda?] was seated
alone in the wild, and the grand scale indicated that thriving Buddhist affairs had
once taken place. Whenever I read its collections, I could do nothing but lament.

Since the establishment of the sacred Song Dynasty, the Feng-Shan ceremony
has been completed, and the Fen-Yin sacrifice has been finished. Then, the law-
preaching Yanhua master Ke Zheng noted the pagoda's tendency to decline and
hoped for its revival for practicing rituals. He turned to dignitaries and let the
heaven (emperor) know of his proposal. Consequently, Kezheng received the
emperor's decree and grant to revive the monastery and pagoda. With Wang
Wen, a Jiang Shilang or assistant teacher in the Hua prefecture, Kezhang shared
the role of head director. He led the laypeople and selected celebrated craftsmen
to participate in the construction of the brick pagoda at the site where light had
radiated throughout history. The pagoda is octagon-shaped and nine-storied,
and is two hundred chi in height. Then, they continued to build structures in the
monastery. □□ presented ten numinous-response relics, as well as the uṣṇīṣa of
the Buddha and relics of holy monks. The relics are placed in nested reliquaries,
which are golden, silver coffins, the Aśoka Stūpa made out of seven jewels, and
the iron casket, in an outward order. On June 18, 1011, rituals were prepared
and a large Zhai ceremony for the whole city was held. Next to the water, the
interment is in hope for eternity.

....

The Yanhua master Kezheng in charge of the pagoda; Administrator of Buddhist
affairs Shanlai; Monk Pulun; Head director Jiang Shilang; Assistant teacher
of Hua Prefecture Wang Wen; his wife Shi Shisiniang; his sons Ning, Xi, Gui,
Zheng ... Monks: Qiji, Wenzhong, Renxiang, Shaozhi ... Relic-donators: Shihu,
Shouzheng, Chonghang ... (NMIA 2015, p. 22)

According to this inscription, the crypt was made as part of the True-Body Pagoda of
the Song Changgan Monastery and dedicated to enshrining the relics presented by Shihu
and others. The entire project was long and complicated. The monk Kezheng first sought
recommendations from influential officials, with which he submitted a proposal to the
emperor at court. After obtaining the emperor's permission, construction began under the
supervision of multiple local officials responsible for religio-cultural affairs. In fact, there
were more people who participated in this project than whose names are engraved on the
stele; more traces can be found of many dedicatory and votive inscriptions on other objects.
In the brief excavation report, the authors provide excerpts from inscriptions on the buried
objects. They consist of over twenty engraved inscriptions on the surface of the Aśoka
stūpa, the dedicatory inscriptions on the bottom of the gold and silver coffins, the ink
inscription on the textiles, and the engraved inscriptions on the bricks. The inscriptions on
the stūpa are especially distinguished by their quantity and rich content, as they intensively
and evenly cover the surface, from the sides and top of the body to the foot of the chattra
(a triple umbrella form on the top of the structure) and the inward-facing sides of the

acroteria. Based on this textual evidence, we can infer additional categories of participants who did not appear in the stele inscription, such as monks from other monasteries (e.g., Siqi from the Chongsheng Monastery 崇聖寺); laypeople grouped by family ties (e.g., Chen Zhihe's 陳知厚 entire family); and individual laity (e.g., Xu Shouzhong 徐守忠 and Liu Yiniang 劉氏一娘).[14] They not only offered financial support for the project, but also generously donated valuable personal belongings. For instance, a bronze mirror tied onto the chattra of the stūpa was donated by the Buddhist disciple Yinwen 印文, according to the ink inscriptions on its surface. Therefore, at least one hundred locals took part in this unusual project, regardless of occupation, gender, or social status.

5. The Analysis of the Seven-Jeweled Aśoka Stūpa

The stūpa is a single-storied, square container, primarily composed of a base, body, and summit. The base is low and square-shaped, with a row of four seated Buddhas in relief on each side. The four reliefs are placed horizontally and evenly spaced, forming an elegant frieze. Above the base is a cube-like container, which is the reliquary body. Each of the body's sides is engraved with a scene from the Jātaka tales, including Prince Mahāsattva giving his body to the hungry tigers, King Śibi saving a dove, King Candraprabha sacrificing his head to a brahmin, and King Sudhira donating his eyes. Besides the Buddhist icons and traditional Chinese patterns, dedicatory inscriptions and auspicious phrases take up the rest of the empty space on the surface. A capital in the shape of a garuda—a golden-winged bird in Buddhist mythology—is welded on each side of the body's edge. The cubic body is topped with a sloping cornice pointing upward. Each side of this cornice is decorated with two reliefs of seated Buddhas, echoing the base's design, but here the reliefs flank a lion-head pattern surrounded by auspicious phrases and votive inscriptions, illuminating the secular feature of this stūpa.

On the top of the cornice, four acroteria in the shape of banana leaves occupy the four corners, whose outward-facing sides are decorated with a total of nineteen narrative scenes chosen from the biography of Buddha. Each of the eight sides is engraved with two or three images, arranged vertically and divided by fine lines. Likewise, the acroteria's inward-facing sides are vertically composed, showing a figure image above and a brief inscription below. Unlike the outward sides, which focus on the Buddha himself, the four inward sides portray a variety of Buddhist icons, including standing and seated buddhas, bodhisattvas, and heavenly guardian kings. Each side is dedicated to a particular figure exclusively. A long pole stands in the center of the top, decorated with a lotus-shaped pedestal. The pole goes up to thread five tiers of disks topped by a flame-shaped orb. Four golden chains, hung with numerous celestial bells, bridge the central pole and the four acroteria at the corners. The base, body, and acroteria are hollow to hold as many relics as possible.

As a reliquary, the design of this Aśoka stūpa adopts the elements of the Aśoka stūpas popular in the Wuyue period. The key features that distinguish the Wuyue stūpas are a single story; a square shape; a three-part composition made of a base, body, and summit; scenes from the Jātaka tales on the body's four sides; and Buddhist icons on the corner acroteria. Although the gilt bronze and gilt iron stūpas slightly vary in the details of their formal design and decoration, the identifying features have been preserved and were also employed in the construction of the Changgan Aśoka stūpa.[15] The Changgan Monastery is in today's Jiangsu Province, a territorial neighbor to Zhejiang Province, the core of the Wuyue Kingdom. The year of its construction, 1011, was within fifty years of the surrender of the Wuyue king and the annexation of his kingdom by the Song. Given this temporal and spatial affinity, the Changgan stūpa is likely a direct legacy of the Wuyue stūpas, and is considered to be an ideal form to worship the bodily relics of the Buddha (Figure 4).

Figure 4. Distribution map of the Wuyue-style stūpas created in the mid-tenth century. (Blue boundaries: administrative prefectures of the Wuyue Kingdom. Red dots: sites of the stūpas. Feature map: (CHGIS 2016)).

The unique form of stūpa was termed as such due to its association with an Indian architectural style alien to the traditions of China. Zhang Yuhuan 張馭寰contends that the form of the Baoqieyin Stūpa (Aśoka Stūpa 寶篋印塔) originated from the stūpas of India and was later influenced by the artistic style of Gandhāra. The form diffused throughout north India, eventually arriving in China (Zhang 2000, pp. 119–20). By examining early Buddhist stūpas in India (third century–first century BCE) and innovative Gandhāran stūpas (first–fifth century CE), Alexander Soper argues that several distinguishing features, namely, the square-shaped body, garuda capitals, and even the four-sided Jātaka tales, all derive from Gandhāran artistic innovations. For instance, the Anda or hemisphere dome is the core component that distinguishes the early Indian stūpas from their descendants. However, after this architectural style spread to northern areas, the dome was gradually transformed into a squarish base and body components (Soper 1940, pp. 652–60; Chen 2006, pp. 63–64). Since the first century CE, the Gandhāran-style stūpas featuring a squarish base gradually spread eastwards to the Tarim basin and then to Central China (Whitefield 2018, pp. 88–89).

During the Han Dynasty (202 BCE–220 CE), the enhanced, bilateral connections between China and Central Asia undoubtedly accelerated the spread of Buddhism and the Gandhāran art to inner China. For instance, King Aśoka dispatched missionary campaigns traveling from India to the Tarim Basin in northwest China, during which they successfully converted lots of local people to Buddhism (Whitefield 2018, p. 83). Likewise, from the second century BCE onward, Chinese people and the government tended to move westward by means of marching, trade, pilgrimage, and so forth.[16] According to *Han shu* (History of the Han Dynasty 漢書), the Han empire established its official connections with the kingdoms to its west after the reign of Emperor Wu漢武帝 (ca. 156–87 BCE).

The number of these kingdoms grew from thirty-six to some fifties in decades (Ban, 1962, p. 3871), which promised safety and reliable supplies to the travelers between China and Central Asia. Consequently, international communication flourished, resulting in frequent exchanges of goods, valuable objects, and agricultural products between China and the West.

Monks were crucial to the eager travelers who traveled in both directions, accelerating the spread of Buddhism across vast areas (Nakamura 1984, pp. 11–12; Dong 2021, pp. 181–84). According to many historical records, these monks not only helped promote Buddhist scriptures and practices (Palumbo 2012), but introduced the Gandhāran style to Chinese Buddhist architecture.[17] For instance, the squarish base and body of the Gandhāran stūpa contributed to the design of Chinese square-shaped stūpas and pagodas, and the four-sided Jātaka tales inspired Chinese art-making during the fifth and sixth centuries (Rhie 2010, p. 362). In Chinese pictorial art, the representation of square stūpas, single- or multi-storied, appeared as early as the Northern Wei (384–534) and Northern Qi (550–577) periods (Wang 2012, pp. 115–16).[18] There is more evidence to strengthen this argument, and an excellent example is the stūpa in Mao County 鄮縣, also considered one of the nineteen Aśoka Stūpas in China.

Legend has it that the Mao County stūpa was discovered when the monk Liu Sahe found it "growing out" of the earth. Liu claimed that the unusual form of this stūpa was similar to those of the foreign kingdom of Khotan (known in Chinese as Yutian 于闐), to be imitated by the Wuyue Aśoka stūpas.[19] More importantly, the Mao County stūpa was transferred to the capital of the Wuyue Kingdom and became the property of Qian Liu 錢鏐(852–932), grandfather of Qian Chu. This episode serves as evidence for scholars who have speculated that the stūpa was a prototype of the miniature Aśoka stūpas mass produced by Qian Chu.[20] In this view, patrons considered the originality and iconography of the reliquary stūpa as verifying its authenticity and superiority. In respect to the Changgan stūpa, the selection of this form was intentional, so that people would associate it with the Aśoka cult and relic worship, establishing the site's place in history.

Furthermore, the Changgan stūpa aspired to the significance of the Wuyue Aśoka stūpas, commissioned by the ruling class as well as ordinary people. In the chronicles of *Fozu tongji* 佛祖統紀, "Wuyue King Qian Chu was naturally disposed to have faith in Buddhism. Admiring Aśoka's building of the stūpas, he [commissioned craftsmen] to manufacture eighty-four thousand [miniature] stūpas using gilt copper and fine iron; inside [each] was deposited the *Baoqieyin xinzhou jing* (Sūtra on the Heart Mantra of the Precious Chest Mudrā 寶篋印心咒經). [The stūpas] were widely disseminated throughout [the kingdom]. This took in all ten years to complete" (Shi Zhiqin, 2012, p. 1018).[21] So far, only around thirty-five of Qian Chu's stūpas have been found, seeming to suggest that the original "eighty-four thousand stūpas" was an exaggeration or a metaphor for devotion. Apart from the silver stūpas dedicated to the bodily relics of Buddha at the Leifeng Pagoda 雷峰塔, most of Qian Chu's stūpas were made exclusively to enshrine Dharma relics, in particular the *Sūtra on the Heart Mantra of the Precious Chest Mudrā*.[22] In the Wuyue Kingdom, the commission of the Aśoka Stūpa was not only a privilege of kings but a trend among the ordinary laity, who seemingly competed with the imperial project in terms of quantity. To date, at least twenty-three replica Aśoka stūpas built by the laity have been discovered. Most temporally and spatially overlap Qian Chu's project during the Ten Kingdoms (902–979) and the Song Dynasty, and most have been found in southeast China. Benefitting from the popularity of this particular form and its techniques, the Changgan Aśoka stūpa, constructed later, can be considered a legacy of the Wuyue stūpas, favored by all social classes. More important, the stūpa's official name—the Seven-Jeweled Aśoka Stūpa—indicates that this is the most venerated stūpa in the *Lotus Sūtra*, and therefore surpasses the value of the Wuyue ones.[23] According to the sūtra, the stūpa is adorned with seven jewels, the hardest and most imperishable materials in the world, radiating, and it miraculously appears in front of the audience. It is not surprising to see that the Changgan stūpa largely fits into its description, and demands a proper treatment of "made offerings . . . reverently worshipping it, holding it in solemn esteem, and singing its praises" (Hurvitz 1976, p. 183).

Interpretation of Distinguished Features from a Secular Perspective

Admittedly, the Changgan stūpa was made in the tradition of Chinese reliquary stūpas, but further questions about its nature require answers. What distinguishes this stūpa from other Aśoka stūpas? What do these features imply? Based on visual characteristics and textual sources, we can conclude that the stūpa has three notable features. First, the scale of the Changgan stūpa is massive, much larger than the Aśoka stūpas of the Wuyue period. According to previous scholarship, the average height of the Aśoka stūpas in the Wuyue period was 30 centimeters, making them portable reliquaries.[24] In contrast, the Changgan stūpa is 120 centimeters tall, four times the size of the Wuyue stūpas, making it less portable. Second, its surface is heavily covered with inscriptions, including both auspicious phrases and dedicatory inscriptions (Figure 5). Among the inscriptions are "Long live the emperor皇帝萬歲", "Praise to important officials for 1000 years 重臣千秋", "Peace for the masses in this world 天下民安", and "Timely wind and rain風調雨順". The dedicatory inscriptions offer a detailed record of benefactors—their names, occupations, and donated objects—that is absent from the Aśoka stūpas commissioned by Wuyue King Qian Chu. Third, as a reliquary stūpa, the Aśoka stūpa is dedicated to bodily or Dharma relics exclusively, in a narrow interpretation of its function, with other offering objects placed outside of it in the hollow body of a stone casket. However, given the considerable space in the body of the Changgan stūpa reliquary, many other relics and offering objects were stuffed inside, violating the defined role of the reliquary stūpa and obfuscating its functionality.

Figure 5. Anonymous. Dedicatory inscriptions on the Changgan stūpa. Courtesy of Zhou Baohua.

The Aśoka stūpas commissioned by Wuyue King Qian Chu appear to be more elegant but stylistically less sophisticated than the Changgan stūpa. With the exception of Buddhist icons and traditional decorative patterns, no ornaments can be seen on its surface. The only engraved text is a short dedicatory inscription from Qian Chu on the bottom (Figure 6): "The king of Wuyue, (Qian) Chu, has reverently made 84,000 precious stūpas as eternal offerings, recorded in 955 or 965".[25] The creators' lack of interest in additional inscriptions is obvious, with no textual information on the surface of the gilt silver Aśoka stūpas at the Leifeng Pagoda; only images are visible. The purpose of creating these Aśoka stūpas, as Shi Zhiru points out, was to stress the Wuyue king's devotion to Buddhism and embody his political legitimacy by repeating King Aśoka's feat of building eighty-four thousand stūpas.[26] As the only owner of merit, Qian Chu did not need to express his religio-political aspirations explicitly. Rather, he reinforced his political position by imitating King Aśoka's act of building and charging certain stūpas with symbolic meaning.

Figure 6. Qian Chu (commissioner). The bottom side of the Aśoka Stūpa, 965. Iron, Huangyan District Museum (Li 2009, p. 38).

The richness of quantity and content, highlighted by the inscriptions on the Changgan stūpa and the offering objects within, may be due to its secular dedication as a counterpoint to the ideology of the ruling class. As mentioned earlier, similar votive inscriptions were found on handwritten scriptures by laypeople of the Tang Dynasty at Dunhuang, seeking the Buddha's blessing for their ancestors, family, and themselves (Yu 2011, pp. 167–68). However, in contrast with these inscriptions—composed, displayed, and taking effect individually—the votive inscriptions of the Changgan stupa, seen in numerous places on its surface as well as on the donated objects inside, show a diversity of gender, provenance, social class, and identity, forming a heterogeneous network that embodied the collective participation of local believers. There are at least thirty inscriptions that can be categorized into two groups. The first are the dedicatory inscriptions, typical on the surface of Aśoka stūpas, recording the donor's name, occupation, and details surrounding the donation. For instance, an inscription on the bottom of the gold coffin states: "Four *liang* of gold, donated by Wang Wen, the head of the community for building the Pagoda of the Changgan Monastery, Jiang Shilang, assistant teacher in the Hua Prefecture, and his wife Shi Shisiniang; three *liang* of gold, donated by Shao [?], Chonghui master in charge of the first Aśoka stūpa".

The second group consists of votive inscriptions, represented by ink inscriptions on the textiles used to wrap the objects. These add the donor's vows to the basic information stated in the dedicatory inscriptions. For example, an inscription on a yellowish-brown silk cloth states:

高郵軍左廂招賢坊弟子荀懷義謹舍水晶杯一隻，碧琉璃杯一隻，白硨磲念珠一串，幸遇皇帝建金陵長幹寺阿育王所造釋迦佛真身舍利塔，下收葬供養舍利。所願劫劫生生長承佛護。時大宋大中祥符三年□月□日，弟子荀□記.

Xun Huai, a Buddhist disciple from Zhaoxian Lane, Junzuo Xiang of Gaoyou, reverently donated a crystal cup, a green-glass cup, and a string of prayer beads made of white shells. Fortunately, I had a chance to contribute to the construction of the True-Body Pagoda, built by King Aśoka to intern the relics for enshrinement and worship, under the emperor's commission. I wish for the eternal protection of the Buddha, in the date [?] of 1010, recorded by the disciple Xun [?]. (Figure 7; NMIA 2015, p. 41)

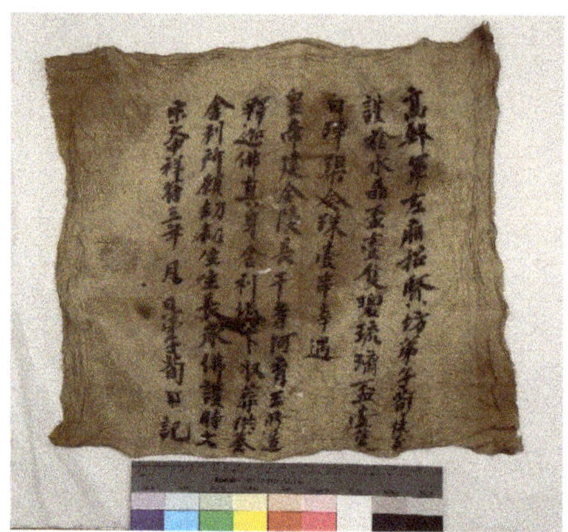

Figure 7. Xun Huai (donor). Silk wrapping, 1011, yellowish-brown, Nanjing Museum. Courtesy of Zhou Baohua.

Other inscriptions include four-character auspicious phrases, such as "Long live the emperor" and a Buddhist poem.

Apparently, the local Buddhist community was inclusive, heterogeneous, and collective, its participants drawn from a wide swath of society, from local officials to monks and ordinary believers. The community was also cohesive and collaborated on this project under the guidance of its leaders, Kezheng and Wang Wen. Since the producers of merits were the entire community, the contributions of every member of the community, including their vows, needed to be clearly documented. This was different from what Qian Chu had done with his stūpas. The contents of the votive inscriptions not only pertained to religious pursuits such as Buddha's protection and the rebirth of the deceased in Buddhist lands, but also revealed several secular hopes, such as good wishes for the emperor and the country's prosperity. These secular desires indicate the diverse expectations and feelings of the common people. Given that utilitarianism and practicality are prominent features of folk Buddhism, it is reasonable to suspect that ordinary people with limited education were not engaged with Buddhist theology. Rather, their emphasis was on Buddhist practices said to produce merit, to be rewarded either in this life or the afterlife.[27] The increasing and consistent participation of ordinary believers in Buddhist affairs accelerated the processes of simplification, socialization, and secularization in folk Buddhism.

Further evidence has been found to explain the uncommon features of the Changgan stūpa by comparing it to other stūpas or Buddhist projects locally commissioned during the Wuyue and Song eras. The imitation Aśoka stūpas, commissioned by Wuyue laity, share with the Changgan stūpa the engraved inscriptions that provide patrons' names and occupations, dates, and the purposes of the commissions. For instance, two bronze Aśoka stūpas were excavated at the site of the Qiyuan Monastery 祇園寺 in Xiaoshan in 1966. A dedicatory inscription carved on the rim of the stūpa body's top states that "the disciple Xia Chenghou and his wife Lin Yiniang, with their entire family, donated all their money to create the two stūpas for true-body relics. In fear of the ample sinful and Dharma hindrance which we pray to eliminate, we hope to undertake the good and reach the pure land in the West; recorded on the 3rd of November, 958" (Figure 8; Chen 2011, p. 34). These inscriptions are similar to those on the Changgan stūpa and offering objects in terms of their locations on the object, their medium, and their contents. The more people who participated in a project, the more diverse their hopes and wishes. Moreover, with more people involved, the roles of the reliquary became more complex, as we know from the Changgan stūpa. In this sense, the reliquary stūpa per se is no longer strictly defined as a container for relic deposits, but a container of laypeople's hopes.

Figure 8. Xia Chenghou (commissioner). The upper side of the body of the Aśoka Stūpa, 958. Bronze, Xiaoshan District Museum (Li 2009, p. 42).

The Changgan stūpa needs to be temporally contextualized from a political and religious perspective so as to facilitate a better explanation of its uncommon features. In the early Song Dynasty, the emperors were often ambivalent about Buddhist affairs and enacted moderate policies towards them in order to maintain social stability and win the support of the people. In its first sixty or seventy years, the Song court also used Buddhism to link China and Central Asia, welcoming Western monks to China to promote Buddhism and translate sūtras. Song Zhenzong, the third emperor of the Song Dynasty, was no exception. He was committed to reconciling Buddhism with Confucian ethics in order to cultivate society.[28] According to the stone stele of the crypt and historical documents, we know that the uṣṇīṣa enshrined in the crypt was a gift from the monk Danapala from Udgana in Northern India. This gift fits with Song Zhenzong's support for Western monks. However, at the same time, the Song court appeared more conservative in its approach to Buddhism, compared to many Tang emperors' devotion to Buddhism. In practice, the Song court intervened more in Buddhist activities and monastic management. For example, the Northern Song emperors inherited the policy of "Request and Grant Plaques" introduced by the Tang emperors to establish their authority in Buddhist affairs, which aimed to control the quantity and scale of monasteries.[29]

According to the stone stele of the crypt, the titles of Changgan Monastery and the True-Body Pagoda were granted by Song Emperor Zhenzong. Their construction and the relic worship subsequently associated with them could not have occurred without the emperor's endorsement. Under these circumstances, any Buddhist activities must have been dependent on political authority and demonstrated obedience to the ruling class. Therefore, it is understandable why the four auspicious phrases, though unrelated to Buddhism, appear prominently on the stūpa.

The first cave of the Gezitou Grotto 閣子頭石窟 was commissioned by laypeople and finished in 1112, near the end of the Northern Song Dynasty. In it, there are similar four-character auspicious phrases, including "Long live the emperor", "Praise ministers for a thousand years", "Stable state and peaceful life", and "Timely wind and rain", (Yuan 1986, pp. 12–13). These secular and political vows reflect the dependence of the local Buddhist community on imperial power, and the compromises they made in its name. They also reflect the extension of folk beliefs to the religious sphere.

Because of its scale, the construction of the crypt and the enshrinement of the uṣṇīṣa must have involved a considerable number of participants. Their success likely lay in the organization and management of the local Buddhist community. Beginning in the Tang Dynasty, with the spread of Buddhism and the urban population growth, relic worship evolved from an elite culture dominated by royalty and the aristocracy to a collective activity accessible to all social classes.[30] Important projects such as the worship of true-body relics demanded the collaboration of many to pool human, material, and financial resources. Since the late Tang, local Buddhist communities, called *Sheyi* 社邑, *Yiyi*邑義, or *Yihui*邑會, had emerged, forming a stable organization in the Northern Song Dynasty. These communities were often centered at a particular Buddhist monastery and engaged in organizing Buddhist activities, celebrating Buddhist festivals, or participating in the construction of Buddhist structures (Gao 2017, pp. 251–52; Teiser 2020, pp. 157–59). Since many communities maintained a high degree of inclusiveness and publicized their activities, membership included ordinary believers, lay Buddhists, and local officials. The inscriptions on the stele and other buried objects specify who held the positions of "head director" and "head of the community." Under their guidance and management, ordinary believers could contribute to and accelerate these projects.

6. Conclusions

Wuyue kings manifested their religio-political aspirations when the role of Aśoka stūpas transitioned from bodily reliquary stūpas to containers for Dharma relics. The Seven-Jeweled Aśoka Stūpa of the Changgan Monastery can be seen as a synthesized product of the heritage of the Aśoka cult and relic worship at this site, the adoption of Wuyue miniature reliquary stūpas, the participation of ordinary believers, and the negotiation between political and religious power. The Changgan stūpa changed its function from a pure reliquary to a materialized embodiment of laypeople's beliefs and expectations. On one hand, the Aśoka stūpa was no longer a depository for a particular type of relic, but a symbol of relic worship with conceptual significance. The three types of relics—the uṣṇīṣa, numinous-response relics, and the relics of holy monks—were simultaneously stowed in the same nested reliquaries. On the other hand, the abundant offering objects, with numerous votive inscriptions, reveal the pragmatic and utilitarian demands of relic worship, displaying the degree of Buddhist secularism in the lower Yangtze River region during the Song Dynasty. With respect to secular features of Song Buddhism, the present-day scholar Ge Zhaoguang 葛兆光 contends that Chinese religions are distinguished by their tendency for pragmatism, which tends to be a psychological comfort, a strategy, or anything besides a pure belief system. Instead of being passive to religious doctrines, the Chinese believers showed remarkable agency by adapting religion to their own interests (Ge 1996, p. 42). Likewise, the Aśoka stūpa of the Changgan Monastery is a reformed reliquary in terms of conception, function, and purpose, and a good representation of secular Buddhism from the Song Dynasty. Through scrutiny of this invaluable reliquary, present-day people can get a glimpse of the artistic innovations and the indigenous practice of relic worship in Nanjing during the Northern Song Dynasty.

Funding: This research received no external funding.

Institutional Review Board Statement: Not applicable.

Informed Consent Statement: Not applicable.

Data Availability Statement: Not applicable.

Conflicts of Interest: The author declares no conflict of interest.

Notes

[1] In recent years, a growing number of scholars have opposed the strict separation of the secular and the sacred in medieval Buddhist practice. Through a close reading of extant materials, they contend that many of the Buddhist practices of laypeople showed a mixture of sacred and secular features, incorporating a variety of factors such as religious imagination, socio-political contexts, and personal interests. See (Copp 2011; Sun 2019; Teiser 2020, pp. 171–72). Sun Yinggang 孫英剛 points out that the boundary between religious and secular texts, monastic and worldly identities, and beliefs and ideologies in medieval China was constantly changing (Sun 2020).

[2] From his study of the family caves at Dunhuang built during the Tang Dynasty, Winston Kyan contends that the boundaries between secular and sacred spaces in these caves are ambiguous. Many murals show a mixture of Buddhist, Confucian, and

Daoist icons, or flaunt the luxurious furnishings and material opulence belonging to this family. In addition, the images depicting ancestor worship directly reflect Confucian ideas, revealing the commissoner's secular considerations. See (Kyan 2010).

3   Onishi Makiko 大西磨希子and Jinhua Chen have demonstrated the ideological meanings behind Emperor Wen's distribution of relics across the country. In particular, the latter argues that this action was not only a device to legitimize the emperor's rule, but also an aid in breaking down racial and cultural barriers to his reunification of the country. See (Chen 2002, p. 42; Onishi 2020).

4   The Edict on Building Stūpas Across the Sui State states that each monastic group should be cautious and diligent and travel taking good care of the relics. Before the groups enter their respective prefectures, ordinary people should have their homes cleaned of all filth. Regardless of belief and gender, people came from all over the city to welcome the returning monks. The prefectural supervisor, governor, and other officials stood along the street and led the team to their destination. The four sections of the people were all arrayed in a solemn manner. See (Yang, 1983, p. 213; Fairbank 1957, pp. 101–2). By analyzing Emperor Wen's call to the whole state to participate in the ritual of relic worship and enshrinment, Zheng Yi 鄭argues that the emperor and nobles were primary beneficaries, despite the involvement of the alleged "all living beings". Therefore, this case is different from the making of the Changgan stūpa, which allowed laypeople to be merit owners. See (Zheng 2016).

5   Wu Hung has summarized the scholarship on Liu Sahe and demonstrated how he became a religious icon in relation to relic worship in medieval China. See (Wu 1996, pp. 32–36). Although Liu's miraculous deed—the discovery of the Buddha's bodily relics beneath the Changgan stūp—is recorded in the *History of Liang Dynasty*, Chen Zhiyuan 陳志遠points out that the mystification of Liu first developed in central China, and then slowly expanded to the southeast of the country. See (Chen 2020).

6   Emperor Wu identified himself as the golden wheel-turning king/cakravartin, following Aśoka's identification as the iron wheel-turning king. See (Endo 2021).

7   Zeng Liping曾立平 has researched the provenance of the bodily relics of the Changgan Monastery and briefly introduces the northern Indian monk Shihu and his donation of the uṣṇīṣa. See (Zeng 2011, p. 70).

8   The Ming gazette *Jinling fancha zhi* (A Record of Jinling Buddhist Monasteries金陵梵志) contains only a record of the history of the site where the Changgan Monastery had been located, but no details on the construction of the True-body Pagoda or the interment of the relics. These had been forgotten by later generations, and thus the crypt remained closed and intact from the Song Dynasty onward. See (Ge, 2007, pp. 459–93).

9   The record of Emperor Yang of Sui's distribution of the relics is also included in *Fayuan zhulin* (The Pearl Forest in the Dharma Park法苑珠林), see (Shi Daoshi, 2003). The record of Li Deyu's distribution is on the stone stele of the crypt of the Ganlu Monastery 甘露寺. A brief description of his actions can be found in (Mao 2009, pp. 212–20).

10   John Strong has carried out a meticulous study of King Aśoka's act of building eighty-four thousand stūpas and its political implications. The king sought to identify himself as a *cakravartin*, the universal king in India culture. Strong also points out that rulers in other countries, such as Japan, emulated this act as a metaphor for their own political legitimacy. See (Strong 1983, pp. 109–25).

11   In a survey of the relic cult practised by rulers in medieval China, Liu Shufen discovered that they sought to legitimize their political authority via this cult. For example, Emperor Wen of the Sui Dynasty ordered a massive project of stūpa building across the country and asked that relics be interred simultaneously on the same day (Liu 2008, pp. 318–22). Shi Zhiru has studied the building of Aśoka stūpas and the acquisition of the Mao County Aśoka stūpa by the Wuyue kings, and concludes that as one of ten kings in China at that time, Qian Chu was anxious to justify and defend his political position by imitating King Aśoka's contribution to Buddhism and expanding his own religio-political influence. See (Shi 2013, pp. 83–109).

12   For detailed information on the crypt of the Song Changgan Monastery, see (NMIA 2015, pp. 4–54).

13   For the entire inscription on the stele, see (NMIA 2015, pp. 14–15).

14   To read the full contents of some of the inscriptions, see (NMIA 2015, pp. 19–48).

15   Most of the gilt bronze stūpas were built in 955 CE, while the gilt iron ones were built in 965. The gap between the production of the two groups of stūpas led to nuanced differences (Chen 2011, pp. 29–30). Lee Seunghye also compared the features of the two groups, and found that major differences lay in the structure and Buddhist iconography. See (Lee 2013, pp. 57–60).

16   As Susan Whitefield points out, Buddhists and merchants established a "symbiotic" relationship as the former spread their religion to foreign areas along the trade routes (Whitefield 2018, p. 85). In another book, Whitefield demonstrates that the development of these trade routes, or the so-called Silk Road, largely resulted from the westward expasion of China since the second century BCE and the rise of the Kushan empire in the first century CE (Whitefield 2015, p. 2). Other scholars emphasize the importance of officials among this group of travelers. Dong Lili argues that soldiers and government officials constituted a large part of the travelers in the early Han Dynasty. One piece of evidence is that the Han emperors commanded military campaigns to combat Huns and therefore protect the western border for years (Dong 2021, pp. 25–26). Besides that, Emperor Wu, in particular, sent his envoy Zhang Qian 張騫 (ca.164–114 BCE) to the western kingdoms twice in order to form military allies. After Zhang's final return in 126 BCE, the Han empire began expanding its control to the west (Hansen 2012, p. 14).

17   By examing the historical records, Xinru Liu stressed that in the first century CE, monks and their merchant patrons brought Buddhist texts and monasteries to China via the Silk Road. See (Liu 2010, pp. 58–59).

[18] However, the four acroteria that often embellished Chinese Aśoka stūpas had no counterparts in ancient India or Gandhāra. Soper suggests, despite a lack of convincing evidence, that this unusual form might have developed from the harmikâ, a square-shaped fence standing on the top of early Indian stūpas. In this regard, Wang Chung-cheng's 王鐘承 speculation may be more persuasive. Wang argues, through a comparision of acroteria and the architectural elements depicted in Han dynasty rubbings, that this distinctive form originated from the Chinese roof decoration on its wings. After being transformed and refined multiple times, the roof decoration lost its original function and became an eye-catching ornament on Chinese Aśoka stūpas. Therefore, acroteria can be seen as a symbol of the sinicization of Indian architecture. See (Wang 2012, pp. 116–17).

[19] For a description of the Mao County stūpa, see (Shi Daoxuan, 1983, p. 404). Shi Zhiru translated the Chinese passage into English. See (Shi 2013, p. 92). Like the episode of Liu's identification of the Changgan stūpa with the King Aśoka stūpas, his identification of the Mao County stūpa is also rife with myth and hyperbole. Scholars such as Wang Chung-cheng question the story's reliability. However, its semi-fictionality does not change the fact that many noble people in medieval China worshiped this reliquary with great devotion. See (Wang 2012, pp. 123–26).

[20] Zhejiang Museum (2008, p. 9). By comparing the images and compositions on several Chinese Aśoka stūpas, Hattori Atsuko suggests that since the Tang Dynasty, the designs of Chinese Aśoka stūpas had more or less been based on the Mao County stūpa, despite nuanced pictorial details. See (Hattori 2011, p. 125).

[21] Shi Zhiru translated the Chinese passage into English. See (Shi 2013, p. 56).

[22] Li Yuxin has collected and arranged information on excavated Aśoka stūpas, especially those built in the Wuyue period, and painstakingly lists many details about most of the excavated or extant Aśoka stūpas. See (Li 2009, pp. 36–41).

[23] John Thompson articulates the construction and veneration of the many-jeweled stūpa described in the *Lotus Sūtra*. As its official name indicates, the Changgan stūpa can be categorized into this group, ranking the highest among all sorts of stūpas. See (Thompson 2008, pp. 126–27).

[24] The average size was calculated by the present author, based on data from excavated Wuyue Aśoka stūpas, and the date taken from Li Yuxin's article. See (Li 2009, pp. 36–41).

[25] For the complete contents of Qian Chu's dedicatory inscription, see (Zhejiang Museum 2008, p. 8).

[26] Shi Zhiru has demonstrated the political motivations behind the Wuyue kings' devotion to Buddhism by studying the Aśoka cult favored by the Qian family, their imitated act of the mass production of stūpas, and their ownership of aged Aśoka stūpas with symbolic meanings. See (Shi 2013, pp. 83–109).

[27] Through an analysis of the production of Buddhist images by local people in southern China during the Tang and Song dynasties, Si Kaiguo summarizes the key characteristics of folk Buddhism as practicality and utility. The social status of the believers defined the quality of these two characteristics, meaning the eclectic and realistic selection of images emphasizes Buddhist beliefs, such as transmigration. See (Si 2013, pp. 210–13).

[28] For more on the Song emperors 'attitudes and policies toward Buddhism, see (Pan 2000, pp. 476–79).

[29] Lee's dissertation argues that political and religious contexts must be considered in order to understand the contents of the dedicatory wishes. Most of the dedicatory wishes were written to protect the living or to ask for a better afterlife for deceased relatives. However, due to the unstable political and social environment of the Southern Song Dynasty, many votive inscriptions began with a blessing for the emperor. See (Lee 2013, p. 213).

[30] Beginning with the introduction of Buddhism in the Eastern Han, relic veneration shifted from an event favored by the upper social classes to a publicly accessible affair. The watershed moment for this change was during the Tang and Song Dynasties. For more on this process, see (Liu 2008, pp. 322–27).

References

Primary Sources

Ban, Gu 班固. 1962. *Han shu* 漢書. Beijing: Zhonghua Book Company.

Ge, Yinliang 葛寅亮. 2007. *Jinling fancha zhi* 金陵梵志. Tianjin: Tianjin renmin chubanshe.

Shen, Yue 沈約. 1983. *Foji xu lue* 佛集序略. In *Taishō shinshū dai zōkyō* 大正新脩大藏經, vol. 52. Edited by Takakusu Junjirō 高楠順次郎, Watanabe Kaikyoku 渡海旭, et al. Taipei: Xinwenfeng Press, pp. 201–2.

Shi, Daoshi 釋道世. 2003. *Fayuan zhulin* 法苑珠林, vol. 3. Beijing: Zhonghua Book Company.

Shi, Daoxuan 釋道宣. 1983. *Ji shenzhou sanbao gantong lu* 集神州三寶感通錄. In *Taishō shinshū dai zōkyō*, vol. 52. Edited by Takakusu Junjirō, Watanabe Kaikyoku et al. Taipei: Xinwenfeng Press, pp. 404–35.

Shi, Huijiao 釋慧皎. 1983. *Gaoseng zhuan* 高僧傳. In *Taishō shinshū dai zōkyō*, vol. 50. Edited by Takakusu Junjirō, Watanabe Kaikyoku et al. Taipei: Xinwenfeng Press, pp. 201–2.

Shi, Zhiqin 釋志磐. 2012. *Fozu tongji jiaozhu* 佛祖統紀校注, vol. 44. Shanghai: Shanghai guji chubanshe.

Yang, Jian 楊堅. 1983. *Sui guoli shelita zhao* 隋國立舍利塔詔. In *Taishō shinshū dai zōkyō*. vol. 52. Edited by Takakusu Junjirō, Watanabe Kaikyoku et al. Taipei: Xinwenfeng Press, pp. 97–363.

Yao, Silian 姚思廉. 1973. *Liang shu* 梁書. Beijing: Zhonghua Book Company.

Secondary Sources

Brose, Benjamin. 2015. *Patrons and Patriarchs Regional Rulers and Chan Monks during the Five Dynasties and Ten Kingdoms*. Honolulu: University of Hawaii Press.

Chen, Jinhua. 2002. Sarira and Scepter. Empress Wu's Political Use of Buddhist Relics. *Journal of International Association of Buddhist Studies* 25: 33–150.

Chen, Lu 陳露. 2006. Cong bamianti fota kan Jiantuoluo yishu zhi dongchuan 從八面體佛塔看犍陀羅藝術之東傳. *Xiyu yanjiu* 西域研究 4: 63–72.

Chen, Ping 陳平. 2011. Bawan siqian ayuwang ta (shang)—Wuyue ayuwang ta shangjie八萬四千阿育王塔(上)—吳越阿育王塔賞介. *Rongbao Zhai* 榮寶齋 1: 22–31.

Chen, Zhiyuan 陳志遠. 2020. Handi sheli chongbai de xingqi—Nanchao fojiao guojiahua zhi yiduan漢地舍利崇拜的興起—南朝佛教國家化之一端. Paper presented at the "Keywords of the Liu-Song Dynasty劉宋關鍵詞國際學術研討會"; Taipei: Institute of Literature and Philosophy, Academia Sinica, December 14.

CHGIS. 2016. Version: 6. (c) Fairbank Center for Chinese Studies of Harvard University and the Center for Historical Geographical Studies at Fudan University.

Choi, Eung-Chon. 2003. Early Korean and Japanese Reliquaries in Relation to Pagoda Architecture. In *Transmitting the Forms of Divinity: Early Buddhist Art from Korea and Japan*. Edited by Naomi Noble Richard. New York: Japan Society, pp. 178–89.

Copp, Paul. 2011. Manuscript Culture as Ritual Culture in Late Medieval Dunhuang: Buddhist Talisman Seals and their Manuals. *Cahiers d'Extrême-Asie* 20: 193–226. [CrossRef]

Dong, Lili 董莉莉. 2021. Sichou Zhi Lu yu Han Wangchao de Xingsheng 絲綢之路與漢王朝的興盛. Ph.D. dissertation, Shandong University, Jinan, China.

Endo, Yusuke 遠藤祐介. 2021. Ryō butei ni oke ru risō teki kōtei zō: Bosatsu kin rinō to shi te no kōtei 梁武帝理想的皇帝像:菩薩金輪王皇帝. *Journal of Institute of Buddhist Culture, Musashino University* 37: 1–32.

Fairbank, John King. 1957. *Chinese Thought and Institutions*. Chicago: University of Chicago Press.

Gao, Jixi 高繼習. 2017. Zhongguo Gudai Sheli Digong Xingzhi Yanjiu 中國古代舍利地宮型制研究. Ph.D. dissertation, Shandong University, Jinan, China.

Ge, Zhaoguang 葛兆光. 1996. Nayige Fojiao, Nayige Daojiao: Tan Zhongguo Zongjiao yu Wenhua Yanjiu Zhongde Yige Wenti 哪一個佛教，哪一個道教-談中國宗教與文化研究中的一個問題. *Dongfang* 東方 3: 39–42.

Hansen, Valerie. 2012. *The Silk Road: A New History*. New York: Oxford University Press.

Hattori, Atsuko 服部敦子. 2011. Aikuō tō no zōryū ni kansuru ichi kōsatsu—bukkyō zuzō no kentō o chūshin ni –阿育王塔造立一考察—教像討中心—. In *Wuyue shenglan guoji xueshu yantaohui lunwen ji* 吳越勝覽國際學術研討會論文集. Beijing: Zhongguo shudian, pp. 107–19.

Hurvitz, Leon. 1976. *Scripture of the Lotus Blossom of the Fine Dharma*. New York: Columbia University Press.

Jao, Tsung-I. 1990. Liu Sahe shiji yu ruixiang tu 劉薩訶事跡與瑞應圖. In *Dunhuang Shiku Yanjiu Guoji Taolunhui Wenji* 敦煌石窟研究國際討論會文集. Shenyang: Liaoning Meishu Chubanshe, pp. 336–49.

Kyan, Winston. 2010. Family Space: Buddhist Materiality and Ancestral Fashioning in Mogao Cave 231. *The Art Bulletin* 92: 61–82. [CrossRef]

Lee, Seunghye. 2013. Framing and Framed: Relics, Reliquaries, and Relic Shrines in Chinese and Korean Buddhist Art from the Tenth to the Fourteenth Centuries. Ph.D. dissertation, University of Chicago, Chicago, IL, USA.

Lee, Seunghye. 2021. What Was in the 'Precious Casket Seal'? Material Culture of the Karaṇḍamudrā Dhāraṇī throughout Medieval Maritime Asia. *Religions* 12: 1–19.

Li, Zhengyu 李正宇. 1999. Tangsong Dunhuang Shisu Fojiao de Jingdian Jiqi Gongyong 唐宋敦煌世俗佛教的經典及其功用. *Journal of Lanzhou Vocational Technical College* 1: 9–36.

Li, Yuxin 黎毓馨. 2009. Ayuwang Ta Shiwu de Faxian yu Chubu Zhengli 阿育王塔實物的發現與初步整理. *Cultural Relics of the East* 31: 36–41.

Liu, Shufen 劉淑芬. 2008. *Zhonggu de Fojiao yu Shehui* 中古的佛教與社會. Shanghai: Shanghai Guji Chubanshe.

Liu, Xinru. 2010. *The Silk Road in World History*. Oxford and New York: Oxford University Press.

Mao, Ying 毛穎. 2009. Zhenjiang Ganlu Si Tangdai Sheli Yimai Zhidu ji Shelizi Yanjiu 鎮江甘露寺唐代舍利瘞埋制度及舍利子研究. *Tangshi luncong* 唐史論叢 1: 212–20.

Nakamura, Hajime 中村元. 1984. *Zhongguo Fojiao Fazhanshi* 中國佛教發展史. Translated by Wan-Chu Yu 余萬居. Taipei: Heavenly Lotus Publishing Co., Ltd., vol. 1.

Nakamura, Hajime 中村元. 2013. 東洋人思惟方法 (*Ways of Thinking of Eastern Peoples: India, China, Tibet, Japan*). Translated by Xu Fuguan 徐復觀. Taipei: Student Book Press.

Nanjing Municipal Institute of Archaeology (NMIA). 2015. Nanjing Da Bao'en Si Yizhi Taji yu Digong Fajue Jianbao 南京大報恩寺遺址塔基與地宮發掘簡報. *Cultural Relics* 5: 4–54.

Onishi, Makiko 大西磨希子. 2020. Sokuten bukō to aiku ō—Gihō nenkan no shari hanpu to 『daiun he sho』 o megu ̂ te 則天武后阿育王—儀鳳年間舍利頒布『大雲經疏』. " *Dunhuang Nianbao* 14: 1–17.

Palumbo, Antonello. 2012. Models of Buddhist Kinship in Medieval China. In *Zhonggu Shidai de Liyi Zongjiao yu Zhidu* 中古時代的禮儀、宗教與制度. Edited by Yu Xin. Shanghai: Shanghai Guji Chubanshe, pp. 287–338.

Pan, Guiming 潘桂明. 2000. *Zhonggu jushi fojiao shi*中國居士佛教史. Beijing: China Social Science Press.

Qi, Haining 祁海寧, and Juping Gong 龔巨平. 2011. 'Jinling Changgansi Zhenshenta Cang Sheli Shihan Ji' Kaoshi ji Xiangguan Wenti '金陵長干寺真身塔藏舍利石函記'考釋及相關問題. *Southeast Culture* 1: 68–75.

Qi, Haining 祁海寧, and Juping Gong 龔巨平. 2012. Beisong Changgansi Gnashengta Digong Xingzhi Chengyin Chutan 北宋長干寺感聖塔地宮形製成因初探. *Southeast Culture* 1: 76–82.

Rhie, Marylin M. 2010. *Early Buddhist Art of China and Central Asia*. Leiden: Brill, vol. 3.

Shang, Lixin 尚麗新. 2007. 'Dunhuang gaoseng' Liu Sahe de shishi yu chuanshuo '敦煌高僧'劉薩訶的史實與傳說. *Journal of Southwest Minzu University (Humanities and Social Science)* 4: 76–82.

Shen, Hsueh-Man. 2019. *Authentic Replicas: Buddhist Art in Medieval China*. Honolulu: University of Hawai'i Press.

Shi, Zhiru. 2013. From Bodily Relic to Dharma Relic Stūpa: Chinese Materialization of the Aśoka Legend in the Wuyue Period. In *India in the Chinese Imagination: Myth, Religion, and Thought*. Edited by John Kieschnick and Meir Shahar. Philadelphia: University of Pennsylvania Press, pp. 83–109.

Si, Kaiguo 司開國. 2013. *Tangsong Shiqi Nanfang Minjian Fojiao Zaoxiang Yishu*唐宋時期南方民間佛教造像藝術. Beijing: China Social Science Press.

Soper, Alexander C. 1940. Japanese Evidence for the History of the Architecture and Iconography of Chinese Buddhism. *Monumenta Serica* 4: 638–79. [CrossRef]

Strong, John. 1983. *The Legend of King Aśoka: A Study and Translation of the Aśokavadana*. Princeton: Princeton University Press.

Sun, Yinggang 孫英剛. 2019. Liangge Chang'an: Tangdai siyuan de zongjiao xinyang yu richang yinshi 兩個長安：唐代寺院的宗教信仰與日常飲食. *Literature, History, and Philosophy*文史哲 4: 38–59.

Sun, Yinggang 孫英剛. 2020. Foguang xia de chaoting: Zhonggu zhengzhishi de zongjiao mian 佛光下的朝廷：中古政治史的宗教面. *Journal of East China Normal University (Humanities and Social Science)* 1: 47–57.

Teiser, Stephen F. 2020. Terms of Friendship: Bylaws for Associations of Buddhist Laywomen in Medieval China. In *At the Shores of the Sky: Asian Studies for Albert Hoffstädt*. Edited by Paul W. Kroll and Jonathan A. Silk. Leiden: Brill, pp. 154–72.

Thompson, John M. 2008. The Tower of Power's Finest Hour: Stupa Construction and Veneration in the Lotus Sutra. *Southeast Review of Asian Studies* 30: 119–36.

Tsiang, Katherine R. 2017. The 'King Aśoka' Type of Stūpa and Its Multivalent Meanings 阿育王式塔所具有的多種意義. *Translated by Wang Pingxian. Dunhuang Research* 2: 22–34.

Wang, Chung-cheng 王鍾承. 2012. Wuyue guowang Qian Hongchu zao ayuwang ta吳越國王錢弘俶造阿育王塔. *National Palace Museum Research Quarterly* 29: 109–78.

Whitefield, Susan. 2015. *Life Along the Silk Road: Second Edition*. Oakland: University of California Press.

Whitefield, Susan. 2018. *Silk, Slaves, and Stupas: Material Culture of the Silk Road*. Oakland: University of California Press.

Wu, Hung. 1996. Rethinking Liu Sahe: The Creation of a Buddhist Saint and the Invention of a 'Miraculous Image'. *Orientations* 27: 32–43.

Yang, Yongquan 楊永泉. 2009. Nanjing zai zhongguo fojiao wenhua zhong de diwei 南京在中國佛教文化中的地位. *Nanjing Journal of Social Sciences* 南京社會科學 2: 7–14.

Yu, Xin 余新. 2011. Personal Fate and the Planets: A Documentary and Iconographical Study of Astrological Divination at Dunhuang, Focusing on the 'Dhāraṇī Talisman for Offerings to Ketu and Mercury, Planetary Deity of the North'. *Cahiers d'Extrême-Asie* 20: 163–90. [CrossRef]

Yuan, Anzhi 員安志. 1986. Shanxi fuxian shiku si kancha baogao 陝西富縣石窟寺勘察報告. *Relics and Museology* 6: 1–18.

Zeng, Liping 曾立平. 2011. Beisong jinling changgan si zhenshen ta digong fajue yu fo dingzhengu sheli chongguang 北宋金陵長干寺真身塔地宮發掘與佛頂真骨舍利重光. *Identification and Appreciation to Cultural Relics* 4: 68–75.

Zhang, Yuhuan 張馭寰. 2000. *Zhongguo ta*中國塔. Taiyuan: Shanxi People's Publishing House.

Zhejiang Museum, ed. 2008. *Dongtu Foguang: Zhejiang Sheng Bowuguan Diancang Daxi* 東土佛光:浙江省博物館典藏大系. Hangzhou: Zhejiang Ancient Books Publishing House.

Zheng, Yi 鄭. 2016. Daocheng sheli: Chongdu Renshou nianjian Suiwendi feng'an fo sheli shijian 道成舍利：重讀仁壽年間隋文帝奉安佛舍利事件. *Art Research* 藝術研究 3: 63–70.

Zheng, Liyong 鄭立勇, and Songguang Lin 林松光. 1996. Shilun zhongguo minzhong de zhuti zongjiao yishi texing 試論中國民眾的主體宗教意識特性. *Studies in World Religions* 3: 11–16, 155–56.

Zhou, Xueqin 周雪芹. 2005. Cong Dunhuang Fayuanwen Kan Tangsong Shiqi Minzhong de Fojiao Xinyang敦煌愿文看唐宋期民的佛教信仰. Master's thesis, Minzu University of China, Beijing, China.

Article

Reconstructing Pure Land Buddhist Architecture in Ancient East Asia

Young-Jae Kim

The Department of Heritage Conservation and Restoration, Korea National University of Cultural Heritage, Buyeo 33115, Korea; kyjandy@nuch.ac.kr

Abstract: Pure land comes from the Indian term "sukha," which means welfare and happiness. However, in East Asia, Buddhism has been associated with the theological concepts of the immortal realm in the bond of death and afterlife. This study reviews detailed conception of Pure Land architecture in Sanskrit literature, as well as Buddhist sutras. The thesis notes that the conceptual explanation of Pure Land architecture, which describes the real world, becomes more concrete over time. Such detailed expression is revealed through the depiction of the transformation tableau. Hence, through Pure Land architecture situated on Earth, this research shows that Buddhist monks and laypeople hope for their own happy and wealthy settlement in the Pure Land. The building's expression of transformation tableaux influences the layout and shape of Buddhist temples built in the mundane real world at that time. Moreover, this study notes that Bulguksa Monastery is a cumulative product of U-shaped central-axis arrangements with courtyards, terraced platforms, high-rise pavilions, and lotus ponds, plus an integrated synthesis of religious behaviors by votaries as a system of rituals. Further, it merges pre-Buddhist practices and other Buddhist subdivisions' notions with Hwaeom thought, in comparison with Hojoji and Byodoin Temples that follow the Pure Land tradition.

Keywords: pure land (sukhavati); Bulguksa Monastery; Buddhist grottoes; transformation tableaux; immoral world

Citation: Kim, Young-Jae. 2021. Reconstructing Pure Land Buddhist Architecture in Ancient East Asia. *Religions* 12: 764. https://doi.org/10.3390/rel12090764

Academic Editor: Shuishan Yu

Received: 10 August 2021
Accepted: 2 September 2021
Published: 14 September 2021

Publisher's Note: MDPI stays neutral with regard to jurisdictional claims in published maps and institutional affiliations.

1. Introduction

In East Asia, the Buddhist Pure Land, which is derived from the Indian term sukha, was intimately concerned with the theological concept of the immortal realm in the bond of death and afterlife. Immortal Taoist paradises and Buddhist Pure Lands have something in common and interact and communicate with each other. In Pure Land tradition, a combination of rituals, such as visualization and meditation, can bring an individual one step closer to the richness of the Pure Land. (Corless 1982; Pas 1987; Xiao 1989; Wu 1992a).

The concept of Pure Land architecture, which is intended to express the real world, became more concrete over time. Such detailed illustration was revealed through the depiction of the transformation tableaux and pictorial spatial representations, which connote narrative moments, events, and places. (Mair 1986; Wu 1992b) The architecture of Maijishan, Xiangtangshan, and Dunhuang Cave Temples depict the Pure Land in the real world. The Buddha sitting at the center of the bodhisattvas, in terraces over a lotus pond, represents a happy configuration. The horseshoe-shaped architectural complex of grand buildings are copies of palaces or great monasteries in the imperial capital. Pure Land consciousness is exhibited as buildings and objects that represent the horizontal courts and ponds of Buddhist monasteries in the real world.

This thesis highlights the detailed conception of Pure Land architecture in understanding Sanskrit literature and the Buddhist sutras,[1] and emphasizes the contribution of Buddhist murals in providing ritual prototypes over time. It also focuses on the roles of symbolic fabrics in monasteries such as Bulguksa, Hojoji, and Byodoin, which represent Pure Land architecture on Earth. In particular, the scattered Bulguksa buildings incorporate elevated flat stone platforms and a lotus-filled pool surrounded by various trees. This study suggests that such an architectural composition stemmed from the Pure Land scriptures and transformation tableaux as a system of rituals that enabled devotees to search for their own happy and comfortable settlement therein.

2. Conceptualization of the Pure Land in Buddhist Literature

The sahaloka (real world) is the Pure Land.[2] The saha world is considered as assorted, ornamented or unornamented, and pure or impure, because all beings remain tied by the laws of causation and are subject to transmigration. However, the saha world can be the land of transformation in the presence of the Buddha, and beings are transformed there within their dwellings. Therefore, the preaching Buddha and stupas (or cetiyas) in the Pure Land allude to the transformative power of the Buddha in purifying this land and preparing people for rebirth (Wong 1998, pp. 67–68; 2008; Tsukamoto 1986, Tsukamoto 1996–1998). Parallel to the nature of stupas and cetiyas, temples with Buddha halls/Buddha pagodas in East Asia were re-designed as new models of Buddhist temples in the real world. They were renamed as a specific building type to depict the Pure Land. Indeed, Pure Land architecture carries the meaning of sacred places or shrines associated with events or stories of meditation, enlightenment, and nirvana. In particular, the term "sukha," notably means welfare and happiness (or comfort). The sukha was discovered in inscriptions in the majority of Buddhist sites, such as Nagarjunakonda, Mathura, and Amaravati (Vogel 1929–1930; Tsukamoto 1996–1998).[3] The inscriptions indicated below are among them.

> For the attainment of welfare and happiness in both the worlds (ubhaya-loka-hita-sukha) and of Nirvana has erected this stone pillar (skambha), in the sixth year of (the reign of) King Siri-Virapurisadata, and the sixth fortnight of the rainy season, the 10th day. From the inscriptions of Nagarjunakonda Sites 1, 5, and 43.[4] (Vogel 1929–1930; Tsukamoto 1996–1998; Macdonell 1929; Apte 1957–1959)

Around Buddhists remain near Nagarjunakonda and Amaravati, it is not unusual to find inscriptions like "ubhaya-loka-hita-sukha" engraved on most ayaka-pillars. Most donor inscriptions on monuments sited by intellectual monastics at primary pilgrimage venues end with the phrase" for the welfare and happiness of all beings." In addition, Nagarjuna, who probably lived in the Purvasaila, Aparasaila, and Caityaka monasteries during the time he wrote the Ratnavali (Walser 2005, pp. 87–88), indicated that we could obtain the "welfare and happiness of all beings" and become great men through a new devotion toward Buddhist images, stupas, and shrines.[5]

Most early sutras in Chinese were published before 220CE, interpreting the "*Sukhavati* (happy land)" in transliteration as the "*xumoti* 須摩提 or 須訶摩提" (Fujita 1970, pp. 141–61; Iwamoto 1978, pp. 57–79). The Banzou sanmei sutra mentions that in the seat of the xumoti, bodhisattvas sat in the center and recited sutras, and all adherents longed for Amitabha. (Taisho 13, no. 418). In the proper sense, the Sukhavati was not a heaven at all; rather, it could be found anywhere on Earth because the sutras said that in xumoti, wealthy merchants had high social standing and possessed a great deal of gold and jewels (Fujita 1970, p. 256; Nakamura 1975, p. 205). In addition, the emergence of the "xumoti" area was closely associated with the occurrence of Amitabha Buddha.[6] In contrast, the Zoroastrian religion was established through the impact of the worship of the sun god because the Pure Land has been described as a "paradise of infinite light" in the Zoroastrian scripture "Avesta" (Tsukamoto 1986, pp. 394–95; Nakamura 1975, p. 204). Paradise literally means enclosure. It designates a demarcated, finite, and protected space from the openness of chao (Pyyhtinen 2014; Kim 2018).

From 220 CE to the early seventh century, a Chinese transliteration of the "Sukhavati" was beginning to be known as the "anle," which implied "comfort and happiness." It was substituted for the "jile" (extreme happiness) and "jingtu,"(pure land), respectively, in the translation of the monk Kumarajiva in the fifth century. It does not refer to Amitabha's land, but rather regards it as purifying the Buddha land.[7] A significant point in the changes in translation regarding the *"Sukhavati"* is that the Chinese transliteration gradually adopts the definition of the "land of welfare and happiness."[8] The correct understanding of the meaning of Pure Land was very significant because such a deliberate attempt helped augment Buddhist authenticity and its mystical power by returning to the original term ostensibly connected with the homeland of Buddhism (Mizuno and Toshio 1941). This is a good representation of the acceptance of foreign culture, and adaptation of important ideas from East Asian countries.

3. Embodiment of the Buddhist Pure Land in Transformation Tableaux

In East Asian culture, the understanding of the Pure Land World was well-expressed in huge depictions, called the *"bianxiang* (transformation tableau)," with rich content and complex compositions at cave temples. They reinterpreted Buddhist stories from the jiangjing (sutra oratory) and guanxiang (visualization) literature produced for ritual participation. (Mair 1986; Wu 1992a, pp. 55–56; 1992b, pp. 111–12). The Buddhist Pure Land's construction on earth was built in concrete ways by devotees of both editorialization and architecturalization. The constructions appeared in paintings, sculptures, and architecture. The figures were composed of buildings, bridges, platforms, and passageways with balustrades, ponds, and many manifested bodies of bodhisattvas and deities. Such depiction was a very creative method and considered a meritorious deed, especially for monks, such as Huiyuan 慧遠 (334–416) in Sichuan, Shandao 善導 (613–681) in Dunhuang, and Zhiyi 智顗 (538–597) in Jiangnan who provided a guideline for the Pure Land's pictorialization (Sponberg and Hardacre 1988, pp. 94–95; Kitagawa 1988; Hay 1999, p. 240). The appearance of the transformation tableau helps people to be reborn in Amitabha's paradise, which is depicted in the main painting.[9] Huiyuan's commentary still emphasized meditating to the Buddha, and thus continued the old tradition of visualizing the Amitabha, *"guanfo,"* in the mind, which was probably based on the Pratyupannasamadhi sutra (Zürcher 1959, pp. 180, 223).[10] However, unlike Huiyuan, his teacher Daoan meditated on Maitreya (Kieschnick 1997, p. 5).[11] Early Chinese monks were well-known for using the meditation techniques of *"bhavana" or "samadhi,"* which consisted of both the meditative absorption of concentration and contemplation through pictures; through this practice, devotees felt that their inner spaces were broadened when they looked at many buildings through the cultivation of samadhi.

On the contrary, Shandao's commentary differed from the older Visualization Sutra in declaring that laypeople, monks, nobles, and commoners had the chance of rebirth in the western Pure Land (Chappell 1977). He confirmed that successfully visualizing the *"sukhavati"* was the key to the attainment of both eternal happiness and wealth (Wu 1992a). Shandao's visualization methods were initially influenced by Tanluan (曇鸞, 476–542), who proposed an easy path through visualization, and whose view consisted of three forms: the merit of manifestation through visualization of the Pure Land, the Buddha, and the bodhisattvas. Tanluan's representation methods were linked with the Taoist visualization of immortal mountains (Corless 1982, pp. 36–45). Nonetheless, Shandao's role was very important. While Tanluan used visualization meditations in Pure Land rituals, Shandao invented the use of illustration for a visualization based on the 16 meditations in the Amitayurdhyana sutra. As a result, the illustration tool for meditation was further extended to the construction of buildings on Earth (Pas 1987; Taisho 47, no 1959).

4. The Immortal Taoist Paradise and the Pure Land in Korean Society

The Dae Hwaeomjong Bulguksa birojanamunsuboheonbosal chanbyeongseo 大華嚴宗佛國寺毘盧遮那文殊普賢像讚幷序 consists of an introduction to and eulogy about the Vairocana Buddha on the left wall, and the Samantabhadra and Manjusiri on its left and right sides at the Gwanghakjang Gangsil 光學藏講室 to memorialize the Silla King Heongang (r. 875~886). Queen Gweon, a concubine of the King, offered the images that led to the statement "Heongang was one of the immortals, while his queen was the manifestation of the bodhisattva." When he came down from a silver palace 鷄林, surprisingly, King Heongang came to the land of Gyerim 鷄林 (Rooster's forest), a legendary site of Silla's founding, to govern Geumseong (City of Gold; present Gyeongju) during the Silla period.[12] Silla society believed that kings came from Taoist heaven and were considered immortal. The merging of rulers and immortals resembled a chakravartin king and an enlightened holy man, because the chakravartin king in the Pali canon is paired with the Buddha as a secular counterpart and conqueror of the universe.[13] He protects his destitute subjects (Zürcher 1959, p. 292). The enshrinement of an Asoka statue at the Silla Hwangnyongsa temple in 549 CE, at the instance of King Jinhung (r. 534–576), was implied to reinforce the power and prestige of the Silla kingship,[14] which was meant to be dedicated to the Buddha, and serve as an allusion to royalty for a ruling world monarch. The political use of Silla Buddhism followed the patterns of the Chinese dynasties of the Northern Wei, Sui, and Tang, in which rulers used Buddhist symbols and ideas, along with Taoism and Confucianism, to support their power.

Silla Buddhism was differentiated from that of Chinese dynasties because the ruling monarch tried to merge the advantages of transcendent immortals, transformative Buddhas, and righteous chakravartin into one frame.[15] Such acts were not limited to the ruling monarchs, but took place among the aristocrats and other lower ranks. Kim Ji-Seong, a true-born noble, commissioned the images of both Maitreya and Amitabha to adorn a monastery, named Gamsansa Monastery, in 716. The inscription mentions that Kim studied Taodejing, and served Zhuangzi and Laozi in his philosophical pursuit and accomplishments.[16] Therefore, pagodas and pavilions are not only an emblem of holy places that signify the Buddha's life stories, but preserve a pavilion for hermits associated with the immortal world. The posthumous settings were a mixture of Taoist dwellings and the Pure Lands (Wright 1948, p. 355; Wu 1992a, p. 135).

5. Real World Representation of a Pure Land at Bulguksa Monastery

Bulguksa 佛國寺, meaning a monastery of the Pure Land, was situated at the base of Mt. Toham, a 745-meter high mountain on the southeastern edge of Gyeongju. It was founded in the first year (742) or the 10th year (752) of King Gyeongdeok's reign (r. 742–765, 景德王) as one typical model among medieval monasteries, with a hall and two pagodas.[17] The late historical records of Korea state that the monastery compound was built by Kim Daeseong, a prime minister of Silla. Kim Daeseong carried out the construction of Bulguksa on the western hillside of Mt. Toham, and Seokbulsa on the western hillside of the same mountain, in honor of his parents in their past and present existence.[18]

Bulguksa has a very impressive plan that represents both the Sakyamuni and Amitabha Pure Lands. The first sphere consists of two elevated Buddha pagodas placed in front of the Daeungjeon (the Great Hero Hall) with five (front facade) by four bays (side). The central bay is much wider than the other bays, It lies on a central north-south axis that devotees enter through the center of Baekwungyo (White Cloud) and Chungwungyo (Blue Cloud) Bridges, plus upper and lower grand terraced platforms, Jahamun Middle Gate (Purple Mist Gate), Daeungjeon Hall, and Museoljeon Lecture Hall (Non-word). This courtyard has two stone pagodas, which were renamed during the reconstruction work of the Joseon period when they came to be called Seokgatap Pagoda (West) for Tathagatha Sakyamuni and Dabotap Pagoda (East) for the abode of Tathagata Prabutaratna, according to the Sadarmapundarika (Lotus) sutra (Taisho 9 no 262). They were originally called the Seoseoktap (West) and Mugujeonggwangtap (Endless Untainted Light or Dongtap (East))

Pagodas according to epigraphical evidence recently restored from four historical records in fragments, which were associated with the restoration campaigns of the pagodas in the Goryeo era (918–1392).[19]

Conversely, another courtyard forms the realm of Amitabha Buddha, which is situated in the western precinct of the Daeungjeon Hall. The ambit has only one building, the Geukrakjeon (Extreme Bliss Hall). Votaries ascend into the present Pure Land utilizing two finely carved stone bridges, the Chilbogyo (Seven-Gems) and Yeonhwagyo (Lotus) Bridges with open terraced platforms, and then enter through the Anyangmun Gate (Comfort and Happiness). The "anyang" means the world of "sukhavati." These "chilbo" and "yeonhwa" are direct references to the sukhavati Pure Land because all sukhavati worlds subsumed the lotus and seven treasures, as referred to in Pure Land sutras.

The tall pavilions, Beomyeongnu (Ugyeongru) Pavilion and Jwagyeongnu Pavilion, were established on both end corners from the front façade, while the "Gupum Yeonji," meaning the lotus pond of the nine ranks, is an oval lotus pond situated below the Cheongwungyo and Baekwungyo Bridges (Figures 1 and 2).

Figure 1. Whole view of Bulguksa after the 1969–73 restoration work (source: Dongailbo, 16 March 1976).

This is related to Hwaeom's idea that several Pure Land concepts coexist and fuse according to the concept of dependent origination, in order to meet realistic demands. These efforts resulted in the construction of great temples to create a unifying plan for harmonizing various doctrines and performing rituals for matching the doctrines. One of these was the Bulguksa. Beopgyedogi chongsurok 法界圖記叢隨錄 records that Kim Daeseong was taught by Pyohun in Hwangboksa Temple (Beopgyedogi Chongsurok 1254), and thus knew the method for practicing the Hwaeomsambonjeong 華嚴三本定 (the three essential concentrations of Hwaeom Samadhi) by realizing the three fundamentals of the undisturbed mind. The highest ideal world in Avatamsaka thought is called the Lotus World, a world of enlightenment. Through all religious behaviors executed within the buildings at Bulguksa Temple as a ritual machine, all distinctions disappear and a state of perfect harmony among all differences is achieved. That is, Minister Kim tried to create an integrated religious tool that could accommodate the various ideologies of Silla society by combining Taoist practices for posthumous rebirth with royalty-and-filial-piety-oriented Confucianism.

Based on this integrated Avatamsaka thought, the visible Sakyamuni Pure Land was identical to the utopian Lotus World, placing Sakyamuni Buddha in a position above the Amitabha Buddha to emphasize the fundamental position of Buddhism. Bulguksa's architecture displays this teaching method with extreme clarity.

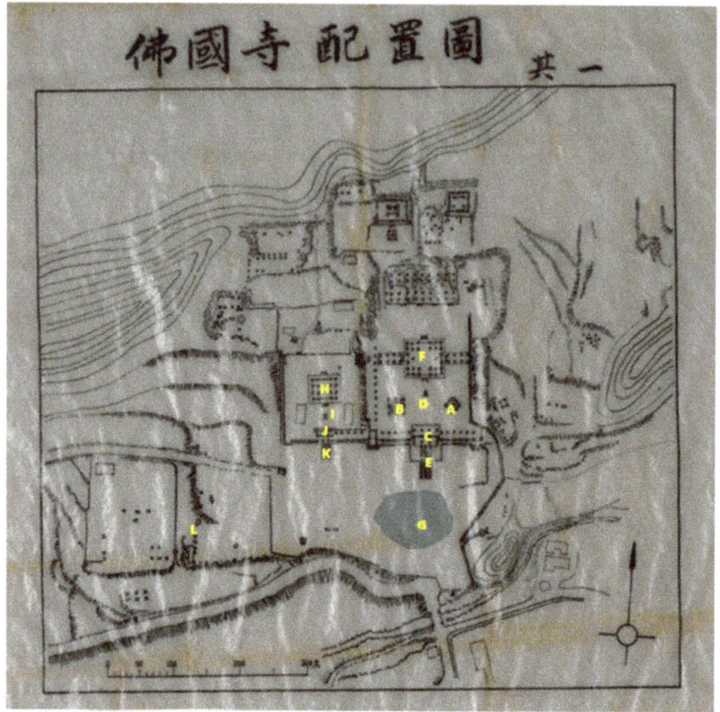

Figure 2. Lotus Pool (Nine Ranks for Rebirth) and Building Remains after 1924 (Gukga girokwon). A: East Pagoda (Dabotap), B: West Pagoda (Seokgatap), C: Gate (Jahamun), D: Worship Stone (Bongnodae), E: Bridges and Platforms (Cheogungyo and Baekun Bridges), F: Mahavira Hall (Daeungjeon Hall), G: Lotus Pool (Pond), H: Amitabha Hall (Geukrakjeon Hall), I: Worship Stone (Bongnodae), J: Gate (Anyangmun) K: Bridges and Platforms (Yeonhwa and Chilbo Bridges), L: Anonymous Building Remains.

5.1. Common Views in Pure Land Scriptures

Primary sutras related to Pure Land were produced from the second century onward. The Kushan monk Lokaraksa translated the Wuliang qingjing pingdengjue jing 無量清淨平等覺經 (Taisho 12, no 361) in the latter half of the second century. A little later, the Kushan Zhi Qian 支謙, a pupil of Lokaraksa, translated the Foshuo amituo sanye sanfosalou guodu rendao jing 佛説阿彌陀三耶三佛薩樓佛檀過度人道經 (Taisho 12, no 362) in the second quarter of the third century. Samghavarman 康僧鎧 translated the Larger Sukhavati vyuha sutra 佛説無量壽經 (Taisho 12, no 360) in 252, and Kumarajiva 鳩摩羅什 translated the Smaller Sukhavati vyuha sutra 佛説阿彌陀經 (Taisho 12, no 366) in around the 400s. Jiangliangyeshe 畺良耶舍 translated the Amitayurdhyana sutra 觀無量壽經 (Taisho 12, no 365) in the fifth century. These sutras share a universal conception that persons who stayed in defiled seeds can go into rebirth before gaining salvation in the Pure Land when they accumulate virtues through alms activities.

Among them, the Wuliang qingjing pingdengjue jing and Foshuo amituo jing mentioned a hermitage called the *"shezai* 舍宅." The Amitabha Buddha was placed inside the lecture hall and hermitage. The concrete expression of the Pure Land appeared in two sutras: the Larger Sukhavati vyuha sutra and the Smaller Sukhavati vyuha sutra. The former records a discourse offered by Bhagavat (Buddha) on Vulture's Peak near Rajagrha in response to his disciple Ananda, "The lecture halls, hermitages, palaces, tall towers, and watchtowers were built on the land. They were adorned by seven jewels with natural consequences, and therefore, multifold pearls that can glow in the dark were covered"

(Taisho 12 no 360). The depictions of buildings on the Pure Land in the Sanskrit version[20] mentioned "gardens, palaces, and pavilions on the land."

The latter, the Smaller Sukhavati vyuha sutra, depicts "lakes, stairs, and pavilions" adorned with the seven jewels made of gold, silver, beryl, and crystal. The Buddha tells Shariputra, "In the land, sukhavati (there are seven rows of balustrades, seven rows of fine nets, and seven rows of arrayed trees; they are all four gems and surround and enclose (the land). The lake bases are strewn with golden sand, and the stairs on the four sides are made of gold, silver, beryl, and crystal. On the land, there are multi-storied pavilions and galleries embellished with gold, silver, beryl, crystal, white coral, red pearl, and diamond" (Taisho 12, no 366). In contrast, the Amitayurdhyana sutra explains sixteen visualization meditations. The beginning of the first visualization is a meditation on the sun. The second envisions that the western region is flooded by pure water, which turns into ice, then into beryl. The flat ground of the Pure Land is made of beryl and is supported by columns made of various jewels. The third and fourth visualizations focus on the lapis lazuli Earth and jeweled trees. The fifth and sixth contemplate the water of eight excellent qualities and the myriad jeweled pavilions on the Pure Land. The seventh contemplates the jeweled lotus throne of Amitabha (Taisho 12, no 365).

It is worth noting that the pavilions (*louge*), including balustrades and ponds enriched with gold, silver, beryl, crystal, white coral, red pearl, and diamond have been commonly and persistently used to adorn buildings. Palaces, lakes, stairs, gardens, and pavilions in the Pure Land were compared in all versions of the scriptures. This suggests a link with Taoist tradition and the building types popularly used at that time in Western areas (Corless 1982, pp. 36–45).

5.2. Architectural Characteristics of Transformation Tableaux

The synthesis of tall pavilions and ponds began to appear in the transformation tableaux in the early sixth century. This combination started in the early sixth century across East Asia, and took place at the following sites: West Pure Land at the Maijishan Cave 127 (Gansu) in the early sixth century (Figure 3), Pure Land depiction with the secular landscape of Wanfosi Stele (Chengdu) of the early sixth century in the Liang era (Figure 4), and the bas-relief of Amitabha Pure Land found in the southern Xiangtangshan Caves 1 and 2 (Hebei) of the sixth century during the Northern Qi period (Katsuki 1992; Li 2007) (Figures 5 and 6).

Figure 3. West Pure Land, Maijishan 127, early-sixth-century Gansu.

Figure 4. Pure Land depiction with secular landscape. The reverse of Wanfosi Stele, early-sixth-century Liang.

Figure 5. West Pure Land of Amitabha, Xiangtangshan Cave 1, sixth-century Northern Qi (Freer Gallery of Art).

Figure 6. West Pure Land of Amitabha, Xiangtangshan Cave 2, sixth-century Northern Qi (Freer Gallery of Art).

The early portrayals of the Pure Land have consisted since the fifth century of two elevated towers or halls on both sides and the Buddha figure flanked by attendants. The pair of towers enhance the centering of the Buddha figure flanked by bodhisattvas, together with many attendants, musicians, apsaras, and reborn beings. In particular, the two bas-reliefs of the Xiangtangshan Cave illustrated paired timber buildings standing on the flat substructure under bracketing sets above columns, with an elevated floor on pile constructions (Figures 5 and 6).

Dunhuang Cave 393 during the Sui period (581–618) presented different patterns. The Buddha figure is naturally located at the center of the painting's composition without any buildings, but a lotus pool is found beneath it. There are beings at the moment of rebirth in the pool, attendant figures, and flying apsaras on both sides (Figure 7). Pure Land's transformation tableaux started to subsume buildings in these illustrations from the Tang period onward, and the number of buildings gradually increased with the appearance of different types. Remarkable development appeared in the following works: Dunhuang Caves 220, 71, 321, 322, 329, 334, 335, 217, and Maijishan Cave 5, all constructed during the early Tang period (618–712).[21]

Figure 7. West Pure Land, Dunhuang Cave 393, Sui.

Among these, the early Tang cave 220 describes a simple great platform with balustrades made of seven jewels, and a lotus pool in which many reborn beings presided. The flat platform where the Amitabha triads stand appears to be floating on the water's surface because the mural fills the wall of the cave. The paired two-story pavilions flank a central hall on both sides, along with covered galleries. Amitabha triads in front of the central hall were placed on a level platform, accompanied by many bodhisattvas and attendants. From Dunhuang Cave 220 onward, the tableaux tended to grow more sophisticated, and the number of buildings gradually increased with the emergence of different types (Figure 8).

Figure 8. Western Pure Land, South Wall, Dunhuang Cave 220, Early Tang (Xiao 1989).

The West Pure Land of the Amitayurdhyana sutra on the north wall of Cave 217 is representative of all transitions during the high-Tang period. All figures in the composition, including architectural elements, are confined to the front of the rear edge of the lotus pool. Many buildings of varying forms are placed on level ground to the rear of the jeweled pool in front of which the Buddha and attendants are placed. The buildings are divided into two types: individual buildings, including a central front hall and four pairs of tall towers; and a rear building complex consisting of a central hall of three bays square with a hip-gable roof, flanked by twin-wing roofed corridors on both sides (Figure 9).

Figure 9. West Pure Land, northern wall, Dunhuang Cave 217, High Tang (Xiao 1989).

The high-rise pavilions called the *"louge* 樓閣*"* were built on the left and right. The paired louge buildings are attached to covered galleries to connect the main hall and have different base storeys. The first is made of decorated bricks that prop up a level masonry platform with balconies, and the other is made of timber columns that support level platforms with banisters surmounted by bracket systems. The "Shigong 釋宮" of the

Erya 爾雅 defines the former as the lou building, narrow and curvedly decorated (Xu 1987, p. 175). The *Shuowen jiezi* 説文解字 defines the lou building as one with multiple storeys, while the ge building has a substructure system (pingzuo) that holds up the level platform on which housing complexes are placed. The two lou buildings have a pole resembling the chattrayashti or chattravali 相輪 (a vertical shaft that protrudes from the top of a pagoda) on top of each pyramidal roof, while the two ge buildings have a pair of chiwei 鴟尾 (owl headed fishtail) on each hip-gable roof with a double row of rafters, which functions as a bell tower with window panel that suggests a belfry and a sutra repository.

Significant changes appeared in the high-Tang cave 172 (Figure 10). To begin with, the central three buildings were grouped on the same platform, further emphasizing them as the center of the architectural composition. The central building at the very front is a five-bay square with a hip-gable roof, and the central bay of the building is much wider than the other bays. The number of buildings along the central axis multiplies, and the figure of the Buddha simultaneously diminishes in scale. This indicates the growing prominence of architectural depiction from the high Tang period onward.

Figure 10. Dunhuang Cave 172: Amitayurdhyana Sutra, south wall, High-Tang (Xiao 1989).

Monastic buildings in the upper half of the illustration correspond closely with the scattered platforms floating on the surface of the lotus pool in the lower half of the mural. From the central line between great platforms with a pond and monastic buildings, and the lotus dais on which the Buddha is placed in front of the main hall with five bays, a one-point perspective was employed. This sense of perspective enables the participants to concentrate on the cultivation of visualization meditation. The cults enable the audience to improve their sense of awe toward the Pure Land. The application of the elevated vantage point in the illustration indicates that the Pure Land was far away from the secular world. Moreover, the well-organized spatial depth in the central group of buildings embraces the votaries in the representation of the paradisiac atmosphere with familiar earthly architectural elements. Thus, this visualization could allow the audience to participate in the bliss of Pure Land in the present world and be quite certain of the possibility of rebirth in the future (Ho 1995, p. 31; Wu 1992a, p. 52).

The building placement in Caves 220, 217, and 172 resembles the main realm of Bulguksa built above the multi-story platforms, as well as the Hoodo of Byodoin sponsored by the Fujiwara family. It is remarkable that high-rise pavilions stand on flat ground surmounted by wooden supports or stone elevated platforms placed along the right and left sides in the transformation tableaux. The elevated towers, flanking the main realm in the center, serve as dwellings for transcendent beings and immortals. Subsidiary buildings on the right and left of the primary buildings in the middle are linked with each other through roofed passageways.

6. Construction of Pure Land Architecture: Bulguksa Monastery and Its Rituals vs. Hojoji and Byodoin Monasteries

The emphasis on open flat platforms and the central hall in the illustration demonstrates a liturgical space representing an increasing share for the entire pure land cult. Therefore, Buddhas, bodhisattvas, heavenly beings, voice-hearers, celestial musicians, and sentient beings not only gathered to dance or play music, but all beings gathered to hear and worship the preaching Buddha. The buildings are connected by stairs, flat bridges, and arched bridges. The pathways between platforms and halls denote the paths followed by the rituals. The flat stone platform played an important role in intimately linking ceremonial buildings and places.

The representation of the Pure Land in the transformation tableaux reflected not an idealized world, but a real one, which appears in pure land monastic complexes, such as Bulguksa Hojoji and Byodoin Monasteries. Of these, Bulguksa Temple provides a prototype of Pure Land architecture because the temple resembles the Pure Land composition in the transformation tableaux. The temple has a main hall with five bays, two elevated towers flanking the main hall on both sides along roofed corridors, western and eastern Buddha pagodas, a lecture hall behind the main hall, a worship stone, and a stone lamp in the middle between the middle gate and main hall, a flight of stairs, arched stone bridges with balustrades, a middle gate, and multi-storyed elevated platforms with banisters, and a lotus pool. The main stone platforms consisted of Baekungyo (blue cloud) and Cheongungyo (white cloud) bridges, which form a long stairway before reaching the Jahamun Gate. The pure-land constructions mounted on the grand platform represent the narratives of the *Amitayurdhyana* and the *Larger Sukhavati vyuha sutras*,[22] in which the main realm is connected with roofed corridors via multi-story platforms (Figure 11).

Figure 11. The upper-lower grand terraced platforms with a central stairway at Bulguksa (**Right**, photo by author) and multi-terraced wooden platform at Dunhuang Cave 148, the Western Amitabha Paradise, Mid Tang (**Left**, source: Xiao Mo).

Accordingly, the multistory stone platforms of Bulguksa Temple are noteworthy in these respects. The level platforms appear in front of the second flight of stairs where devotees climb a flight of stairs at the first arched bridge. The first-level platform is wider than the others, which have enough room for performance and playing musical instruments. An arched gate was naturally formed underneath the second flight of stairs and second arched bridge. The grand platform was constructed by combining the pile-up structure with the framed one. Capital blocks that are almost square-shaped do not resemble the kind of capital block that supports columns. The outlines show that hidden transverse beams and penetrating tie beams exist behind the walls. The stone platform shapes a framework joined by mortises and tenons. It is derived from a Chuandou framework composed of columns and tie beams. Thus, the stone framework results in a pile construction (Figure 12).

Daeungjeon and Geukrakjeon Halls were reconstructed in the 18th century, but their foundation and stairways were built in the first foundation during the eighth century. The buildings were revived in an old location. The distance from the vicinity of the worship stone to the foundation of Daeungjeon Hall is 7.07 m. The distance from the principal image in the Hall to the end of the foundation is 8.45 m. When the two distances are combined, the entire distance from the principal image to the worship stone is 15.52 m (Figure 13).

Figure 12. Terraced platforms at Bulguksa Temple and wooden platform with pile construction at Dunhuang Cave 172, the Western Amitabha Paradise, Mid Tang (source: Xiao Mo).

The distance between a worshiper standing in front of the worship stone and the principal image is approximately 15.5 m (Mitsumori 1999). The current appearance shows that the stone lamp obstructs the devotees' field of vision toward the principal image because it is located in front of the worship stone. This implies that the method of present-day worship is different from that of the Silla era. Unfortunately, although there is no information about daily rituals in the courtyard during the Silla period, other countries have records of their daily activities during the same period. For example, the Nanhaiji guinei fachuan 南海寄歸內法傳 by the monk Yijing (635–713) records, "One sitting down, one's feet touch the ground," and "devotees walk three times around a pagoda, offering incense and flowers. They all kneel down."(Yijing and Takakusu 1970, pp. 123–24). Thus, the rituals were not intended to allow devotees to see the principal image, rather these devotees prayed to images because the process of worship was more significant in the sangha community of the time.

In the case of Daeungjeon Hall, the distance between a devotee and a principal image is approximately 15.5 m, and that between the base stones of the middle gate and the worship stone is 12.7 m. In the case of rain and snow, if devotees worship the principal image from the middle gate, the distance between the principal image and the devotee standing in the middle gate and corridor would be more than 30 m, which is the proper distance to provide a perspective view for extreme concentration and one-way vocal communication during the ritual performance (Hall 1966, pp. 42–43) (Figure 13).

In special cases, it should be considered that worship existed from the middle gate and corridor (including intermediate corridors to connect the main hall and the east-west gates or corridors on both sides), although there was a long distance from the principal image and devotees. In contrast, it is possible to cultivate stupa worship from the corridor and the middle gate. To understand this relationship, a few measurements were performed. In particular, the ratio of the foundation's height of the main hall to that of the two pagodas has a significant meaning. The eastern pagoda's foundation is 1.97 m high, while the main hall's foundation is 1.66 m high. Their relationship offers an equivalent height hierarchy. When the image's dais is more than approximately 1.6 m high, the total height from the image dais to the ground level of the main hall is approximately 3.13 meters, which is comparable to those of a lion statue sitting on the foundation of the eastern pagoda and the first story of the western pagoda, 3.28 m and 3.43 m each (Table 1).

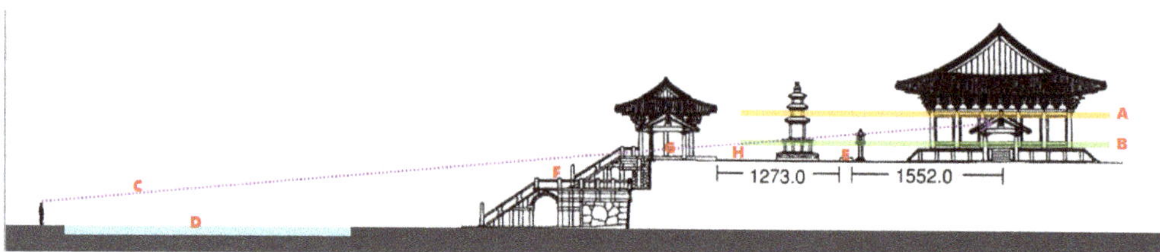

Figure 13. Ritual Concept at Bulguksa Temple (Kim 2011). A: Height of Buddha Image; Location of Sarira Casket; B: Height of the West Pagoda's Foundation; Height of the Main Hall's Foundation; C: Human Angle (10 to 15 degrees) from the Lotus Pool (Nine Ranks for Rebirth); D: Lotus Pool (Pond); E: Worship Stone (Bongnodae); F: Flat Stone Platforms; G: Gate (Jahamun); H: Flat Platform with Enclosure Pure Land.

The western pagoda's foundation is 2.44 m tall, and the entire height from the image dais to the ground level of the main hall is approximately 3 m. The location of the inner chamber considered is 5.11 m high. The Buddha image enshrined in the Dauengjeon Hall sits about 1.66 m high, and the total height from the image to the ground level of the main hall is 4.79 m. In comparison with the location of the inner chamber in the western pagoda standing at 5.11 m, the hierarchical relationship of the buildings is almost in an equivalent position with an error range of 0.31 m. Meanwhile, in the eastern pagoda, the top side of the substructure supporting the pavilion is 4.59 m high. Moreover, the hierarchical link between the two buildings is almost in a corresponding position with an error range of 0.20 m (Figure 13, Table 1). The equivalent relationship is applied to other sites in the same period as the monastic buildings at Bulguksa Temple: such a hierarchical approach has something in common with the decisions regarding the building layout of Hwangny-ongsa, Sacheonwangsa, Gameunsa, Geodonsa, and Mangdeoksa Buddhist Monasteries. In particular, in Hwangnyongsa Monastery, the image pedestal height corresponds to that of the foundation platform for a wooden pagoda. In the Geodonsa Temple site, the three-story stone pagoda with a three-stepped foundation platform is comparable to an image pedestal in the main hall. The height of the image pedestal was identified with that of a three-stepped foundation platform (Han 2003) (Table 1).

The equivalent hierarchy between the two pagodas and the Daeungjeon Hall in Bulguksa Temple, and other Buddhist sites in the same period or afterward, demonstrates the important notion that devotees regarded the pagodas and the main hall as a total frame. They understood them in a narrative frame that constructs the Buddha's life (Kim 2011).

The miniature shrine (Geumdang) stored in the West Pagoda (Seokgatap) chamber is hierarchically identified with the Buddha image of the main hall, and the horizontal level between the Buddha sculpture and miniature golden hall inside the pagoda needs to be considered to understand the relationship between the two pagodas and the main hall. The Bulguksa seoseoktapjungsu hyeongjigi noted that the western Buddha pagoda stores a golden hall (sarira casket), a gilt-bronze lotus throne, 47 sariras, flavorings, 39 beads, and 15 miniature votive stupas in a heavenly chamber built inside the pagoda's body at the second story in the period of its foundation. Even though the report was written in the early 11th century, the name "Geumdang" represents the symbolic meaning of Buddha's stupas and sarira. The Geumdang Hall's container for placement of the Buddha's sarira indicates that stupas were considered places where the Buddha resides, and sariras were regarded as the living presence of the Buddha.

There are no records concerning Pure Land rituals in Bulguksa Monastery. However, the records of Pure Land's cults in the ancient Japanese monasteries of Hojoji and Byodoin are comparable to those of Bulguksa in architectural portrayals.

Table 1. Comparison in Height between the Pagoda's Foundation and the Main Hall's Foundation (Measurement unit: meters).

Height (H)	Height of the Pagoda's Foundation: A	Height of the Main Hall's Foundation: B	Height Difference between A and B	
Bunhwangsa 634 CE	1.50	unknown	unknown	
Hwangnyongsa 645 CE	1.60	1.10	0.50 (A/B = 1.45)	Buddha dais: 0.5.
Sacheongwangsa 679 CE	1.50	1.15	0.35 (A/B = 1.30)	
Gameunsa 682 CE	2.58	1.90	0.70 (A/B = 1.35)	Height of underground channels: 0.60
Mangdeoksa 684 CE (or 679 CE)	1.26	0.95	0.31 (A/B = 1.32)	
Bulguksa East Pagoda (EP) 742 CE	1.97	1.53	0.44 (A/B = 1.29)	Height(E) to the platform above column network from the ground level: 4.59
Bulguksa West Pagoda (WP) 742 CE	2.44	Image (H): 1.66 Dais (H): 1.60 Total height: 4.79	0.91 (A/B = 1.59)	Height(W) to the heavenly chamber from the ground level: 5.11

6.1. Monastery Hojoji and Its Buddhist Rituals

The Hojoji Monastery法成寺was developed in the process of converting Fujiwara Michinaga's 藤原道長 (966–1028) home into a Pure Land monastery on the western bank of the Kamo River in Kyoto (Sugiyama 1981, pp. 87–93; Ota 2010, pp. 149–77, 193–203) (Figure 14). The Eiga Monogatari recorded there describes an entire plan for establishing the monastic complex of Hojoji by superimposing mountains, digging a pond, and planting trees in a row, and for the form of the main buildings, sanctums, and the manner of ornamentation of a golden hall. The Amitabha hall has a tile-roof facing east, which is flanked by subsidiary buildings, such as pavilions, attached to long corridors on both sides. All doors were covered in golden leaves. The Yakushi Hall was built on the western side of the link with the Amitabha hall, in terms of the form and ornament of the building; i.e., the hall was identified with the appearance of the Amitabha hall (Matsumura and Yutaka 1965, pp. 152–54; Shimizu 1986a, 1986b).

According to Eiga Monogatari, a lotus pond in front of the Amitabha hall was fully completed on 22 March 1020. Laypeople enjoyed boating in the pond during the ritual, while carrying a carriage. All buildings (east, west, north, and south), including a belfry and sutra tower, forming shadows on the pond. The landscape provided the appearance of 3000 worlds (Matsumura and Yutaka 1965, pp. 445–46; Minamoto 1965). All halls were designed with gold leaf ornaments that cast shadows on the pond. This implies that all building locations were close to the lotus pond. The layout of Hojoji was intended to represent the 3000 worlds in mutual associations with all the shrine's halls and ponds. To ensure the visual effect of the pond's siting, the landscape around the pond was considered with several points in mind: sand flashed like a crystal, water from the pond became clear and pure, and a Buddhist image floated in harmony with the multicolored lotus flower. Buddha halls in the north, south, east, and west, a sutra tower, and a belfry tower were projected onto the pond. Treasure trees were used to hang the multicolored fishing nets along the pond. A bridge was decorated with seven treasures and ships were embellished with various treasures. There were imitations of a peacock and a pelican parrot on the central island. In addition, during a 10,000-lamp festival that took place around the pond on 10 March 1023, treasure trees were erected, and artificial multicolored lotus flowers and Buddha images were floated in the pond (Matsumura and Yutaka 1965, pp. 108–10; Shimizu 1986a). During the performance of the ritual, the combination of the buildings, the pond with shadows cast in the daylight, and the lights at night provided an important visual effect and presented an unusual scene, different from the mundane world. The landscape produced an impressive state, like a pure land.

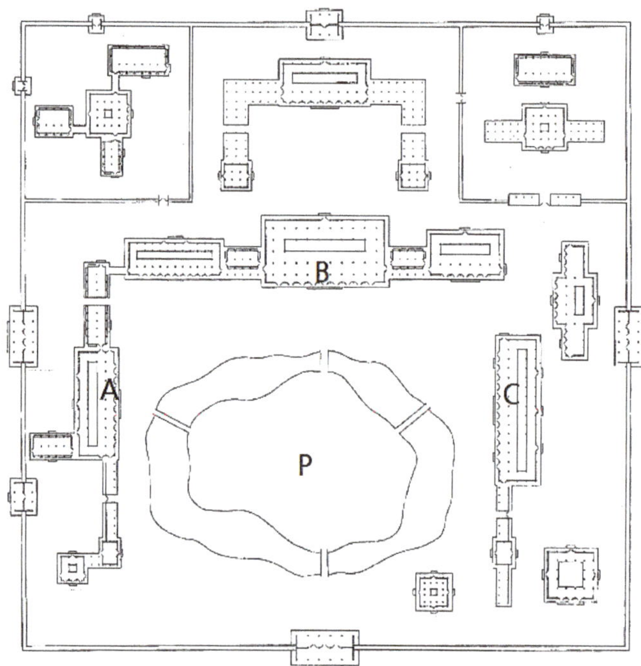

Figure 14. The first construction of the Monastery Hojoji. P: A platform for rituals on an island, a lotus pond, C: Yakushi Hall, A: Amitabha Hall, B: Main Hall.

6.2. Byodoin Monastery and its Buddhist Rituals

The Phoenix Hall, Hoodo, and Amidado are situated on the western bank of the Uji River in the scenic town of Uji, not far from the ancient Heian capital, Kyoto. The Phoenix Hall at Byodoin 平等院 consists of a central hall to enshrine an image of Amitabha, with intermediate corridors to its left and right, and a rear corridor extending behind it. The rear corridor is seven bays long by one bay wide. The central hall does not have narrow aisles around the core sanctum of Hoodo, and a roof enclosure is directly added to the main building. The hall was built on the central island (Nakashima) at Byodoin (Tatara et al. 1998).

The Byodoin precinct is situated along the Uji River (Figure 15) and has been regarded as the Pure Land. On 4 March 1053, an offering service for the Phoenix Hall was held according to the Fuso Ryakki 扶桑略記. In the record at the time of this memorial service, "Yorimichi built Byodo-in. A tall Buddhist statue 4.6 m high was enshrined therein." (Koen [1169] 1965; Ota 1988, pp. 94–112). Concerning the performance of the offering, there are no records about the location where visitors were seated and how long dances and music were performed during the cult—although information about the dance and music was reported in the Bugaku yoroku 舞楽要録 (Hanawa 1951). However, during a memorial service for a new pagoda on 25 October 1061, Sadaie asonki 定家朝臣記, another piece of text stated that there was an inner altar on a long couch that was extremely splendid and utterly beautiful. There were 20 seats for the devotees. Under the floors of the building, there were other seats for consecration ceremonies installed on the southeast, northeast, northwest, and southwest corners of the new pagoda. Along narrow aisles and corridors, there were seats for the visitors. The use of the aisles and corridors was akin to the example of the Hojoji Monastery (Kurokawa and Tsunoda 1969; Ota 1988, pp. 94–112).

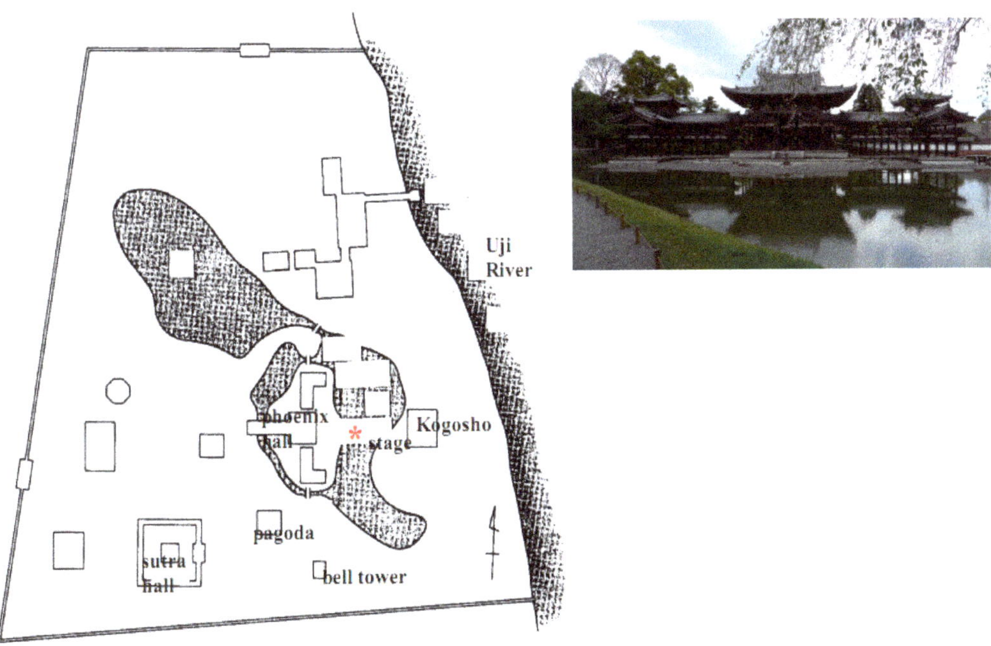

Figure 15. Byodo-in, the Phoenix Hall and Little Imperial Palace (Kogosho).

On 25 October 1061, the garden site and Amitabha hall were described as unusual and outstanding in the Fuso Ryakki, and a sukhavati was also represented. The text added that the Phoenix Hall was "conducive to meditative reflection on sukhavati" Moreover, the Uji River is considered part of the long river previously covered with reeds which leads groups to a world representing the highest complete understanding of truth, currently 'gone to the beyond' (Koen [1169] 1965). They speak to the psychological condition of the visitors who reached the Byodoin Monastery, utilizing the Uji River from the Heian capital.

Chuyuki, on 21 September in 1118, explained 10 kinds of memorial services performed by Fujiwara Yorimichi, reporting "lotus, waterfowls, trees, sand-made swans, and so on were made and then installed to float on the pond, and the hill of the water no longer had any space. The Kogosho 小御所 (Little Imperial Palace) in the eastern side was for an empress dowager, and the Saiin 斎院 was an imperial palace for a daughter of empress dowager. Some women served the imperial palace. A temporary house with four bays stood for high-ranking officials on the western side, and the one facing the eastern aisle was used for visitors at Byodoin Monastery, and tent structures were established in the eastern garden."(Fujiwara 1965). The text states that Kogosho was located on the eastern bank of the lake across the Amitabha hall. The palace was situated at a vantage point so that the rituals could be observed. It was also the place where Yorimichi and his heirs gathered to observe the garden. The text also implies that there was a stage in the pond in front of the

Amitabha hall, and the entire territory around the pond was filled with imitations, such as flowers and waterfowls. In addition, boats might float on the pond to create a stage (Motonaka 1994, p. 250).

6.3. Re-Interpreting Buddhist Rituals at Bulguksa Temple

As seen in the votive events at Hojoji and Byodoin Monasteries, Bulguksa Monastery created similar spectacles during the Pure Land cults along with the grand terraced platforms and around the Yeonji Lotus Pond on the ground. The Jahamun (golden purple) and Anyangmun (peace enhancing) gates, stairways with banisters, and four bridges are situated on the terraced stone platforms at Bulguksa. The main platform with balustrades is located in front of the Lotus Pond measuring 39.5 m from east to west and 25.5 m from north to south. The second-storyed platform with balustrades has a stage on which dances and music were played during a ritual (Munhwagongbobu 1976, pp. 63, 69). Additionally, a small arched gate on the second-storied platform might be used for circumambulation rituals. The devotees marched in procession toward two Buddha pagodas through two worship stones and two stone lanterns, and the Buddha Hall at the end. The processions might also appear along a narrow path with balustrades installed on the open terraced platform, via a square-shaped stage on the second platform and cloud-shaped pillars of Beomyeongnu (Figures 16 and 17).

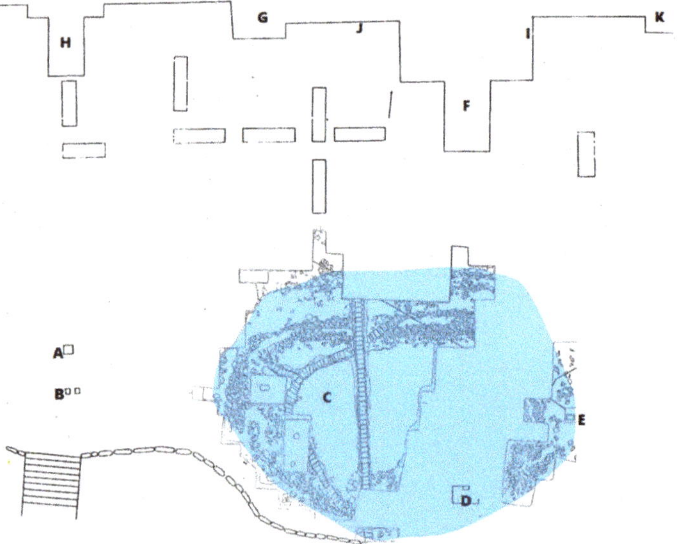

Figure 16. The Archeological Findings of Lotus Pool and Surrounding Building Configuration (Kim 2011). A: flagpole support, B: Sarira Stupa, C: Lotus Pool, D: Previous Main Gate, E: Bulguksa Stele, F: Cheongungyo and Baekungyo Bridges, G: Beomyeongnu (Ugyeongnu) Pavilion, H: Yeonhwagyo and Chilbogyo Bridges, I: Multi-storied Grand Stone Platforms, J: Pathways of Grand Stone Platforms, K: Jwagyeongnu Pavilion.

Figure 17. The territory of Lotus Pool represented by researchers in front of Baekungyo Bridge (Gyoeonghyang Sinmun, 8 November 1994).

Likewise, donors who stood in open-roofed corridors watched the spectacles that happened in a lotus pool of nine ranks for posthumous rebirth. This indicates that the symbolic organization of the nine levels or three types of rebirth was easily accepted by ruling clans to common persons in Silla society. The open roofed corridors are located in two places: on platforms with balustrades to an east-west axis, and in the inner courtyard. The roofed corridors open without any wooden-framed walls in the two photos of the Sekino Tadashi archive, although they are closed between the main courtyards and stone platforms at present (Tokyoteikokudaigaku Kokadaigaku 1904; Gukgagirokwon 1918–1924) (Figures 18 and 19). In addition, the remains of several buildings were found on the western side under the grand-terraced platforms (Figure 2). The buildings might be associated with the lotus pond on the east-west axis, as Kogosho was located at a vantage point to watch the rituals on the east-west axis with the Amitabha hall.

Figure 18. Sekino's photo on the Bulguksa Monastery's Front Buildings (Tokyoteikokudaigaku Kokadaigaku 1904).

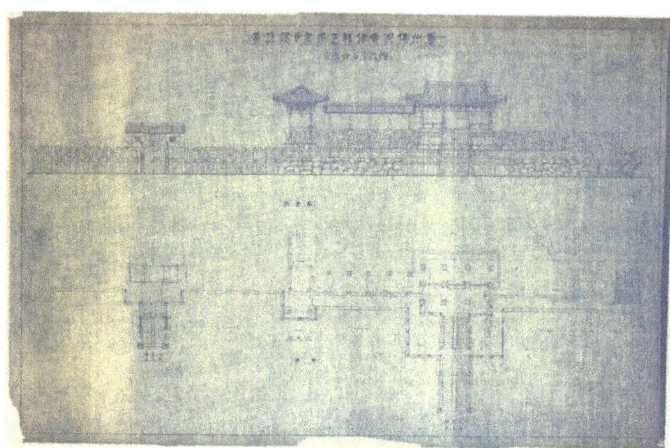

Figure 19. Restoration Plan (a line drawing between 1918 and 1924) on the basis of Sekino's photo (Gukgagirokwon).

In the past, the pond at Bulguksa might be filled with numerous imitations of lotuses, waterfowls, trees, and swans made of paper or sand during the festival, compared to the appearance of the lotus pool in the transformation tableaux of monastery caves. In particular, the Pure Land murals at Dunhuang Caves 217 and 172 in High-Tang consist of three buildings on a level platform at the center of the painting (Figures 9 and 10). The front central hall has five bays, like the Daeungjeon Hall at Bulguksa Monastery, and the wider central bay of the front building resembles the central bay of Daeungjeon Hall. The mural composition focuses on the lotus pool in front of the main hall using a one-point perspective method. This sense of perspective occurs through the difference in the distance between the lotus pool and buildings. The former is located in front of the devotees, while the latter is located far away from them.

Similarly, the application of the elevated vantage point appears in Bulguksa architecture. An impressive perspective appears in the combination of the lotus pond and elevated grand platform with main buildings situated at a very long distance. The elevated platforms maximize a sense of perspective and emphasize the Pure Land surrounded by splendid buildings spread over a long distance. When Pure Land cults take place on the lotus pool or around it, devotees look at the grand-terraced platform and buildings seen through the open-roofed corridors. The positions on which devotees stand around the lotus pool make it possible for them to feel a sense of awe toward the Pure Land. Such application of the elevated vantage point indicates that the Pure Land is far away from the secular world. The intended spatial depth in the central group of buildings on the elevated platforms is the result of embracing the devotees in the paradisiac atmosphere. Such an atmosphere makes the devotees feel the bliss of the Pure Land (Figures 13 and 16).

7. Conclusions

The early concepts of the Pure Land as "sukhavati" in the Chinese translation, such as the "anle" and "jile," which signify "extreme happiness," become more concrete and powerful through images for visualization and meditation rituals, in the mixture of a supernatural transcendent being in Taoism with a chakravartin king in Buddhism.

Furthermore, the Pure Lands as the abodes of miscellaneous deities correspond to the lands of Taoist hermits in their quest for immortality, and the cosmic order that symbolizes the authority of rulers. Images of paradise appear in the buildings (e.g., platforms, pile-built constructions called "galan," high-rise pavilions, pagodas), landscape (e.g., mountains, clouds, ponds), and simultaneously through the architectural representation of Buddhist temples.

Together with these ceremonial constructions, the rituals and samadhi practice towards the Pure Land served as powerful tools to dignify and strengthen the Buddha Land's position. Cults and practices as tools controlled the behavior of laypeople and monks. They satisfied their psychological and emotional needs in the law of dependent origination (Pratityasamutpada) between buildings and devotees.

The Monastery Bulguksa is a ritual apparatus. The architectural representation of a Pure Land that subsumes lotus ponds, courtyards, and open terraced platforms with banisters as stages for dancers and musicians is further emphasized. It provides an application of the elevated vantage point from a perspective and focuses on ritualized practices associated with meditation and votive events linked to pure land ceremonies.

As a result, based on Buddhist narratives, such framed story structures of the Pure Land continue to produce an integrated and pure world far away from the secular one, and toward the sacred places of key events in the Buddha's life.

Funding: This research was supported by Basic Science Research Program through the National Research Foundation of Korea (NRF) funded by the Ministry of Education (No.2021R1I1A3059811).

Conflicts of Interest: The authors declare no conflict of interest.

Notes

1. Four significant scriptures related with the Pure land tradition follow the Wuliang qingjing pingdengjue jing (Taisho12, no 361), the Foshuo amituo sanye sanfosalou guodu rendao jing (Taisho12, no 362), the Larger Sukhavati vyuha sutra (Taisho 12, no 360), and the Amitayurdhyana sutra (Taisho12, no 365). These Pure Land-related scriptures were translated from the Sanskrit version into the classical Chinese version, which influenced the establishment of Pure Land architecture in East Asia. These documents will be dealt with in detail in Section 5.1.

2. The Pure Land is interpreted as follows: Skt. Buddhaksetra; Ch. shatu, fotu, foguotu, foshentu; Kor. ch'altu, bulto, bulgukto, pulsint'o. The Pure Land is also translated into "jingtu" and "jile" for Sinification. It implies that the real world had been part of the existence of the living Buddha specifically in terms of sacred sites associated with key events in the Buddha's life story; these places became important for traveling to get the merit (Bharati 1963, pp. 135–36).

3. The author referred to Tsukamoto's two books. Tsukamoto re-organized the afore-mentioned inscriptions of other sites in India, including those of Nagarjunakonda sites (site 5 and site 43), and then compiled all Indian Buddhist inscriptions which were translated by Vogel, Sircar, Sarkar, Narasimha, Rama, Shizutani, and Sadakata. The first version of Nagarjunakonda inscriptions translated to English can be identified in Vogel's works. (Vogel 1929–1930; Sircar 1963–1964; Tsukamoto 1986, Tsukamoto 1996–1998; Rama 1967; Shizutani 1979; Sadakata 1994).

4. In particular, the well preserved text came from ayaka-pillar B5 (one of five pillars usually erected on the four cardinal directions), namely the fifth pillar on the south side of the great Buddhist Stupa at Site 1, Nagarjunakonda. This noted the gift of a stone pillar by the Mahadevi(Queen) Rudradharabhatarika, King Siri-Virapurisadata's daughter from Ujjeni (Skt. Ujjayini), while the Mahachetiya (great stupa) was raised by the ladies, the Mahatalavaris, Chamtisitinika of (the family of) the Pukiyas. (Vogel 1929–1930, p. 19).

5. Nagarjuna wrote, "through proper honoring of stupas, you will become a Universal Monarch. Your glorious hands and feet marked with (a design of) wheels. Through the practices there are fame and happiness here, there is no fear now or at the point of death, in the next life happiness flourishes, therefore always observe the practices." (Nagarjuna, Taisho 32 no 1656) Thus, we should note that the welfare and happiness of the entire world mean both the present world and posthumous future world as labeled on the inscriptions. Devotees at Nagarjunakonda and other areas have dreamt of the representation of the Pure Land or heavenly world in the real world from early Buddhist time. These epigraphical and literary proofs indicate that in ages past the monuments were crucial instruments for obtaining merit-transferring and merit-making devotees, and ultimately, for fulfilling Pure Land architecture.

6. The Amitabha not only means "measureless light," but also "measureless wisdom." In the Chinese translation, the *jingtu, jile,* and *qingjing* have a similar context. They refer to a pure, a no-ado, and an *anle* pleasure place, or an ideal nirvana.

7. The *"qingjing"* is similar in meaning to the *"anle"* and *"jingtu,"* which are frequently used in the *Wuliangshoujing lun* and Tan Luan's *Wuliangshoujing lunzhu.* The term *"jingtu"* representing Amitabha's religion is derived from the name of the Buddha in the *Pingdeng juejing* and *Wuliang qingjing.*

8. It was expressed in the text of the *Ratnavali* and the inscriptions of Nagarjunakonda (Walser 2005).

9. Afterwards, the ritual method was consistently followed by descendent monks. Xuanzang, a Tang monk, mediated upon the Bodhisattva Maitreya. He turned all his thoughts to the Heaven of the blessed one, praying ardently that he might be reborn there to pay homage to the Bodhisattva. His boat was once attacked by pirates who attempted to kill him as a sacrificial offering to the ferocious Sivaite goddess Durga, while a Silla monk, Wonhyo, proposed "there are three grades for people to cultivate the

visualization, and the highest grades of people are those who either cultivate the Samadhi of Buddha visualization or repentance as their method of practice. In their present body, they will succeed in seeing Maitreya. According to the quality of their mind, the image they see will either be great or small." (Sponberg and Hardacre 1988, pp. 94–95; Kitagawa 1988).

[10] Lokasema's translated texts in the second-century state the Buddha of the Ten Directions, "if one's heart is focused on Amitabha one will be reborn in *sukhavati*, the Western Pure Land presided over by Amitabha." Additionally, Huiyuan established a Buddhist center at Lushan, Jiangxi, and another at Xiangyang, Hubei. He described a miraculous sculpture at Xiangyang and a painting of the "shadow" or "reflection" of the Buddha (*foyingxiang*) at Lushan. The statue and picture might be related to meditation or visualization practices in the locales.

[11] When Tanjie fell seriously ill, Daoan chanted continuously, the name of Maitreya Buddha never leaving his lips. Zhisheng (Taoan's disciple), who waited on him in his illness, asked him why he did not want to be reborn in the Heaven of Peaceful Response (i.e., Amitabha's paradise, sukhavati). Tanjie replied, "Together with the Reverend (Daoan) and eight others, I have vowed to be reborn in Tusita (i.e., Maitreya's paradise). The Reverend, Daoan, and the others have already been born there, but I have not. That is why I have this wish" (Kieschnick 1997, p. 5).

[12] "*Dae Hwaeom Bulguksa birojanabulmun munsu bohyeon bosal chanbyeongseo* is listed in the *Bulguksa gogum changgi*佛國寺古今歷代記.

[13] The conquests of Asoka were realized through the imperial ideas of India. Through his strong espousal of the *sangha* communities, as well as the construction of 84,000 stupas all over the *Jambudvipa*, he was both a transformative body of the Buddha and a cakravartin (Strong 1989, p. 117). The cakravartin legends became an archetypal example of the Buddhist kingship, and were spread among several hagiographical works.

[14] A similar record that the ruling class built a monument to protect his subjects appeared in an inscription that was manufactured for the repair works of the nine-story pagoda of Hwangnyongsa in the Silla period. The inscription writes, "Hitherto (The construction of the pagoda) has led to the peaceful and happy life of sovereign and subject 君臣安樂至今賴之." As in the *Hwangnyongsagucheungmoktap chaljubongi* 皇龍寺九層木塔刹柱本記. in the 11th year (871) of Silla King Gyeongmun, the repair work began on a nine-story wooden pagoda at Hwangryongsa Monastery. In this process, Park Geo-Mul recorded the construction of the wooden Buddhist pagoda and repair process from 871 (the 11th regnal year of King Gyeongdeok) to 872 CE.

[15] Korean Buddhism did not show political upheavals and heavily depended on the personal preferences of rulers, while at least four anti-Buddhist campaigns occurred in China in 446 (Northern Wei), 557 (Northern Zhou), 845 (Tang), and 955 (Later Zhou) in the contemporaneous period.

[16] See the *Gamsansa Seokjomireukbosaripsang Josanggi* 甘山寺石造彌勒菩薩立像造像記. The inscription consists of 21 rows and 391 characters, which have been left on these images. The Taoist idea shows a teaching about the various disciplines for achieving "perfection" by becoming one with the unplanned rhythms of the universe called "the way" or "Tao," in association with the faith of ancient East Asians who desired to become hermits in mountains or tall buildings after death (Gamsansa Seokjomireukbosaripsang Josanggi 720).

[17] Recently, some scholars have argued that the completion of the main territory of a Buddha hall and Buddha pagodas had already been done before 742 CE, as determined by epigraphical evidence such as *Bulguksa mugujeonggwangtap jungsugi* 佛國寺無垢淨光塔重修記 (Bulguksa Mugujeonggwangtopjoongsugi 1024) and *Bulguksa seoseoktap jungsu hyeongjigi* 佛國寺西石塔重修形止記 (Bulguksa Seoseoktapjungsuhyeongjigi 1038).

[18] Kim Daeseong and his father, Kim Munryang, of *Samgukyusai*, and Kim Daejeong and Kim Munryang of *Samguksagi* are the same person. Kim Munryang served as Prime Minister from 706–711 under King Seongdeok, and Kim Dae jeong served as King Gyeongdeok from 745–750.

[19] *Bulguksa mugujeonggwangtap jungsugi*, *Bulguksa seoseok-tap jungsu hyeongjigi*, and *Bulguksa jungsu bosimyeong gongjungsomyoeonggi* show that the original names of the pagodas were not Tathagata Prabutaratna and Tathagata Sakyamuni Pagodas at the time of the first construction. (Cheon 1996).

[20] The *larger Sukhavativyuha sutra* (The Sutra on the Buddha of Eternal Life), translated from the Sanskrit by F. Max Mueller, edited by Richard St. Clair. Available on: https://www.nanputuo.com/npten/html/201203/2816073173499.html (accessed on 10 June 2021).

[21] In order to make an argument about historic shifts in style at Dunhuang, I utilize the reference scheme developed by the Dunhuang Research Academy. The number within the parentheses pertains to the number of caves allocated in the following periods: early-Tang 618–704 (40), high-Tang 705–80 (81), mid-Tang 781–847 (46), late-Tang 848–906 (60), Five Dynasties 907–59, Song 960–1035, and Western Xia (Xi xia) 1036–1226 (Xiao 1989, p. 30).

[22] In particular, the lotus pond is a reminder of the "nine levels of rebirth" in the *Amitayurdhyana sutra* or the "three types of persons to be reborn" in the *Larger Sukhavati vyuha sutra*.

References

Primary Sources

Larger Sukhavati Vyuha Sutra 佛説無量壽經. Taisho 12, no. 360.

Wuliang Qingjingpingdengjuejing 無量清淨平等覺經. Taisho 12, no. 361,

Foshuo Amituo Sanye Sanfosalou Guodu Rendao Jing 佛説阿彌陀三耶三佛薩樓佛檀過度人道經. Taisho 12, no. 362.

Amitayurdhyana Sutra 佛説觀無量壽佛經. Taisho 12, no. 365.

Amitayurdhyana Sutra (Smaller Sukhavati vyuha sutra) 佛説阿彌陀經. Taisho 12, no. 366.

Pratyupannasamadhi Sutra Banzhou Sanmei Jing 般舟三昧經. Taisho 13, no. 418

Suvarnaprabhasa sutra 金光明經 (*Golden Light Sutra*). Taisho 16, no. 663.

Guannian Amituofo Xiangbai Sanmei Gonde Famen Jing 觀念阿彌陀佛相海三昧功德法門 (*Meritorious Methods of Contemplating Amitābha Buddha*). Taisho 47, no. 1959.

Saddharmapundarika 妙法蓮華經 (*Lotus Sutra*). Taisho 9 no 262.

Nagarjuna, Paramārtha. *Ratnavali* 寶行王正論 (*Precious Garland*), Taisho 32 no 1656.

Dae Hwaeomjong Bulguksa Birojana Munsu Boheon Bosal Chanbyeongseo 大華嚴宗佛國寺毘盧遮那文殊普賢像讚并序.

(Gamsansa Seokjomireukbosaripsang Josanggi 720) *Gamsansa Seokjomireukbosaripsang Josanggi, 720*, 甘山寺石造彌勒菩薩立像造像記 (*the Dedicatory Inscriptions (Josanggi) on the Amitabha Buddha and the Maitreya Bodhisattva statues of Gamsansa Temple*).

(Ilyeon 1281) Ilyeon, Samguk Yusa. 1281. 三國遺事 (*Memorabilia of the Three Kingdoms*).

(Kim 1145) Kim, Busik. 1145. Samguk Sagi 三國史記 (*History of the Three Kingdoms*).

(Hwaram Doeun 1740) Hwaram Doeun. 1740. Bulguksa Gogeum Changgi 佛國寺古今創記 (*Architectural History of Bulguksa*).

(Beopgyedogi Chongsurok 1254) Beopgyedogi Chongsurok. 1254. 法界圖記叢隨錄 (*Collected Essentials of Records on the Dharma Diagram*).

(Koen [1169] 1965) Koen, Katsumi Kuroita. 1965. *Fuso Ryakki* Teio Hennenki 扶桑略記 帝王編年記. Tokyo: Yoshikawa Kobunkan. First published 1169.

(Matsumura and Yutaka 1965) Matsumura, Hiroji, and Yutaka Yamanaka. 1965. *Eiga Monogatari* 榮花物語. Nippon koten bungaku taikei 76, Tokyo: Iwanami Shoten, a story from 1028 to 1107.

(Fujiwara 1965) Fujiwara, Munetada. 1965. *Chuyuki* 中右記, Zoho Shiryo Taisei Kankokai, Kyoto: Rinsen Shoten, Fujiwara's diary written from 1087 to 1138.

(Bulguksa Jungsubosimyeong Gongjungsomyoeonggi 1038) Bulguksa Jungsubosimyeong Gongjungsomyoeonggi. 1038. 佛國寺塔重修布施名公衆僧小名記 (*Document on the Name of Almsgiving from Laypeople and Monks of Bulguksa*).

(Bulguksa Mugujeonggwangtopjoongsugi 1024) Bulguksa Mugujeonggwangtopjoongsugi. 1024. 佛國寺無垢淨光塔重修記 (*Repair Report of Rasmivimalavisuddhaprabhadharani Cetiya (Endless Untainted Light Pagoda) of Bulguksa Monastery*).

(Bulguksa Seoseoktapjungsuhyeongjigi 1038) Bulguksa Seoseoktapjungsuhyeongjigi. 1038. 佛國寺西石塔重修形止記 (*Repair Report of the West Stone Pagoda of Bulguksa Monastery*).

(Hwanglyongsa Gucheungmogtab Geumdongchaljubongi 871) *Hwanglyongsa Gucheungmogtab Geumdongchaljubongi* 皇龍寺九層木塔利柱本記. 871. (Basic Annals concerning the Pillar of the Nine-story Pagoda of Hwangnyong-sa).

(Gukgagirokwon 1918–1924) Gukgagirokwon 1918–1924, *Restoration Plans of Monastery Bulguksa*, Gukgagirokwon(National Archive of Korea).

Anonymous, n.d. Western Paradise of the Buddha Amitabha(Northern Qi dynasty, 550-577), Freer Gallery of Art, Smithsonian Institution, Washington, DC. Available online: https://asia.si.edu/object/F1921.2/ (accessed on 26 June 2021).

Secondary Sources

Apte, Vaman Shivaram. 1957–1959. *A Revised and Enlarged Edition of Prin. V. S. Apte's The Practical Sanskrit-English Dictionary*. Poona: Prasad Prakashan, vol. 3.

Bharati, Agehananda. 1963. Pilgrimage in the Indian Tradition. *History of Religions* 3: 135–67. [CrossRef]

Chappell, David W. 1977. Chinese Buddhist Interpretations of the Pure Lands. In *Buddhist and Taoist Studies I*. Edited by Michael Saso and David Chappell. Honolulu: University Press of Hawaii.

Cheon, Haebong. 1996. Mugujeonggwangdaranigyeong 無垢淨光大陀羅尼經, Hanguk Minjok Munhwa Daebaekgwasajeon. Available online: http://encykorea.aks.ac.kr/Contents/Item/E0018943 (accessed on 20 August 2021).

Corless, Roger J. 1982. *T'an-Luan: Taoist Sage and Buddhist Bodhisattva. Buddhist and Taoist Practice in Medieval Chinese Society*. Edited by David W. Chappell. Buddhist and Taoist Studies 2. Honolulu: University of Hawaii Press, pp. 36–45.

Fujita, Kotatsu. 1970. *Genshi Jodo Shiso No Kenkyu*. Tokyo: Iwanami Shoten, pp. 141–61.

Hall, Edward Twitchell. 1966. *The Hidden Dimension*. New York: Doubleday.

Han, Jeongho. 2003. Silla seoktabui ijunggidan balsaengwonine daehan gochal (The Origination Cause of Double Deck Foundation of the Silla Stone Pagodas). *Silla Munhwa-je Haksul Balpyohoe Nonmunjib* 24: 61–96.

Hanawa, Hokiichi. 1951. *Zoku Gunshoruiju*. Tokyo: Zoku Gunshoruiju Kanseikai, vol. 19.

Hay, John. 1999. Questions of influence in Chinese art history. *Res: Anthropology and Aesthetics* 35: 240–62. [CrossRef]

Ho, Poay-Peng. 1995. Paradise on earth: Architectural depiction in pure land illustrations of high Tang caves at Dunhuang. *Oriental Art* 41: 22–32.

Iwamoto, Yutaka. 1978. *Bukkyo Setsuwa no Densho to Shinko*. Tokyo: Kaimei Shoin, pp. 57–79.

Katsuki, Genichiro. 1992. Tonko Mogaoku dai 220ku amida joudohenzo zu ko. *Bukkyo Geijutsu* 202: 67–92.

Kieschnick, John. 1997. *The Eminent Monk: Buddhist Ideals in Medieval Chinese Hagiography*. Honolulu: University of Hawaii Press.

Kim, Young Jae. 2011. Architectural Representation of the Pure Land: Constructing the Cosmopolitan Temple Complex From Nagarjunakonda to Bulguksa. Ph.D. dissertation, University of Pennsylvania, Philadelphia, PA, USA.

Kim, Young Jae. 2018. The Origin and Development of the Pavilion at the Contact Points of the East and the West–Between Asia Minor and Indian Subcontinent. *The Eastern Art* 38: 34–61. [CrossRef]

Kitagawa, Joseph M. 1988. "The Many Faces of Maitreya". In *Maitreya, the Future Buddha*. Cambridge: Cambridge University Press, pp. 7–22.

Kurokawa, Harumura, and Bun'ei Tsunoda. 1969. *Sadaie ason ki, Rekidai Zanketsu Nikki*. Kyōto-shi: Rinsen Shoten.

Li, Baijin. 2007. *Tang Fengjian Zhuyingzao*. Beijing: Zhongguo Jianzhu Gongye Chubanshe.

Macdonell, Arthur Anthony. 1929. *A Practical Sanskrit Dictionary with Transliteration, Accentuation, and Etymological Analysis Throughout*. London: Oxford University Press.

Mair, Victor H. 1986. Records of Transformation Tableaux (pienhsiang). *T'oung Pao* 72: 3–43. [CrossRef]

Minamoto, Tsuneyori. 1965. *Sakeiki左経記, Zoho ShiryoTaisei Kankokai 6*. Kyoto: Rinsen Shoten.

Mitsumori, Masashi. 1999. Kodai bukkyou jiin no reihai kuukan to reihai seki. In *Bukkyo Bijutsu No Kenkyu*. Kyoto: Jishosha, pp. 185–212.

Mizuno, Seichi, and Nagahiro Toshio. 1941. *Kanan Rakuyo Ryumon Sekkutsu No Kenkyu (Longmen Caves' Study in Luoyang, Henan)*. Tokyo: Zauho kankoka.

Motonaka, Makoto. 1994. *Nippon Kodai No Teien to Keikan*. Tokyo: Yoshikawako Bunkan.

Munhwagongbobu, Munhwajaegwalliguk. 1976. *Bulguksa Bokwongongsa bogoseo (Restoration Work Report of Bulguksa)*. Gyeongju: Gyeongju City Government.

Nakamura, Hajime. 1975. Jodo Kyoten no Seiritsu Nendai to Chiiki. In *Jodo Sanbukyo*. Tokyo: Iwanami Shoten.

Ota, Hirotaro. 1988. *Byodo-in Taikan: Dai 1 Kan Kenchiku*. Tokyo: Iwanami Shoten.

Ota, Seiroku. 2010. *Shindenzukuri No Kenkyu. Shinsoban*. Tokyo: Yoshikawa Kobunkan, pp. 149–77, 193–203.

Pas, Julian. 1987. Dimensions in the Life and Thought of Shan-tao(613–681). In *Buddhist and Taoist Practice in Medieval Chinese Society*. Edited by David W. Chappell. Honolulu: University of Hawaii, University of Hawaii Press, pp. 65–84.

Pyyhtinen, Olli. 2014. *The Gift and Its Paradoxes: Beyond Mauss*. Farnham: Ashgate Publishing, Ltd.

Rama, Rao. 1967. *Inscriptions of the Andhradesa*. 2 vols. Sri Venkatesvara University Historical Series, No. 5. Tirupati: Sri Venkatesvara University.

Sadakata, Akira. 1994. Tōkai Daigaku kiyō Bungakubu 61 (Bulletin of the Faculty of Literature of Tokai University). Available online: https://searchworks.stanford.edu/view/4227370 (accessed on 10 June 2021).

Shimizu, Hiroshi. 1986a. Hojo-ji no garan to sono seikaku. *Nippon Kenchiku Gakkai Keikaku Kei Ronbun Houkoku Shu* 363: 158–67.

Shimizu, Hiroshi. 1986b. Byodo-in no garan to sono seikaku. *Tokyokougei Daigaku Kougakubu Kiyou* 8: 65–72.

Shizutani, Masao. 1979. *Indo Bukkyō Himei Mokuroku (Catalog of Indian Buddha Monument Inscription)*. Kyōto: Heirakuji Shoten.

Sircar, D. C. 1963–1964. More Inscriptions from Nagrjunakonda. *Epigraphia Indica* XXXV: 1–36.

Sponberg, Alan, and Helen Hardacre, eds. 1988. *Maitreya, the Future Buddha*. Cambridge and New York: Cambridge University Press.

Sugiyama, Shinzo. 1981. *Inka Kenchiku No Kenkyu*. Tokyo: Yoshikawa Machi Koubun Kan.

Tatara, Miharu, Andreas Hamacher, and Shirai Hikoe. 1998. Jodo teien no kuukan kousei nikansuru kousatsu: Hojo-ji oyobi Byodo-in o jirei toshite (Consideration on the spatial composition of the Pure Land Garden: Hojoji Temple and Byodoin Temple). *Chiba Daigaku Engei Gakubu Gakujutsu Houkoku* 52: 33–42.

Tokyoteikokudaigaku Kokadaigaku. 1904. *Kankoku kenchiku chōsa hōkoku韓國建築調査報告 (Research Report of Korean Architecture)*. Tokyo: Tokyo Teikoku Daigaku.

Tsukamoto, Keisho. 1986. *Hokekyo No Seiritsu to Haikei: Indo Bunka to Daijo Bukkyo 法華経の成立と 背景: インド文化と大乗仏教 (Formation and the Background of the Lotus Sutra: Indian Culture and Mahayanist Buddhism)*. Tokyo: Kosei.

Tsukamoto, Keisho. 1996–1998. *Indo Bukkyo Himei no kenkyu インド仏教碑銘の研究 (The Study of the Buddhism Epigraph)*. 2 vols. Kyoto: Heirakuji.

Vogel, J. P. 1929–1930. No. 1 Prakrit Inscriptions from a Buddhist Site at Nagarjunikonda. In *Epigraphia Indica, XX ed*. Hiranand Shastri and New Delhi: Archaeological Survey of India, pp. 1–37.

Walser, Joseph. 2005. *Nagarjuna in Context: Mahayana Buddhism and Early Indian Culture*. Columbia: Columbia University Press.

Wong, Dorothy C. 1998. Four Sichuan Buddhist Steles and the Beginnings of Pure Land Imagery in China. *Archives of Asian Art* 51: 56–79.

Wong, Dorothy C. 2008. The Mapping of Sacred Space Images of Buddhist Cosmographies in Medieval China. In *The Journey of Maps and Images on the Silk Road*. Edited by Andreas Kaplony and Philippe Foret. Leiden: E. J. Brill, pp. 66–95.

Wright, Arthur. 1948. Fo-t'u-teng, a Biography. *Harvard Journal of Asiatic Studies* 11: 321–71. [CrossRef]

Wu, Hung. 1992a. Reborn in Paradise: A Case Study of Dunhuang Sutra Painting and its Religious, Ritual and Artistic Context. *Orientations* 23: 55–56.
Wu, Hung. 1992b. What is Bianxiang? On the Relationship between Dunhuang Art and Dunhuang Literature. *Harvard Journal of Asiatic Studies* 52: 111–92.
Xiao, Mo. 1989. *Dunhuang Jianzhu Yanjiu (Research on Dunhuang Architecture)*. Beijing: Wenwu Chubanshe.
Xu, Chaohua. 1987. *Er Ya Jin Zhu* 爾雅今注. Tianjin: Nan Kai Da Xue Chu Ban She.
Yijing, I.-Tsing, and J. Takakusu. 1970. *A Record of the Buddhist Religion as Practiced in India and the Malay Archipelago*. Oxford: The Clarendon Press.
Zürcher, Erik. 1959. *The Buddhist Conquest of China*. Leiden: E. J. Brill.

Article

The Background of Stone Pagoda Construction in Ancient Japan

Asei Satō

Faculty of Humanities and Culture, University of Shiga Prefecture, 2500 Hassaka-cho, Hikone 522-8533, Japan; sato.a@shc.usp.ac.jp

Abstract: In this study stone pagodas from ancient Japan (7th to 9th centuries) were analyzed. The findings show that there are some apparently influenced by the Korean Peninsula and two other types. While there are examples of the former type that are large and serve as temple buildings, the latter are located in mountain forest temples. I am of the opinion that stone pagodas were important mechanisms that made possible the existence of mountain forest temples as Mahayana precepts-based transgression repentance (*keka* 悔過) training sites that complemented flatland temples. This use of stone pagodas is different than China and Korea, which treated both wooden and stone pagodas in the same way. Moreover, ideas regarding Mahayana precepts-based transgression repentance were introduced from China, and I hold that the increase in stone pagodas at mountain forest temples corresponds to the Sinicization of Japanese Buddhism.

Keywords: stone pagodas; multistory pagodas; Ishidōji 石塔寺; Okamasu Ishindō 岡益石堂; Roku-tanji 鹿谷寺 temple ruins; mountain forest temples (*sanrin jiin* 山林寺院); transgression repentance (*keka* 悔過); Mahayana precepts

Citation: Satō, Asei. 2021. The Background of Stone Pagoda Construction in Ancient Japan. *Religions* 12: 1001. https://doi.org/10.3390/rel12111001

Academic Editors: Shuishan Yu and Aibin Yan

Received: 30 August 2021
Accepted: 9 November 2021
Published: 15 November 2021

Publisher's Note: MDPI stays neutral with regard to jurisdictional claims in published maps and institutional affiliations.

1. Introduction

While there are many examples of stonework in Japan, the most familiar are pagodas and Buddhist figures. Stonework pieces from the latter half of the 13th century and later are particularly prevalent throughout Japan, and have drawn the attention of many researchers and lay historians. However, there are few stone pagodas from the Heian period (794–1185) and earlier, in terms of both type and absolute number. Therefore, the background to their construction and historical characteristics were unclear, and research that brought together examples of these stone pagodas was necessary. While prominent Japanese stonework scholar Kawakatsu Masatarō published his famous work *Nihon sekizai kōgeishi* 日本石材工芸史 (*The History of Japanese Stone Craftwork*), covering the background of 7th and 8th century stonework creation and providing an overview of existing materials from the time, he also noted that there was insufficient knowledge of their details due to lack of materials (Kawakatsu 1957). In recent years, Sagawa Shin'ichi has pointed out that many stone pagodas predating the 9th century exist in places that were primarily used for mountain forest religious training, offering a new perspective for the field (Sagawa 2021). However, there is still a gap in the literature concerning the background of stone pagoda construction during that time. While scholars have discussed ancient stone pagodas' genealogies and dates with a focus on their morphological similarities to stone pagodas in China and Korea, Japan's stone pagodas are different from those pagodas in size, basic form, and construction techniques, making such discussions not very meaningful. Therefore, it is necessary to discuss the places where they were located, as Sagawa does. In this paper, I will focus on the nature of such places and stone pagodas' functions from this perspective.

2. Unusual Ancient Stone Pagodas

2.1. Ishidōji's Three-Story Pagoda

The three-story pagoda at Ishidōji 石塔寺 temple in the city of Higashi Ōmi in Shiga Prefecture is one of Japan's ancient-style stone pagodas. It features a thin, tall body measuring 7.4 m in height, with a wide roof and thin eaves. While the wheels at its top (called *sōrin* 相輪) were replaced at some point, the other parts appear to be original. This stone pagoda also has an unusual appearance. In fact, there are no similar examples in Japan. As it resembles a Baekje pagoda from the Korean Peninsula, some scholars have connected it to an entry in the *Nihon shoki* 日本書紀 from the eighth year of Emperor Tenchi's reign (669) that describes settlers from abroad entering the Gamō District (see below), and estimate that the pagoda was constructed during the second half of the 7th century (Kawakatsu 1957, etc.). However, there are no stone pagodas with the same shape on the Korean Peninsula from that time. The most similar stone pagoda is located in Zhang ha ri, Chungcheongnam-do, South Korea, and is thought to have been built between the 10th and 12th centuries. While based on this estimate, it has been argued that Ishidōji's pagoda is from the Heian period (Nomura 1985), I believe it is from the 8th century for the reasons below.

Of all the stone pagodas in South Korea, the Zhang ha ri pagoda certainly resembles the Ishidōji pagoda in several ways. However, there are actually many differences, including the shape of the roof, body construction, and shapes of the areas above and below the eaves (Figure 1). These differences make the use of resemblance as the sole basis for dating problematic. Here, an issue of particular note is the groove carved out below the eaves at the top story of the Ishidōji pagoda, which was designed to prevent rainwater from running down the body (Figure 2). While these types of grooves are often found in Korean stone pagodas, none of the pagodas from the ancient period or middle ages in Japan handle rainwater in this way. It is highly probable that this technique was introduced by someone who worked with stone on the Korean Peninsula. While relations with the Korean Peninsula changed greatly from the 8th century onwards and there was contact via trade, the migration of foreign groups, including those with technical skills, completely ceased, and it is unclear why stoneworkers would have traveled from the Korean Peninsula to Japan to build peninsula-style layered pagodas during the Heian period. For this reason, it is reasonable to assume that the Ishidōji pagoda was built by people from the Baekje Kingdom who entered Japan during the latter half of the 7th century or their successors. This is also clearly supported by the above-discussed historical record. The pagoda likely differs in shape from those on the Korean Peninsula because these settlers did not bring precise stone pagoda blueprints, but instead worked from broad mental outlines.

The aforementioned *Nihon shoki* entry from the eighth year of Emperor Tenchi's reign (669) states, "About 700 men and women move to the Ōmi Gamō District" (男女七百餘人遷居近江國蒲生郡). This textual passage indicates that a group of people from the Korean Peninsula settled around Ishidōji, where researchers have unearthed a "Kotō-style" roof tile with the same pattern as one from the abandoned temple Kabataji 綺田寺 (Gamōmachi Kokusai Shinzen Kyōkai 2000). Ogawasara Yoshihiko and Hishida Tetsuo argue that this roof tile reflects a new style introduced by Baekje immigrants (Ogasawara 1992; Hishida 2013). It appears that the Ishidōji stone pagoda is part of a temple in which such immigrants were involved. While it is not easy to determine when it was built, we can assume that it was before the 9th century, when this immigrant community dissolved.

2.2. Okamasu Ishindō

Further evidence that Japanese pagodas from the 7th century were built by settlers with roots in the Korean Peninsula can be found in the Okamasu Ishindō 岡益石堂. The Okamasu Ishindō is located in Kokufuchō, Tottori City, and was also strongly influenced by Korean stone pagodas. It is made entirely of green tuff breccia. The center of its cut stone *danjōzumi kidan* 壇上積基壇 style podium (6 m per side) has somewhat of a bulge (Figure 3). It contains a circular pillar with a central stand, the bottom of which features an ornamental double lotus petal motif (some believe this is actually a palmette). The structure is surrounded by a stone wall measuring 4.3 m in width and 2.1 m in height (Kasano et al. 1999) and which is engraved with the seal of the Ikeda family, who were the Tottori domain lords during the Edo period (1603–1868). It therefore appears that the Ikeda family repaired the wall during that time. However, as Mizuno Masayoshi points out, these repairs were limited to the stone walls (Mizuno 2006).

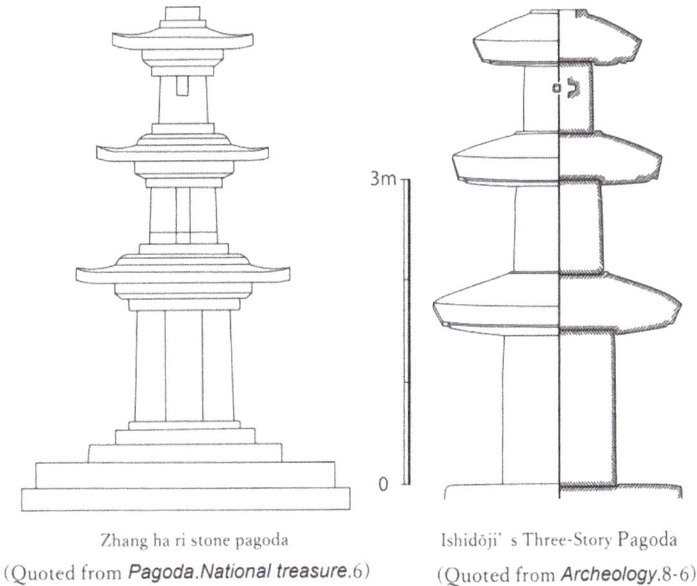

(Quoted from *Pagoda.National treasure*.6)   (Quoted from *Archeology*.8-6)

Figure 1. Morphological comparison of Zang ha ri Pagoda and Ishidōji Pagoda.

Figure 2. The groove below the eaves at the top story of the Ishidōji pagoda.

Figure 3. Drawing of Okamasu Ishindō (based on (Kasano et al. 1999), with modifications by the author).

The top of the central circular pillar is a piece of a multistory pagoda. The same pieces are also found at the neighboring temples Daiun'in 大雲院 and Okamasuji 岡益寺. Sagawa Shin'ichi posits that these were originally part of a five-story pagoda (Sagawa 2006) that he believes can be dated to the 9th century, based on a comparison with ancient multistory pagodas extant in Japan. Mizuno holds that the multistory pagoda, stone wall, and circular column form a set, and has reconstructed them by placing the multistory pagoda on top of a base with a stone wall. Noting that a tuff fragment, probably material for the pagoda, was unearthed from a ground layer dated to the 7th century, Mizuno also argues that the stone wall and other features were built during that time. He believes that the top part of the pagoda was initially wooden, but was changed to stone during the 9th century. Here, I will just mention that currently these two theories of Sagawa and Mizuno exist regarding the time when this pagoda was built.

The Okamasu Ishindō pagoda is unique in Japan, with no similar examples anywhere in the country. Certain characteristics do not have genealogies in Japan and clearly demonstrate the influence of peninsular stonework, including the ornamental double lotus petals on the bottom of the circular pillar's flower-shaped support (*ukebana* 請花) and the entasis pillar structure. In this regard, we should look to the peninsula for similar examples. However, at present, I will refrain from pursuing its prototypes.

This pagoda was built where the eastern pagoda of the temple's golden hall was located, and we can assume it was created as one of the temple's buildings.

The western part of Tottori Prefecture, where Okamasu Ishindō is located, contains Korean Peninsula-style relics. It is likely that settler clans from the Korean Peninsula were involved in the construction of both Okamasuji and Ishindō.

As we have seen, stone pagodas influenced by the Korean Peninsula tend to be large and form part of a temple complex. This is similar to how large stone pagodas were, like wooden pagodas, built as temple buildings on the Korean Peninsula. However, the only stone pagodas in Japan with such peninsular influences are the two discussed above. Other stone pagodas are clearly different. Below, I will review the characteristics of such stone pagodas from the eighth and ninth centuries.

3. Other Ancient Stone Pagodas

Table 1 shows a list of multistory pagodas from the 8th to 9th centuries in Japan (Table 1, Figures 4 and 5). In Japan, there are no examples of stone pagodas with inscriptions or unearthed artifacts from the 900s. Stone pagodas began to recede in the 10th century and then were absent up through the 12th century. Included in the table's stone pagodas are ones that, like wooden pagodas, have a deeply contoured shape (5–10, 13–17) and ones with wooden pagoda structures, such as hip rafters (*sumigi* 隅木) on eave undersides and descending ridges (*kudarimune* 降棟) on eaves (3, 4, 5, 7, 8, 10, 13–16). A ceramic pagoda held by Shōsōin 正倉院 from the 8th century is similar. Also, at the Enichiji 恵日寺 temple pagoda was a pot apparently from the 9th century. Furthermore, Iwanaga Shōzō points out that ancient-style pagodas' roofs are slanted the same amount and that this slant is different from stone pagodas from the 12th century onwards (Iwanaga 2001). While I ask the readers to refer to my below reference list for information on the dating of individual ones, we can see these pagodas as all being from the 8th or 9th centuries. During this time, the number of stone pagodas in Japan greatly increased. They are notable for being relatively small, around 2 m in height. Regarding location, several were found at temples deep in the forest mountains, including Ōganji 応願寺 temple's multistory pagoda fragment (6), Tōnomori's 塔ノ森 hexagonal 30-story pagoda (7), Murōji 室生寺 temple's wish-granting jewel pagoda (10), Rokutanji temple ruins' 13-story pagoda (11), Iwaya's 岩屋 three-story pagoda (12), Iwatakiyama's 岩滝山 multistory pagoda (13), and Shusshakuji 出釈迦寺 temple's multistory pagoda (15). On the other hand, Kannonji temple's three-story pagoda (5) and Ryūfukuji 龍福寺 temple's five-story pagoda (9) were found in valleys at the bases of mountains, and can be seen as parts of mountain forest temples. Excluding that of Rokutanji, these stone pagodas are small, around 2 m in height. Further, many of the temple buildings connected to them do not exhibit clear monastery style (*garan* 伽藍; Skt. *saṃghârāma*) structures. Ōganji's multistory pagoda (6) and Tōnomori's hexagonal 13-story pagoda (7) are found at mountain forest training sites east of Tōdaiji 東大寺 temple and Kōfukuji 興福寺 temple. As Sagawa Shin'ichi points out, they appear to have been created as pagodas for mountain forest temples that formed sets with flatland temples (Sagawa 2021). Moreover, their shapes and locations are clearly different from those of the many wooden pagodas built during this time.

Table 1. Multistory Pagodas from the 7th to 9th Centuries in Japan.

№	Name of Pagoda	Period	Location	Type of Location	References	Remarks
1	Ishidōji's three-story pagoda	latter half of the 7th century	Shiga Prefecture's city of Higashi Ōmi	Hills	(Gamōmachi Kokusai Shinzen Kyōkai 2000)	
2	Okamasu Ishindō	7–9th century	Tottori Prefecture's city of Tottori	Hills	(Inaba Man'yō Rekishikan 2006)	
3	Yamagami's multistory pagoda	801	Gunma Prefecture's city of Kiriu	Hills	(Honma 2021)	
4	Hōda pagoda	9th century	Nagano Prefecture's city of Nagano	Level ground	(Fukuzawa 2002; Takei 2021)	
5	Kannonji's three-story pagoda	End of the 8th century	Kyoto Prefecture's city of Kyōtanabe	Hills		Original position unknown
6	Ōganji Temple's multistory pagoda fragment	End of the 9th century	Nara Prefecture's city of Nara	Mountain	(Shimizu 1984)	
7	Tōnomori's hexagonal thirty-story pagoda	latter half of the 8th century	Nara Prefecture's city of Nara	Mountain	(Shimizu 1984)	

Table 1. *Cont.*

№	Name of Pagoda	Period	Location	Type of Location	References	Remarks
8	Zutō's multistory pagoda fragment	8 or 9th century	Nara Prefecture's city of Nara	Level ground	(Iwanaga 2001)	
9	Ryūfukuji temple's five-story pagoda	751	Nara Prefecture's town of Asuka	Hills	(Kokuritsu Rekishi Minzoku Hakubutsukan 1997)	Estimated position from inscription
10	Murōji temple's wish-granting jewel pagoda	End of the 8th century	Nara Prefecture's city of Uda	Mountain	(Kishi and Masao 1955)	
11	Rokutanji's thirteen-story pagoda	Middle of the 8th century	Osaka Prefecture's town of Taishi	Mountain	(Taketani 1989; Yamamoto 1993)	
12	Iwaya's three-story pagoda	8th century	Osaka Prefecture's town of Taishi	Mountain	(Yamamoto 1993)	
13	Iwatakiyama's multistory pagoda	8 or 9th century	Okayama Prefecture's city of Kurashiki	Hills	(Mori 1994)	
14	Zentsuji Tanjōin's multistory pagoda	9th century	Kagawa Prefecture's city of Zentsuji	Level ground	(Matsuda and Kaibe 2009)	
15	Shusshakaji's multistory pagoda	9th century	Kagawa Prefecture's city of Zentsūji	Mountain	(Matsuda and Kaibe 2009)	
16	Sanuki Kokufu excavated multistory pagoda	8th century	Kagawa Prefecture's city of Sakaide	Level ground	(Matsuda and Kaibe 2009)	
17	Enichiji's multistory pagoda	9th century	Fukushima Prefecture's town of Bandai	Mountain	(Sagawa 2018)	

Inscribed on Ryūfukuji's five-story pagoda in Nara Prefecture's village of Asuka, we find the year Tenpyō-shōhō 3 (751) and the name of Princess Takeno (Takeno-ō 竹野王; female royalty connected to Prince Nagaya or Nagaya-ō) (Figures 6 and 7). It states, "South of Asakaze and North at Tōnomine . . . " (朝風南葬談武之峯北). A hill approximately 1 km to the northwest is still referred to as Asakaze. It therefore appears that this stone pagoda was originally located on top of this hill. One of Prince Nagaya's wooden tablets (*mokkan* 木簡) contains the following: "Prince Takeno hired people for 3.6 L of rice and dispatched them to a mountain temple" (竹野王子山寺遣雇人米二升γγγ; Nagaya-ō Mokkan No.1829). Apparently, Prince Takeno built this stone pagoda for his own memorialization and to, as parishioner of this mountain temple, support the overall development of its sangha. While the primary aim of this pagoda's construction was the repose of Prince Takeno after death, secondarily it functioned as the sangha's core, symbolizing Ryūfukuji's status as a mountain temple. The next section expands on this issue with a discussion of Rokutanji (Taishi Town, Osaka Prefecture).

Figure 4. Stone pagodas in ancient Japan.

Figure 5. Stone Pagoda distribution map.

Figure 6. Ryūfukuji's five-story pagoda (photo).

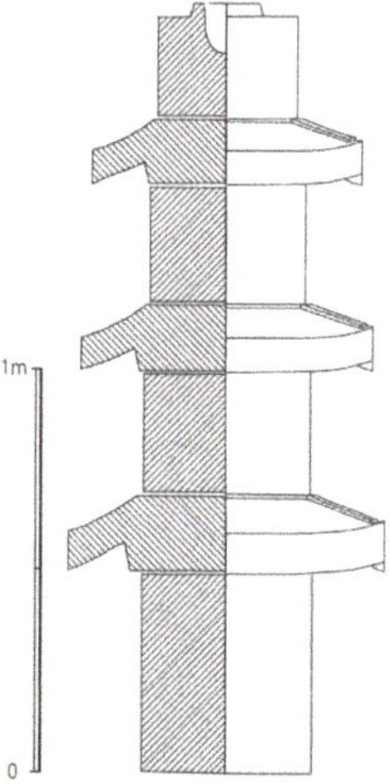

Figure 7. Drawing of Ryūfukuji's five-story pagoda (1/20 Sakurai and Fumihiko 1989).

4. The 13-Story Pagoda at the Rokutanji Ruins and the Background of Its Creation

As discussed in the previous section, clear patterns and tendencies in construction and usage emerge by inspecting numerous 8th century stone pagodas in Japan. Close consideration of one specific 8th century pagoda, located at the Rokutanji ruins, brings even further to light important aspects of the 8th century pagodas and their background influences.

The Rokutanji ruins are located at an altitude of about 280 m to the west of Mt. Nijō, which separates Yamato and Kawachi (Figure 8). Its ground consists of a 400 m² flat area created by quarrying. Compared to other mountain forest temples from its time, Rokutanji is unusual due to its stone 13-story pagoda, which is located in front of a cave containing three engraved *nyorai* 如来 (Skt. *tathāgata*). This stone pagoda is 5.1 m in height, with a base width of 1.5 m (Figure 9). It was created by shaping a tuff outcrop into a stone pillar, then adding features (Yamamoto 1993). There is a hole at the top of the base. The relative positioning of the stone pagoda and stone cave indicate that it was probably a single pagoda/golden hall (金堂) temple with *nyorai* as its main object of veneration (Taketani 1989). The closely connected eaves, wide body, and the shape of the hole of this 13-story pagoda are not found elsewhere in Japan. The stone pagoda was certainly built in the mid-eighth century; it is made from an outcrop carved into a stone pillar, and only relics from the mid-eighth century exist around it.

Figure 8. The Rokutanji site.

Figure 9. 3D model of Rokutanji's 13-story pagoda (provided by Honma Takehito).

Rokutanji's main object of veneration was likely the *nyorai* triad that is line-carved into the stone cave (Figure 10). These *nyorai* figures are thought to be either a Miroku 弥勒 (Skt. *Maitreya*) triad (Fujisawa 1985), or Amida 阿弥陀 (Skt. *Amitâbha*), Shakamuni 釈迦牟尼 (Skt. *Śākyamuni*), and Yakushi 薬師 (Skt. *Bhaiṣajyaguru*) (Nishimura 1942).

Taketani previously published a detailed measured drawing of these *nyorai* (Figure 2) (Taketani 1989). Comparing it with the figures on site, I found that the north figure clearly forms a meditation mudra (*jōin* 定印), and that the central figure probably forms a wish-granting mudra (*yogan'in* 与願印). By contrast, the southern figure's mudra is unclear; having observed it in person, I believe it is a Miroku. These three figures are the buddhas of the three dimensions of time (*sanzebutsu* 三世仏); that is, the past, present and future. It is highly likely that transgression repentance at this site entailed the recitation of buddhas' names based on the *Sangōsanzenbutsumyōkyō* 三劫三千仏名経. Based on earthenware found around the ruins, it should be noted that Rokutanji appears to have been built during the second quarter of the 8th century, around 745. Generally, rituals that involve recitations of the names of buddhas and transgression repentance are referred to as *butsumyōe*. Based on the *Nihon kiryaku* 日本紀略, such a ritual is thought to have first appeared in Tenchō 7 (830) (Takei 1980). Therefore, the Buddhist service conducted here would not have been a full-fledged *butsumyōe* 仏名会. However, from the time that Buddhism was transmitted to China, sutras on the names of the Buddhas existed there, and gained wide popularity during the Tang Dynasty (at the latest) (Yamaguchi 2018). In addition, Shōsōin sutra copying records indicate that such sutras existed in Japan during the first half of the 8th century. There are also many popular-level sutras on the names of the buddhas that are not found in the *Kaiyuan lu* 開元録 (Jp. *Kaigen roku*). It is entirely possible that primitive buddha name recitation-based transgression repentance was conducted at Rokutanji.

Figure 10. Nyorai figures inside the caves (Taketani 1989).

In addition, the figures comprising Rokutanji's main object of veneration (Amida, Shakamuni, and Miroku) are different from the buddhas of the three dimensions of time generally found in Japan (Amida, Shakamuni, and Yakushi). This combination appeared in China during the Sui dynasty, and was popular during the Tang dynasty (Hagiwara 2007). Here, as well as in the above-discussed existence of buddha name-based transgression repentance, we can identify a strong influence from Chinese Buddhism. In this context, it is important to note a point made by Inoue Kaoru regarding the appearance of Rokutanji's 13-story pagoda. With the strong influence of Tang culture during the Nara period in mind, Inoue notes that Rokutanji's pagoda resembles the Small Wild Goose Pagoda (Xiaoyan ta 小雁塔) at Jianfu 薦福寺 temple in China's Xi'an (Figure 11: Inoue 1982). In light of my above examination, we should fully consider this possibility. With that said, there are no technical similarities; only a mental image of a pagoda was introduced to Japan. I also noted this with regard to Ishidōji's three-layer pagoda. Almost all other stone pagodas in Japan were, in the same way, built only based on an image and with completely different techniques from the peninsula. These pagodas were not supported by the full-fledged transfer of techniques using blueprints.

Figure 11. Xiaoyan ta Pagoda (provided by Yamaguchi Hiroyuki).

5. The Adoption of Stone Pagodas in Ancient Japan

The preceding sections have examined the development of stone pagodas in Japan. Some were strongly influenced by the Korean Peninsula, and some were not. The Ishidōji three-story pagoda is a stonework made by settlers from Baekje or their descendants. Due to its distinctive scale, it appears to have been made in imitation of pagodas that were built as temple buildings. Similarly, excavations have shown that Okamasu Ishindō overlaps with the building of the abandoned Okamasuji (Tottori-ken Maizō Bunkazai Sentā 2000); it appears to have functioned as a temple building (Mizuno 2006). These Korean Peninsula-type stone pagodas are different from other stone pagodas in terms of their large size and status as temple buildings, and have strong settler clan monument characteristics.

By contrast, stone pagodas not strongly influenced by the Korean Peninsula include many measuring approximately 2 m in height. The majority of these are found in mountain forests, like Ryūfukuji's pagoda, and some were built as mountain temple equipment, so to speak. Around Ryūfukuji, we can reconstruct an early *butsumyōe* ritual space in which transgressions were repented to the buddhas three times. It can be assumed that many of these mountain forest stone pagodas were used for these transgression repentance rituals.

In the past, scholars believed 8th-century mountain forest temples were private Buddhist hubs "of the people" that stood in opposition to state Buddhism and, in light of *ritsuryō* 律令 rules, sometimes even illegal (Futaba 1957). However, it eventually became clear that mountain forest temples and government temples actually formed sets, as pointed out by Sonoda Kōyū, who argues that government temple priests traveled between government temples and mountain forest temples in order to acquire *jinenchi* 自然智, or knowledge with which individuals are originally equipped, yet currently unaware of (Sonoda 1981). In recent years, Kikuchi Hiroki has also highlighted cases in which lay male practitioners engaged in *jōgyō* (浄行), or religious training, to become monks, arguing that this practice originated in Chinese scripture and was transgression repentance (Skt. *poṣadha*, Jp. *fusatsu* 布薩／*keka* 悔過) in essence based on the Mahayana precepts. He holds that while at flatland temples monks had to observe the complete precepts (based on the Theravāda four-part *Vinaya*), at mountain forest temples they maintained Mahayana precepts based on the likes of the much simpler "three categories of pure precepts" (Skt. *trividhāni śīlāni*, Jp *sanju jōkai* 三聚浄戒), and engaged in active mountain forest religious training characterized by more freedom in daily life (Kikuchi 2020). In this way, mountain forest temples in Japan were religious training sites that belonged to flatland temples, and at them transgressions were repented and diverse activities engaged in. Why, then, were stone pagodas installed in such places?

In ancient Japan, there were no stone pagodas without peninsular influences built as temple buildings. Wooden and stone pagodas had different roles. Mahayana scriptures often advocate building pagodas and making offerings. The *Lotus Sutra* is a major example of this. Its "Expedient Means" and "Spiritual Powers" chapters state that to build a pagoda is both a virtuous act that opens the path to buddhahood, as well as that where a pagoda is built—whether in forests, under trees, on mountains, or in valleys—will become a site for religious training and enlightenment. In East Asia, where Mahayana Buddhism spread, the influence of the *Lotus Sutra* has been consistently strong. In Japan, importance was particularly attached to the "Expedient Means" and "Spirituals Powers" chapters (they were considered one of the sutra's four essential chapters), and the sutra would widely spread amongst not only priests but also laypeople. Based on these facts, it appears that at mountain forest temples, which were the Mahayana precept-based training sites of flatland temples, first a pagoda was built to symbolize their status as such. In other words, we should see these stone pagodas as having been built on small mountains to establish places for Buddhist training. While, of course, in Japan large wooden pagodas were built at mountain forest temples that served as hubs, I believe that these small stone pagodas were installed at religious training sites where it was not possible to build large temple complexes. In this sense, it appears that in Japan stone pagodas were entirely replacements for wooden pagodas.

Looking at the background of such repentance systems, religious beliefs and practices related to the buddhas of the past, present, and future, and ideas regarding pagoda construction and the establishment of religious training sites, we can clearly see the influence of Chinese Buddhism. In the case of Rokutanji, it appears that a repentance ritual existed which predates the standardization of Buddhism that followed the establishment of a national precepts platform when Jianzhen 鑑真 arrived in Japan. This ritual was based on diverse groups of sutras continually introduced to Japan beginning in the 7th century (e.g., the group of sutras brought by Dōshō 道昭[1]). Excluding the time after Buddhism's introduction to Japan, Sinification advanced greatly in the seventh century, and it appears that this is a manifestation of this phenomenon.

Yamamoto Jun's research has found that transgression repentance rituals, which were introduced with the transmission of Buddhism to Japan, transitioned from repentance by individuals in the 7th century to rituals directed toward specific objects of veneration in the 8th century. Then, in the 9th century, he argues, these prevailing practices were again replaced by new esoteric Buddhist rituals, including the *butsumyōe*, before becoming obsolete (Yamamoto 2018). Due to the old transgression repentance rituals held at stone pagoda-centered mountain forest temples shifting to esoteric Buddhist rituals from the 9th century onwards, the symbols of mountain forest temples changed from small stone pagodas to esoteric mandalas, buddhas, bodhisattvas, and deities. The decline in stone pagodas from the tenth century onwards was probably a result of this.

6. Conclusions

In this paper, I argued the following. First, I noted that ancient stone pagodas in Japan fall under three types, one of which is pagodas exhibiting the influence of the Korean Peninsula. The two types without Korean Peninsular influences are small or medium-sized multistory pagodas, and the small ones were primarily located in mountain forest temples. Second, I pointed out the possibility that all Korean Peninsula-influenced pagodas were built as temple buildings and the other types as symbols to establish mountain forest temples as sites of religious training, as well as that the decline in stone pagodas coincided with the shift to of esoteric Buddhist rituals. Japanese Buddhism was established by taking in Buddhist culture from the Korean Peninsula, and then would became increasingly inclined toward Chinese Buddhism as envoys were sent to China. While one cause of this was changes in foreign relations during the latter half of the 7th century that led to a confrontational relationship with the peninsula, this orientation toward China remained unchanged even after relations improved. This paper's findings basically match the process of Japanese Buddhism maturing as it incorporated new information on Chinese Buddhism. The development of stone pagodas matches the development of ancient Japanese Buddhism.

As noted at the beginning of this article, Kawakatsu Masatarō points out that there are few stone pagodas from this period, and of those that do exist, only several can be dated with certainty. In order to clarify the background of pagoda construction from this limited information, it is necessary to decipher the ideas regarding the spaces in which pagodas were placed. I hope this paper will serve as a springboard for further advancements, based on the relationship between mountain forest temples and ancient Buddhism, in our understanding of stone pagodas.

Funding: This work was supported by JSPS KAKENHI (Grant Number JP21H04359, JP19K01107 and JP19H00536).

Institutional Review Board Statement: Not applicable.

Informed Consent Statement: Not applicable.

Acknowledgments: I thank Kikuchi Hiroki (Historiographical Institute, University of Tokyo) for his considerable guidance, as well as Kitani Satoshi and Sagawa Shin'ichi for their assistance, in drafting this manuscript. I would also like to express my gratitude to Yan Aibin for suggesting that I write this paper.

Conflicts of Interest: The author declares no conflict of interest.

Note

[1] An entry for the third month of Monmu 4 (700; 56th year of the sexagenary cycle) in *Zoku Nihonki* 続日本紀 states that many of the scriptures brought by Dōshō were stored at Zen'in 禅院 in the Heijō-kyō's Sakyō area, which was the predecessor of Tōnan Zenin 東南禅院.

References

Fujisawa, Kazuo 藤澤一夫. 1985. Santai Mirokuzō o honzon to suru Nihon kodai jiin: Kin Samuyuon sensei no kakō o ga shite 三体弥勒像を本尊とする日本古代寺院－金三龍先生の華甲を賀して－ [Japan's ancient temples with Miroku triads as the main object of veneration: Celebrating the sixty-first birthday of Kim Samryong-sensei]. In *Han'guk munhwa wa wŏn Pulgyo sasang: Munsan Kim Sam-nyong Paksa hwagap kinyŏm* 韓國 文化 와 圓佛教思想: 文山 金 三龍 博士 華甲 紀念 [*Korean Culture and Won Buddhist Thought: Marking the Sixty-First Birthday of the Prolific Writer Dr. Kim Sanryong*]. Iri-si: Wŏn'gwang Taehakkyo Ch'ulp'anbu, pp. 1–26.

Fukuzawa, Kunio 福澤邦夫. 2002. Shinano Shinonoi shozai kosekitō no yōshiki 信濃篠井所在古石塔の様式 [The styles of ancient stone pagodas in Shinano's Shinonoi]. In *Fujisawa Kazuo sensei sotsuju kinen ronbunshū* 藤澤一夫先生卒寿記念論文集 [*Article Collection Marking the Ninetieth Birthday of Fujisawa Kazuo-Sensei*]. Nara: Tezukayama Daigaku Kōkogaku Kenkyūjo, pp. 452–62.

Futaba, Kenkō 二葉憲香. 1957. Taika no kaishin no shūkyō kōzo. 大化改新の宗教構造 [The Religious structure of the Taika Reforms]. *Bukkyō shigaku.* 教史 [*Buddhism History*] 6: 171–98.

Gamōmachi Kokusai Shinzen Kyōkai 蒲生町際親善協会, ed. 2000. *Ishidōji sanjū sekitō no rūtsu o saguru* 石塔寺三重石塔のルーツを探る [*Exploring the Roots of Ishidōji's Three-Story Stone Pagoda*]. Hikone: Sanraizu Shuppan.

Hagiwara, Hajime 萩原哉. 2007. Sanzebutsu no zōzō: Shōzan sekkutsu dai 3 gō kutsu no sanbutsu o chūshin to shite 三世仏の造像－鐘山石窟第3窟の三仏を中心として [Visualizing the buddhas of the three times: The three buddhas at Zhongshan Cave No. 3]. *Indo bukkyōgaku kenkyū* 印度仏教学研究 [*Indian and Buddhist Studies*] 56: 489–86. [CrossRef]

Hishida, Tetsuo 菱田哲郎. 2013. Hakusuki no e igō Nihon no bukkyō jiin ni mirareru Kudara imin no eikyō 白村江以後日本の仏教寺院に見られる百済遺民の影響 [The influence of Baekje survivors seen in Buddhist temples of Japan from Baekgang onwards]. *Dongyang misulsahak* 동양미술사학 [*Journal of Asian art History*] 2: 32–75.

Honma, Takehito 本間岳人. 2021. Yamagami tajūtō shōkō 山上多重塔小考 [Some thoughts on the Yamagami multi-story pagoda]. *Hibiki* 日引 [*Hibiki*] 16: 29–48.

Inaba Man'yō Rekishikan 因幡万葉歴史館, ed. 2006. *Okamasu Ishindō: Ima, Toikakeru Mono* 岡益石堂いま、問いかけるもの [*Okamasu Ishindō: What It Asks Us Now*]. Tottori: Inaba Man'yō Rekishikan.

Inoue, Kaoru 井上薫. 1982. Izumi no kuni Ōnodera dotō no genryū 和泉国大野寺土塔の源流 [The origins of the earthen pagoda of Ōnodera temple in Izumi Province]. *Bunkazai gakuhō* 文化財学報 [*Cultural Properties Academic Bulletin*] 1: 1–11.

Iwanaga, Shōzō 岩永省三, ed. 2001. *Shiseki zutō hakkutsu chōsa hōkoku* 史跡頭塔発掘調査報告 [*Report of the Zutō Historical Site Excavation*]. Nara: Nara Kokuritsu Bunkazai Kenkyūsho 奈良国立文化財研究所 [Nara National Research Institute for Cultural Properties].

Kasano, Tsuyoshi 笠野 毅, Fukuo Masahiko 福尾正彦, Nakahara Hitoshi 中原斉, Yamamasu Masayoshi 山枡雅美, and Tsugawa Hitomi 津川ひとみ. 1999. *Ubenoryōbo sankō chinai 'Okamasu no Ishindō' no hozon shori chōsa hōkoku* 宇倍野陵墓参考地内「岡益の石堂」の保存処理・調査報告 [*Report on Preservation Measures for and Survey of Ubeno Imperial Mausoleum Reference Site's 'Okamasu Ishindō'*]. *Shoryōbu kiyō* 書陵部紀要 [*Archives and Mausolea Department Bulletin*]. Tokyo: Archives and Mausolea Department, p. 50.

Kawakatsu, Masatarō 川勝政太郎. 1957. *Nihon sekizai kōgeishi* 日本石材工芸史 [*The History of Japanese Stone Craftwork*]. Kyōto: Sōgeisha.

Kikuchi, Hiroki 菊地大樹. 2020. *Nihonjin to yama no shūkyō* 日本人と山の宗教 [*Japanese People and Mountain Religion*]. Tōkyō: Kōdansha Gendai Shinsho.

Kishi, Kumakichi 岸熊吉, and Suenaga Masao 末永雅雄. 1955. *Uda-gun Murō-Mura Murōji Nyoimine Shutsudo ibutsu* [*Artifacts Unearthed from Mt. Nyoi at Murōji temple in Uda District's Murō Village*]. *Nara-ken Shiseki Meishō Tennen Kinenbutsu chōsa shōhō* 奈良県史跡名勝天然記念物調査抄報 5 [*Report on Nara Prefecture Historical Site and Famous Natural Monument Surveys*] 5. Nara: Naraken Kyōiku Iinkai奈良県教育委員会 [Nara Prefectural Board of Education].

Kokuritsu Rekishi Minzoku Hakubutsukan 国立歴史民俗博物館. 1997. *Kodai no hi* 古代の碑 [*Ancient Period Monuments*]. Chiba: National Museum of Japanese History.

Matsuda, Tomoyoshi 松田朝由, and Hiroshi Kaibe 海邉博史. 2009. Shikoku no kodai sōtō 四国の古代層塔 [Shikoku's ancient multi-story pagodas]. *Kagawa kōko* 香川考古 [*Kagawa Archaeology*] 11: 75–94.

Mizuno, Masayoshi 水野正好. 2006. *Okamasu no Ishindō kara sugata naki tō o okangaeru* 岡益の石堂から姿なき塔を考える [*Thinking about Absent Pagodas from Okamasu's Ishindō*]. Tottori: Inaba Man'yō Rekishikan, pp. 48–66.

Mori, Akira 森 章. 1994. Iwatakiyama sanjū sekitō: Tsūshō 'Shakutō-sama' no seisaku nendai 岩滝山三重石塔－通称＜釈塔様＞の制作年代 [The time of creation of the three-story pagoda of Iwatakiyama, commonly known as the Śākyamuni pagoda]. *Shiseki to bijutsu* 史迹と美術 [*Historical Sites and Art*] 650: 396–407.

Nishimura, Tei 西村貞. 1942. *Nara no sekibutsu* 奈良の石仏 [*Nara's Stone Buddhist Figures*]. Ōsaka: Zenkoku Shobō.

Nomura, Takashi 野村隆. 1985. Ōmi Ishidōji sanjū sekitō no sōritsu nendai 近江石塔寺三重石塔の造立年代 [The time of the creation of Ōmi Ishidō's three-story stone pagoda]. *Shiseki to bijutsu* 史迹と美術 [*Historical Sites and Art*] 55: 342–53.

Ogasawara, Yoshihiko 小笠原好彦. 1992. Ōmi no bukkyō bunka 近江の仏教文化 [Ōmi's Buddhist culture]. In *Kodai o kangaeru: Ōmi* 古代を考える近江 [*Thinking about the Ancient Period: Ōmi*]. Edited by Mizuno Masayoshi 水野正好. Tokyo: Yoshikawa Kōbunkan, pp. 156–72.

Sagawa, Shin'ichi 狭川真一. 2006. *Okamasu Ishindō to kodai no sekitō* 岡益石堂と古代の石塔 [*Okamasu Ishindō and Ancient Stone Pagodas*]. Tottori: Inaba Man'yō Rekishikan, pp. 88–116.

Sagawa, Shin'ichi 狭川真一. 2018. Enichiji-den Tokuitsu-byō gojū sekitō no saifukugen 慧日寺伝徳一廟五重石塔の再復原 [Re-restoration of the Enichiji temple Tokuitsu mausoleum five-story stone pagoda]. In *Gankōji Bunkazai Kenkyūsho kenkyū hōkoku*元興寺文化財研究所研究報告 [*Gangoji Institute for Research of Cultural Property Research Report*]2017. Nara: Gangoji Institute for Research of Cultural Property, pp. 73–80.

Sagawa, Shin'ichi 狭川真一. 2021. Kodai no sekizōbutsu 古代の石造物 [Ancient stonework]. *Hibiki* 日引 [*Hibiki*] 16: 3–16.

Sakurai, Toshio, and Fujisawa Fumihiko. 1989. Basic Planning of the Stone Pagoda at Ryufukji-Temple in Asuka. *Journal of the Faculty of Science and Technology, Kinki University* 25: 251–59.

Shimizu, Toshiaki 清水俊明. 1984. *Nara-ken shi* 奈良史 [*History of Nara Prefecture*]. vol. 7. *Sekizō bijutsu* 石造美術 [*Stone Art*]. Tōkyō: Meicho Shuppan.

Sonoda, Kōyū 園田香融. 1981. *Heian bukkyō no kenkyū* 平安仏教の研究 [*Research on Heian Buddhism*]. Kyoto: Hōzōkan.

Takei, Akio 竹居明男. 1980. Butsumyōe ni kansuru sho mondai: Jū seiki matsu made no dōkō (jō) 仏名会に関する諸問題－十世紀末までの動向－（上） [Various issues related to buddha name recitation rituals: Developments up through the end of the tenth century (part 1)]. *Jinbungaku* 人文学 [*Studies in humanities, Doshisha University*] 135: 21–46.

Takei, Narumi 武井成実. 2021. Chūbu chihō no kodai sekizōbutsu: Nagano-shi Shinonoi shozai no sekizō tasōtō [Ancient stonework of the Chūbu region: A multi-story stone pagoda in Nagano city's Shinonoi. *Hibiki* 日引 [*Hibiki*] 16: 49–54.

Taketani, Toshio 竹谷俊夫. 1989. Kawachi Rokutanji ato shutsudo no ibutsu 河内鹿谷寺址出土の遺物 [Artifacts unearthed from the site of Rokutanji temple in Kawachi]. *Kobunka dansō* 古文化談叢 [*Plethora of Ancient Culture Topics*] 20: 85–94.

Tottori-ken Maizō Bunkazai Sentā 鳥取県埋蔵文化財センター. 2000. *Okamasu haiji* 岡益廃寺. [*Okamasu Abandoned temple*]. Kokufu Town (Tottori Prefecture): Tottori Prefecture Embedded Cultural Properties Center.

Yamaguchi, Masateru 山口正晃. 2018. "Chūgoku bukkyō" no kakuritsu to butsumyōkyō 「中国仏教」の確立と仏名経 [The establishment of "Chinese Buddhism" and buddhas' names sutras]. *Kansai Daigaku Tōzaigakujutsu Kenkyūsho kiyō* 関西大学東西学術研究所紀要 [*Bulletin of Kansai University's Institute of Oriental and Occidental Studies*] 51: 233–59.

Yamamoto, Yoshitaka 山本義孝. 1993. Nijōsanroku no sekkutsu jiin: Rokutanji, Iwaya no shiryōka to haikei 二上山麓の石窟寺院鹿谷寺・岩屋の資料化と背景－ [Stone cave temples at the foot of Mt. Nijō: Turning Rokutanji and Iwaya into materials and their background]. *Shigaku ronsō* 史学論叢 [*Historical studies review of Beppu University*] 23: 17–41.

Yamamoto, Jun 山本潤. 2018. "Kodai ni okeru keka no juyō no hen'yō: Zange kara shūzen e" 古代における悔過の受容と容－懺悔から修善へ—— [The adoption and transformation of transgression repentance in ancient times: From repentance to the cultivation of goodness]. *Bukkyōshi kenkyū* 仏教史研究 [*Buddhism History Studies of Ryukoku University*] 56: 26–54.

Article

Rethinking the Proportional Design Principles of Timber-Framed Buddhist Buildings in the Goryeo Era

Ju-Hwan Cha and Young-Jae Kim *

The Department of Heritage Conservation and Restoration, Korea National University of Cultural Heritage, Buyeo 33115, Korea; cjhsww@naver.com
* Correspondence: kyjandy@nuch.ac.kr

Abstract: This study examines how the wooden architecture of the Goryeo Dynasty in Korea evolved in an original way while incorporating Chinese architectural principles. For the Goryeo Era's timber-framed buildings, eave purlin height was determined according to $\sqrt{2}$H times the eave column height (H), while the eave column height influenced the proportional location of each purlin, determined by the $\sqrt{2}$H times decrease rate in the cross-section. Thus, eave column height was proportionately connected to a geometric sequence with a common ratio of $\sqrt{2}$H. This technical approach, achieved using an L-square ruler and a drawing compass, contributed to determining eave purlin and ridge post placement, bracket system height, and outermost bay width. This study notes that the practical works were consistently preserved in East Asian Buddhist architecture, in that a universal rule of proportion was applied to buildings constructed during the Tang–Song and the Goryeo Dynasties, surmounting differences in local construction methods. These design principles were a vestige of socio-cultural exchange on the East Asian continent and a minimal step toward the establishment of structurally safe framed buildings.

Keywords: Goryeo Buddhist architecture; *Zhoubi Suanjing*; *Jiuzhang Suanshu*; *Yingzao Fashi*; proportional principle; reconstruction of timber-framed buildings

Citation: Cha, Ju-Hwan, and Young-Jae Kim. 2021. Rethinking the Proportional Design Principles of Timber-Framed Buddhist Buildings in the Goryeo Era. *Religions* 12: 985. https://doi.org/10.3390/rel12110985

Academic Editors: Shuishan Yu and Aibin Yan

Received: 21 July 2021
Accepted: 5 November 2021
Published: 10 November 2021

Publisher's Note: MDPI stays neutral with regard to jurisdictional claims in published maps and institutional affiliations.

1. Introduction

This study shows that a universal construction method exists and confirms that it has been used in regions of East Asia, such as Korea and China, for a long time. The universal design principle was combined with and adapted to vernacular exiting methods over time. This study notes that the wooden-framed buildings in the Goryeo Era in Korea were transformed according to the local architectural environment based on the universal framework of obtaining the proportional principle and *yeongjocheok* (construction measurement unit) by making circles and squares using an L-square ruler and a drawing compass. Accordingly, in comparison with ancient Chinese buildings in the same period, this study confirms that Goryeo architecture demonstrates certain commonalities and differences, sharing common universality and adjusting differences in regional construction methods.

In both the East and the West, ancient people applied certain architectural objects and real numbers in building design. Concerns regarding the fundamentals of beauty and proportionality provide parallels between Western classical orders following the Vitruvian module (Vitruvius Pollio and Morgan 1960; Stevens 1990; Kim 2016) and East Asian classical works on mathematics, such as the *Zhoubi Suanjing* 周髀算經 (*The Arithmetical Classic of the Gnomon and the Circular Paths of Heaven*) (Anonymous n.d.c.; Cullen 2007) and the *Jiuzhang Suanshu* 九章算術 (*The Nine Chapters on the Mathematical Art*) (Anonymous n.d.a.; Guo 2009). The *Jiuzhang Suanshu* was composed through the interpretive efforts of several generations of scholars from the tenth through the second century BCE, followed by Liu Hui (225–295 CE), who edited and published the work in 263 CE, during the Wei Era[1] (Cha and Kim 2019). This book greatly influenced mathematical logic in Korea and Japan[2] and contributed to philosophical contemplation and cultural formation in ancient East Asia. As in the traditions of Western classical architecture, certain proportional systems were used to create line drawings of wooden buildings utilizing L-square rulers and compasses.

To prevent collapse, ancient wooden buildings in East Asia were constructed following a ground plan with a rectangular contour. In buildings composed of columns, bracket sets, and a roof, the columns were the most essential, and they served as a standard reference to adjust each component's location.[3] The use of a regular cubit system matched the arrangement and setting of the internal frame structure. Wooden architecture requires large trees to be cut and polished into components with regular dimensions that can be used to erect large timber-framed buildings by assembling columns, beams, and purlins using mortise and tenon joints.[4] To make the desired timber sections, craftspeople cut massive trees according to a specific ratio. Using the standard cubit system to produce regular components and maintain structural stability is vital for this process. However, the "standard" cubit system would often differ, even within the same period and region (Yun 1975). This shows that while there was a universal way to build a stable structure, there was also a different cubit system employed depending on the region and period (Kim 2011).

Hence, to delve into the universal and vernacular ways practiced in the Korean Peninsula, this study examined nine buildings in South and North Korea from the mid-to-late Goryeo Era: Bongjeongsa Monastery's Geungnakjeon Hall (鳳停寺 極樂殿, twelfth or thirteenth century) in Andong; Buseoksa Monastery's Muryangsujeon (浮石寺 無量壽殿, eleventh or thirteenth century) and Josadang Halls (祖師堂, 1377) in Yeongju, Seongbulsa Monastery's Eungjinjeon Hall (成佛寺 極樂殿, 1327); Bogwangjeon Hall at Simwonsa Monastery (心源寺 普光殿, fourteenth century) in Yeontan; Sudeoksa Monastery's Daeungjeon Hall (修寺 大雄殿, 1308) in Yesan; Seongbulsa Monastery's Geungnakjeon Hall (成佛寺 極樂殿, 1374), Eunhaesa Geojoam Hermitage's Yeongsanjeon Hall (銀海寺 居祖庵 靈山殿, 1375) in Yeongcheon; and Imyeonggwan Guesthouse's Sammun Gate (臨瀛館 三門, fourteenth century) in Gangneung (Figure 1).

Figure 1. A Map depicting the location of the target buildings in the Korean Peninsula.

Seven of the buildings followed the *jusimpo* style (柱心包, bracket complexes placed directly on column heads), and two followed the *dapo* style (多包, intercolumnar bracket complexes) (Figure 2). Bongjeongsa Monastery's Geungnakjeon Hall preserved the mid-Goryeo building's appearance between the twelfth and thirteenth centuries (Cultural Heritage of Administration (hereafter CHA) 2003, p. 102), while Buseoksa Monastery's Muryangsujeon Hall retained its distinctive architectural style and techniques between the eleventh and thirteenth centuries (CHA 2002, p. 64; Cha and Kim 2019) (Table 1).

Figure 2. The distinction of *Jusimpo* (**left**) and *Dapo* (**right**) styles. (Left: Muryangsujeon Hall at Buseoksa Monastery, Plate 18627, courtesy of National Museum of Korea; right: Bogwangjeon Hall at Simwonsa Monastery, Plate 18195, courtesy of National Museum of Korea).

Table 1. Goryeo-era's buildings in the Korean Peninsula.

Building Name	Construction Period	Bracket Set's Style	References
Geungnakjeon Hall at Bongjeongsa Monastery	Replacing roof tiles in 1363Built between the twelfth and thirteenth centuries in that roof tile repairs are usually replaced at intervals of 120–150 years	*Jusimpo*	(CHA 2003, p. 102)
Muryangsujeon Hall at Buseoksa Monastery	Repairing in 1376Built between the eleventh and thirteenth centuries due to many structural techniques related to the eleventh-century construction	*Jusimpo*	(CHA 2002, p. 64)
Daeungjeon Hall at Sudeoksa Monastery	1308, first construction	*Jusimpo*	(CHA 2005a, p. 78)
Eungjinjeon Hall at Seongbulsa Monastery	1327, first construction	*Dapo*	(NRICH 1998, p. 39)
Geungnakjeon Hall at Seongbulsa Monastery	1374, first constructionExtended outermost bays at front façade in the sixteenth century	*Jusimpo*	(NRICH 1998, p. 39)
Yeongsanjeon Hall at Geojoam Hermitage, Eunhaesa Monastery	1375, first construction	*Jusimpo*	(CHA 2004, p. 59)
Josadang Hall at Buseoksa Monastery	1377, first construction	*Jusimpo*	(CHA 2005b, p. 63)
Bogwangjeon Hall at Simwonsa Monastery	Rebuilding in 1374The fourteenth century	*Dapo*	(NRICH 1998, p. 36)
Sammun Gate at Imyeonggwan Guesthouse in Gangneung City	The fourteenth century	*Jusimpo*	(CHA 2004, p. 140)

Dapo: inter-columnar bracket set; *jusimpo*: column top bracket set.

This study aimed to elucidate the proportional design principles of $\sqrt{2}$ times the eave purlin height by examining wood-frame buildings in ancient Korea and China. Ancient wood-frame architecture in East Asia relied on columns, beams, and purlins to form a basic structure and determine a plan for the buildings' construction. Critical factors regulating a building's scale were the column height and purlin location. This study methodically and accurately analyzed the geometry used in nine examples of Goryeo architecture in Korea.

Essentially, this study examined how ancient Chinese architectural culture influenced the evolution of building principles in ancient Korea. Accordingly, architectural design principles in Goryeo Era buildings were analyzed, following which Chinese buildings were examined to focus on aspects relevant to these principles. This study identified similarities and differences through comparisons with Korean and Chinese buildings that shared the same design principles since the eighth century. As a common point, the positions of the purlins were determined by applying a proportionality of $\sqrt{2}$ times the eave pillar height. Such a universal point recognized that Goryeo architecture inherited the rules of ancient Chinese architecture. Differences can be found in the methods used to determine purlin location, purlin size, and the presence of column purlins. Thus, this study indicates the process of adapting as a universal method accepts the construction method of local architecture on the Korean Peninsula and is assimilated into vernacular buildings, sharing existing construction techniques thereon.

2. Preliminary Study of the Proportional System of Traditional Wood-Frame Architecture in East Asia

A few previous studies have been conducted on the proportional systems of traditional wood-frame buildings in Korea. Sim and Kim (1983) analyzed the proportional systems of the bracket elements in *dapo* intercolumnar bracket complexes, identifying formal similarities among the bracket systems of wooden buildings in Korea, China, and Japan, despite disparities caused by the lack of agreement on the width of the *cai* (材, a length unit in the *cai-fen* 材分 carpentry system). Kim (2003) addressed the proportional system of the front and side façades of buildings with intercolumnar bracket systems, including purlin location, roof type, and floor area. Kim (2008) examined the adoption of a cubit system based on an average of 31.29 cm and wooden components with proportions of 5:3, 9:5, and 3:5 used to construct the Daeungjeon Hall at the Sudeoksa Monastery.[5] In a study on the Muryangsujeon Hall at the Buseoksa Monastery, Kim (2014) posited that an eleventh-century building method was used, based on the composition principle of the plan and cross-section and repetition of wooden components identical to a cubit system (30.3 cm), deduced from the actual measurements. Cha and Kim (2019) discussed mathematical proportional systems for the overall structural design of wooden architecture, identifying the $\sqrt{2}$ ratio system for arranging wooden components in an internal framed structure and the relationship between outer bay width and eave column height. Comparing construction methods in the Song Era with Korean wooden structures, Bak and Mun (1992) noted other details that indicated local differences, such as the width of wooden components, proportions of the front and side façades, and roof types.

Among the research from China and Japan on the proportional system of $\sqrt{2}H$ times the eave column height for the proportion of interior timber-framed buildings, Chen (1981) found that eave column height determined the dimensions of the middle roof purlins and the queen posts from the ground level. Meanwhile, Fu's (2001) research was more advanced, revealing that the height of the eave columns along the cross-section had a noticeable proportional link to the overall length of the frontal view on a ground plan along the longitudinal direction, corresponding to standard proportions in the application of mid-rise buildings with two or more above-ground stories (Figure 3). Wang (2011a) and Chen (1981) conducted significant studies on the height of eave columns and purlins, showing the feasibility of the $\sqrt{2}$ times proportional system[6] (Figure 4).

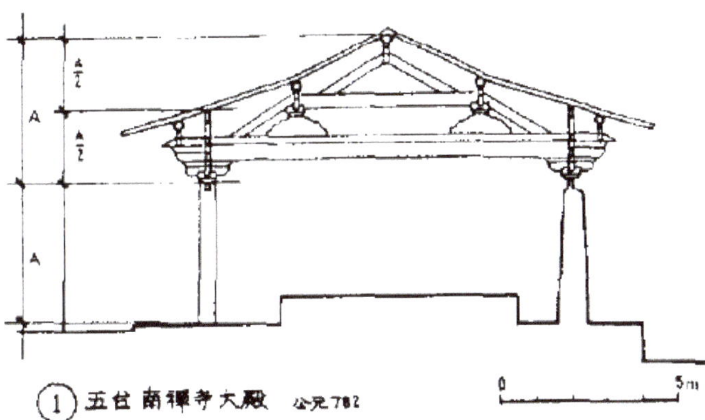

Figure 3. The main hall of Nanchansi Monastery and Fu Xinian's proportional interpretation, Adapted from Fu (2001, p.108).

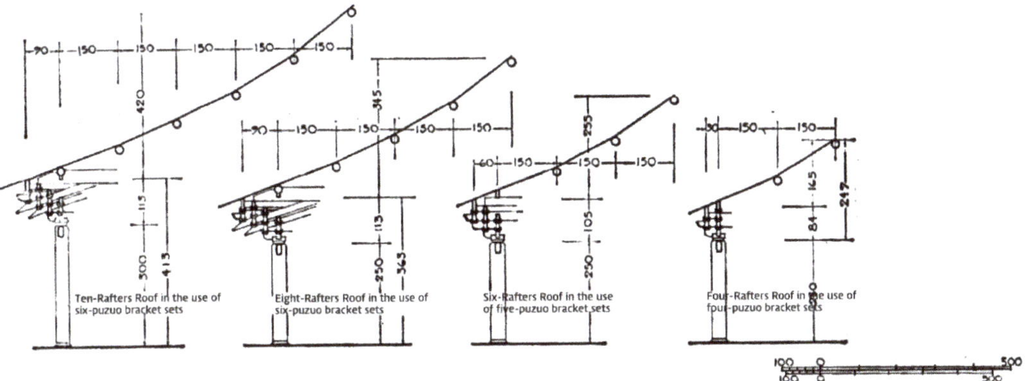

Figure 4. A proportional concept on the roof purlins location according to the eave columns' height, Adapted from Chen (1981, p. 4).

Yoneda conducted significant research on the proportional system of framed stone elements during the Japanese colonial period at the Seokguram Buddhist Grotto[7] (Yoneda 1944; Yoneda and Sin 1976). Although Chinese and Japanese scholars have provided important information regarding the proportional system of ancient wooden buildings, there are still no definitive answers to the questions that concern the purlin locations relying on the proportional link of $\sqrt{2}$ H times the height of the eave columns in the cross-section. The $\sqrt{2}$ H ratio is key to restoration studies. The current study provides a better understanding of Korean monuments by comparing them with ancient wooden structures in East Asia.

Most studies after the 2000s have examined the proportionality of aesthetic perfection in East Asia, but there are some records of the design of wood-frame buildings using proportional systems or accurate information regarding the manuals for line drawings and measurements (Fu 2001; Wang 2011a, 2011b; Wang 2017a, 2017b, 2017c, 2018a, 2018b, 2019; Wang et al. 2021a). In particular, the study of Guixiang Wang (2011a) is noteworthy in that the arithmetic of the square root of two from the ancient Chinese mathematical texts contributed to determining the 1:$\sqrt{2}$ ratio of the heights of the eave columns (exterior columns or perimeter columns, named *yanzhu* [檐柱] or *weizhu* [外柱]) and eave purlins (Wang 2011a, 2011b; Cha and Kim 2019).

Wang Nan's research confirms that the "circle–square map" (*yuanfangtu*) and "square–circle map" (*fangyuantu*) were commonly applied during the design of ancient capital cities and timber-framed buildings for the utilization of geometric rules; similar techniques for structural composition appeared in the Buddhist halls, pavilions, and pagodas of the Tang, Liao, and Qing dynasties (Wang 2017a, 2017b, 2017c, 2018a, 2018b, 2019). Based on these achievements, Wang et al. (2021a)'s recent study concerning the wooden frame pagoda in Ying County, Shanxi Province, notes that during the design of Buddhist buildings in dynastic China, statues and their surrounding spaces were designed in a uniform style, allowing a harmonious proportion between the statues and the neighboring inner spaces on every floor. He indicates that the most frequently used proportions for the organization of interior spaces and placement of statues inside the pagoda are $\sqrt{2}$:1, 3:2, or 5:3 (or 8:5), and 9:5, which are compatible with the 1.5:1 ratio, approximately. The results show that a uniform measurement system was employed to determine the building scale between the spaces and sculptures in the Buddhist Wooden Pagoda (Wang et al. 2021a).

The application of the $\sqrt{2}$ H ratio in deciding the placement and the length of individual members as a manual for building design spread to the Korean Peninsula around the seventh century and had an important influence as a building design principle on the Korean Peninsula until modern times (Cha and Kim 2019). Likewise, a morphological approach on multi-storied brick pagodas of the Tang Dynasty implies the use of proportion and modules that took other wooden structures as a reference (Wang et al. 2021b). This designates that a universal construction method existed and that it has been preserved in East Asia, for example, in countries such as Korea and China, for a long time (Kim 2011; Kim and Park 2017).

3. Research Method

To build a traditional wood-frame structure, builders placed plinth stones on a platform and then situated the columns on the platform. They positioned transverse beams beneath purlins to establish the building framework. They completed the basic wooden framework using timber-framed interior structures with rafters above. Roofs were shaped by two diagonally slanted struts whose heads were joined together and topped by a bearing block in the synthesis of a top beam with an upper roof purlin. The wooden structures relied on columns, beams, and purlins as their basic components, the location and size of which, and the method used to connect each part, determined the building scale.

The following settings were used to describe the concepts of squares, circles, and geometric models.

First, in this study, "(H)" means the eave column height of a wooden building and serves as a standard reference for the overall composition and scale of the timber-framed structure.

Second, the circle–square (圓方圖) and the square–circle (方圓圖) diagrams include the principle for making a circular form using a square (Figure 5), according to the *Zhoubi Suanjing* and *Jiuzhang Suanshu*. The unremitting and repetitive overlap of squares within circles and circles within squares shows that there were regular rules to calculate the length comparable to geometric sequences in modern mathematics. The proportional principle determines each diagonal line length made through the superimposition of squares and circles outwards or inwards, resting on the length ratio of verticality and horizontality (Figure 6). According to the "Cuifen" (衰分, Distribution by Proportion) of the *Jiuzhang Suanshu*, this book introduced the problem regarding "*nuzhi*" (women who weave textiles) to explain the principle as a geometric progression.[8] The *nuzhi* show a ratio based on 2 in the sequence 1, 2, 4, 8, 16, etc.[9] (Figure 7). The ancient Chinese knew the geometric law that was created through the repetitive combination of circles and squares for such a consistent system of proportions. Through this principle, a method of weaving was developed. The same calculation method was also used in the *juzhe* ((first) raise (the total height) and (then) break (for individual heights of purlins)) method of the *Yingzao Fashi* (*State Building Standards*) of the Song Dynasty (Li 1100).

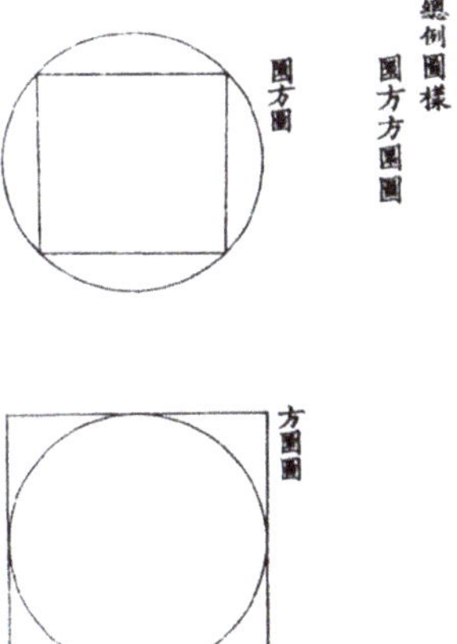

Figure 5. Song's *Yingzao Fashi* circle–square diagram (upper), square–circle diagram (bottom).

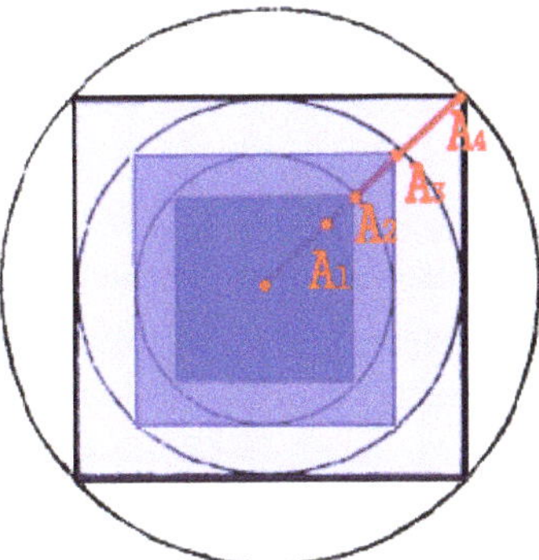

Figure 6. Superimposing the circle–square and the square–circle maps.

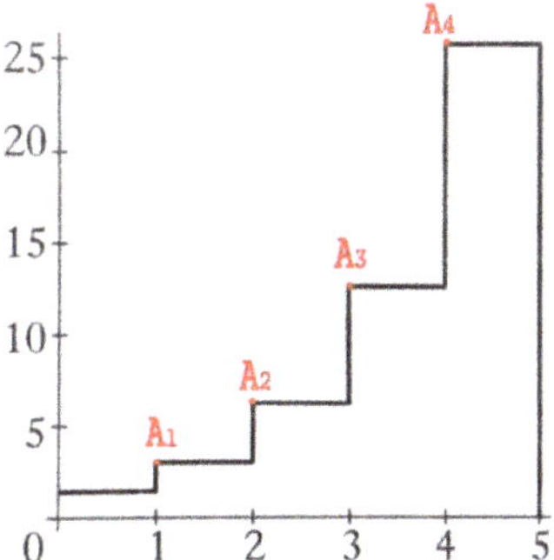

Figure 7. The *Jiuzhang Suanshu*: about problem four on the proportional distribution (*Cuifen*), a graph for the solution on "weaving woman as skillful weaver".

Third, assuming that the roof curve from the eave purlins to the ridge purlin becomes steeper in the wooden framework, the purlins would be situated in a gradually higher position, with decreasing height ratios of 1/10, 1/20, 1/40, and 1/80, respectively, according to the *juzhe* method for roof curvature in the *Yingzao Fashi*, juan 5, "Major Carpentry System." The ridge purlin height must be determined first by basically considering the main crossbeam length, and then other purlins' heights can be determined to make the roof curvature by placing rafters on the roof purlins (Li 1100). Consequently, this shows that the geometric rule had been persistently used to shape a curved roof in ancient timber-framed buildings, conforming to the location and arrangement of all purlins (Figure 8).

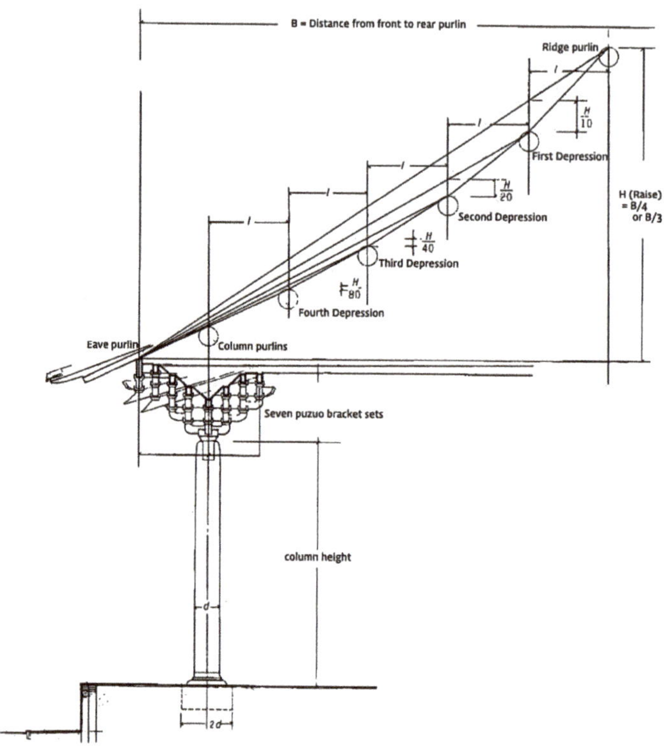

Figure 8. *Ceyang* (側樣, Section), *Chicun* (尺寸, Feet and Inches) and *Juzhe* (舉折, Raising and Depression) *Methods* in Song's *Yingzao Fashi*, Adapted from Pan and He (2005, p. 51).

Fourth, regarding the second and third assumptions, $\sqrt{2}$ H times the eave column height is served in the cross-section of the wooden structure as a reference standard for the geometric concept. Therefore, in successive order, the numerical values are as follows: 2H, $\sqrt{2}$ H, H, $\frac{H}{\sqrt{2}}$, $\frac{H}{2}$, $\frac{\sqrt{2}H}{4}$, and $\frac{H}{4}$ in regular sequence, implying a geometric sequence in which each term after the first is $\frac{H}{\sqrt{2}}$ times the previous term. The geometric series with a ratio of $\frac{H}{\sqrt{2}}$ makes it possible to organize interior timber-framed structures by applying a proportional principle.[10] The geometric concept is applied as the reference standard for eave column height, depending on the drawing of squares and circles that used a drawing compass and carpenter's square, and is manipulated according to the proportion of $\sqrt{2}$ H times the eave column height (Figures 9 and 10).

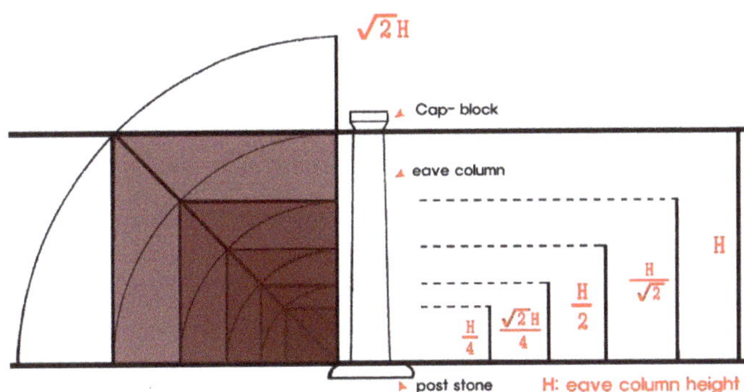

Figure 9. The geometrical diagram based on $\sqrt{2}\,H$ times the eave columns' height (H).

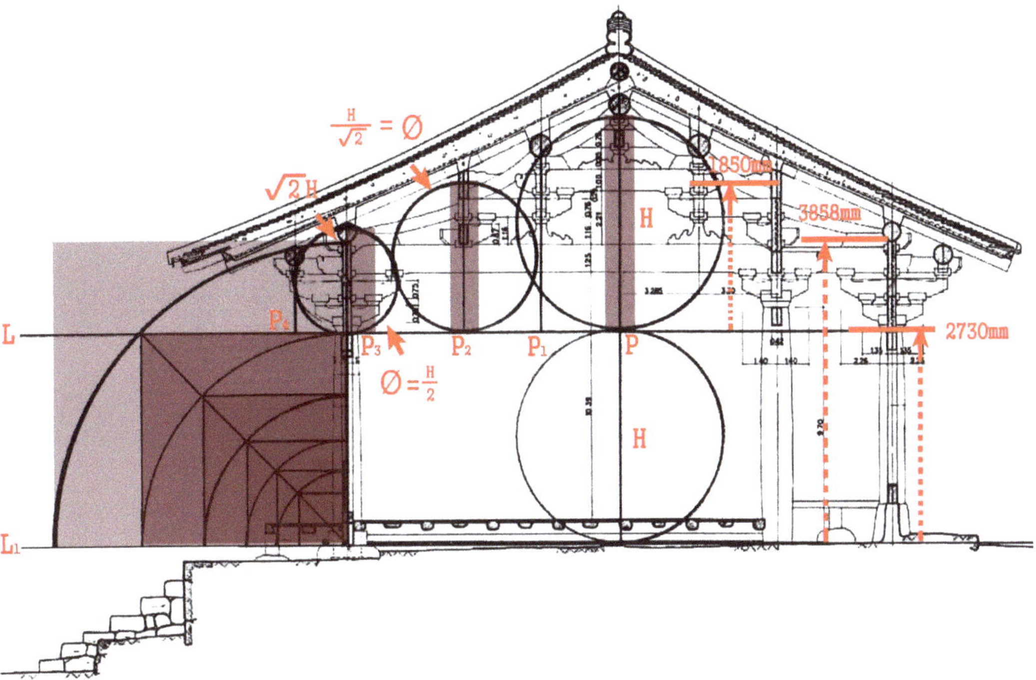

Figure 10. The Geungnakjeon Hall of Bongjeongsa Temple (before the repair work in 1973). proportional analysis of the sectional elevation (H = eave column height, Ø = circle diameter), Adapted from CHA (2003, p. 389).

Further, this study established a standard line for a detailed comparison of the purlin location and arrangement between each building using the following abbreviations: (H) is the eave column height, (L) is the horizontal standard line that touches the upper side of the front and rear eave columns, (L₁) is the horizontal line meeting the upper side of the front and rear plinth stones, and (P) is the intersection between (L) and the vertical downward line from the centerline of the ridge purlin. Where (L) connects to the upper side of the front and rear eave columns, the vertical downward lines from the upper-middle roof purlins, middle roof purlins, lower-middle roof purlins, and column purlins are identified as P_1, P_2, P_3, and P_4, respectively. As noted above, we analyzed the proportional rapport between the location and distance of the wooden components in timber-framed structures (Figure 9).

4. Interpretation of the Proportional System in Nine Goryeo Era Timber-Framed Buildings

4.1. Associations between the Eave Column Height (H) and $\sqrt{2}H$

The line drawing of the Bongjeongsa Monastery's Geungnakjeon Hall reveals a notable point. Achieved using a square with a circle created with compasses and L-square rulers, the $\sqrt{2}$ H times the eave column height from the top plinth stones to the top eave columns is equal to the height of the column purlins. In the Muryangsujeon Hall at the Buseoksa Temple, the ratio is entirely different. Although the builders used a method similar to that used for Bongjeongsa's Geungnakjeon Hall, the eave column height of the Muryangsujeon Hall is $\sqrt{2}$ H times the height from the top of the plinth stones to the top of the eave purlins, not the column purlins. Thus, the column purlin height was $\sqrt{2}$ H times the eave column height (Figures 10 and 11). The same method was used for other Buddhist monastery buildings, such as the Daeungjeon Hall at Sudeoksa, Josadang Hall at Buseoksa, Geojoam Hermitage's Yeongsanjeon Hall at Eunhaesa, and Sammun Gate at Imyeonggwan Guesthouse (Figures 12–18). For these buildings, the height from the top of the plinth stones to the top of the column purlins was employed as $\sqrt{2}$ H× the eave column height.

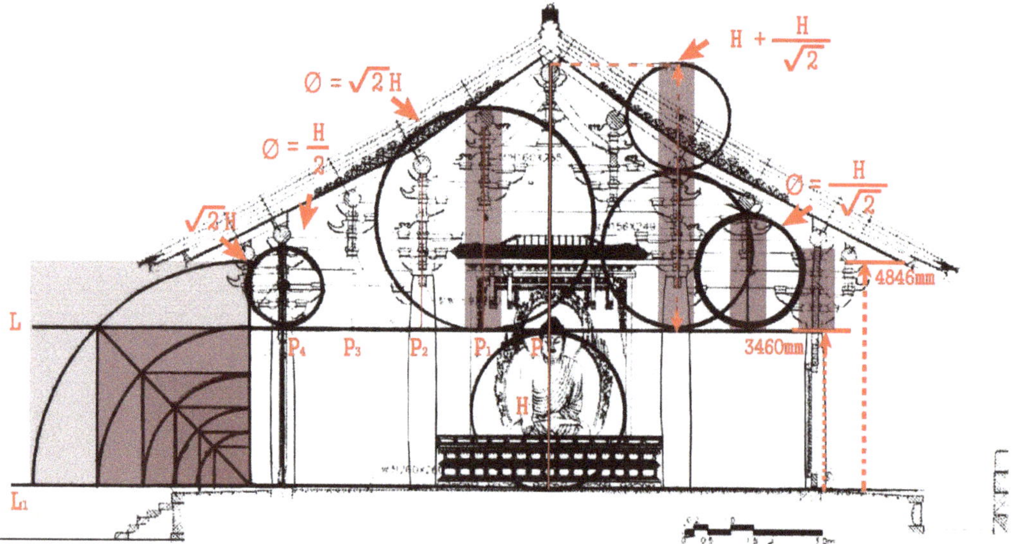

Figure 11. Muryangsujeon Hall of Buseoksa Monastery. Proportional analysis of the sectional elevation (H = eave column height, Ø = circle diameter), Adapted from CHA (2002, p. 133).

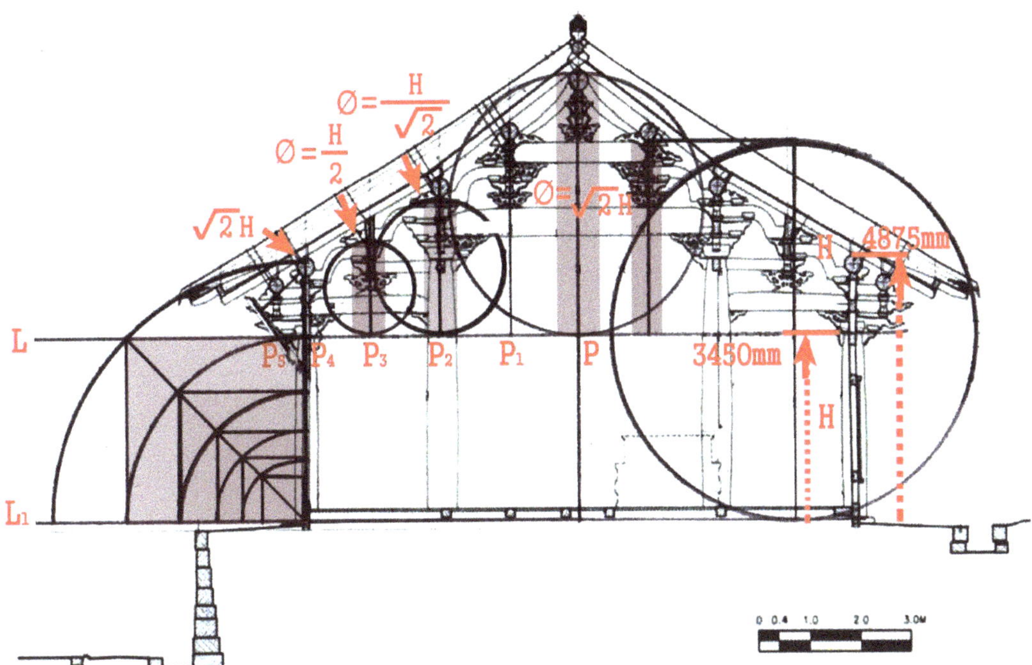

Figure 12. Daeungjeon Hall of Sudeoksa Monastery. Proportional analysis of the sectional elevation (H = eave column height, Ø = circle diameter), Adapted from CHA (2005a, p. 78).

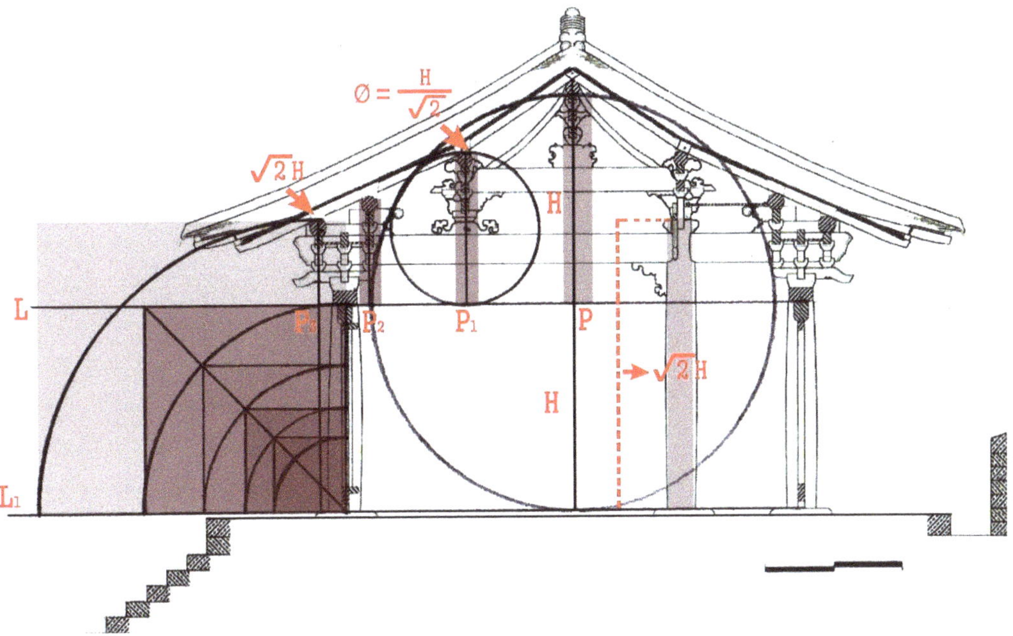

Figure 13. Eungjinjeon Hall of Seongbulsa Monastery. Proportional analysis of the sectional elevation (H = eave column height, Ø = circle diameter), Adapted from CHA (NRICH 1998, p. 42).

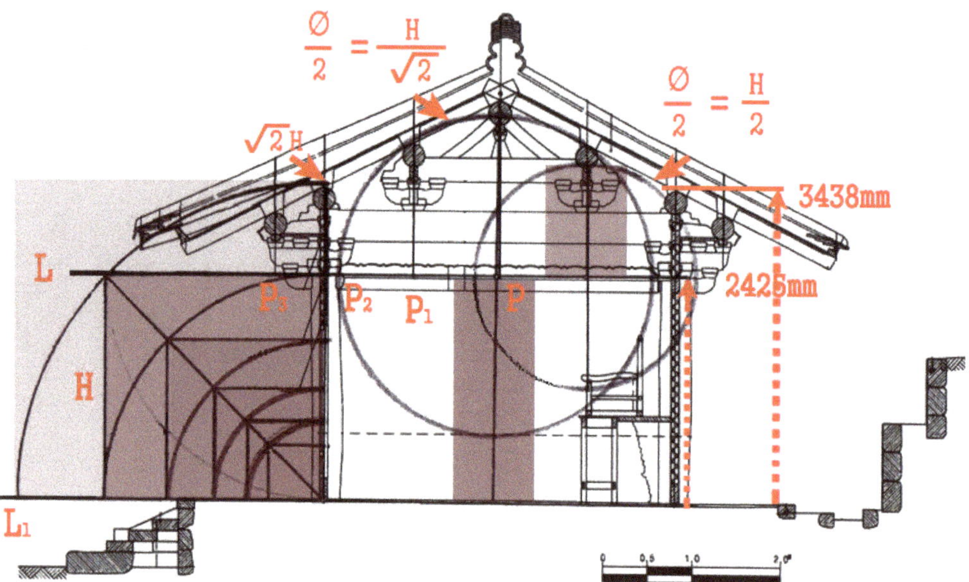

Figure 14. Josadang Hall of Buseoksa Monastery. Proportional analysis of the sectional elevation (H = eave column height, Ø = circle diameter), Adapted from CHA (2005b, p. 274).

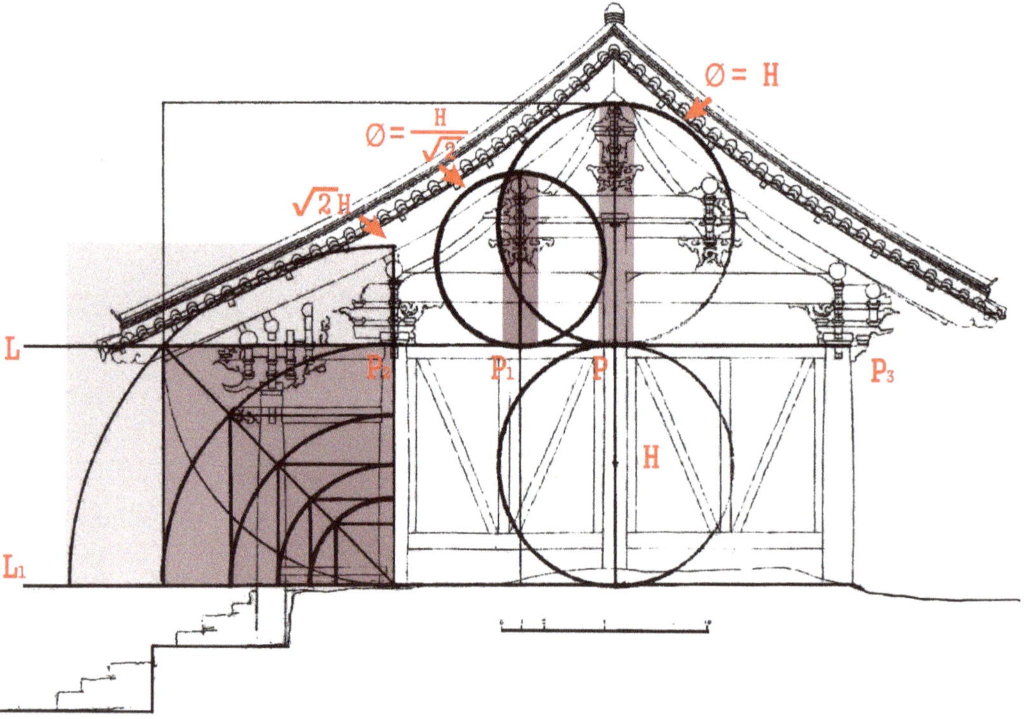

Figure 15. Geungnakjeon Hall of Seongbulsa Monastery. Proportional analysis of the sectional elevation (H = eave column height, Ø = circle diameter), Adapted from CHA (NRICH 1998, p. 38).

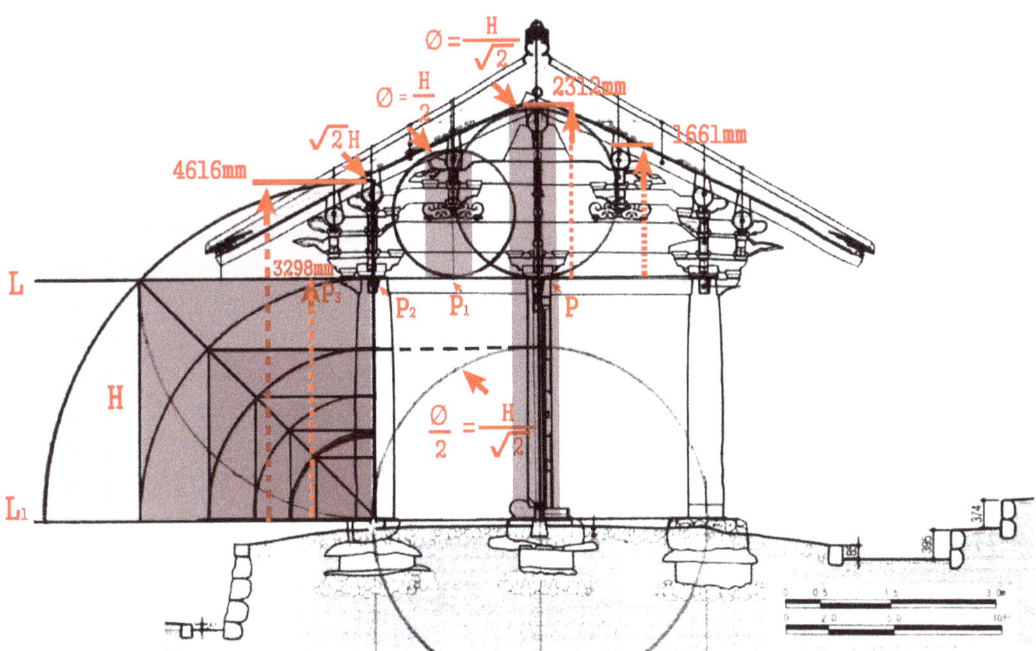

Figure 16. Sammun Gate of Imyeonggwan Guesthouse. Proportional analysis of the sectional elevation (H = eave column height, Ø = circle diameter), Adapted from CHA (2004, p. 442).

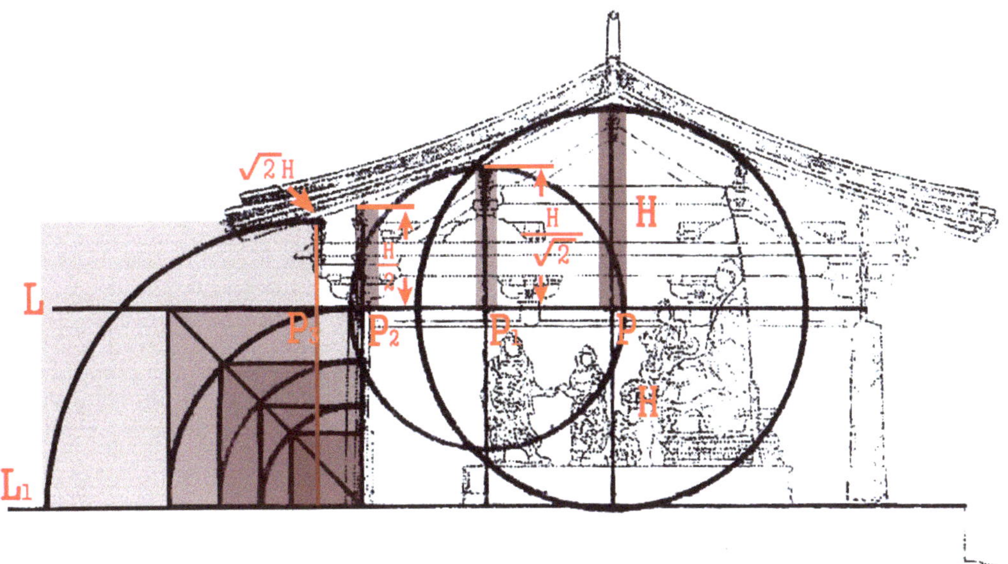

Figure 17. Main Hall of Monastery Nanchansi. Proportional analysis of the sectional elevation (H = eave column height, Ø = circle diameter), Adapted from Wang et al. (2011, p. 143).

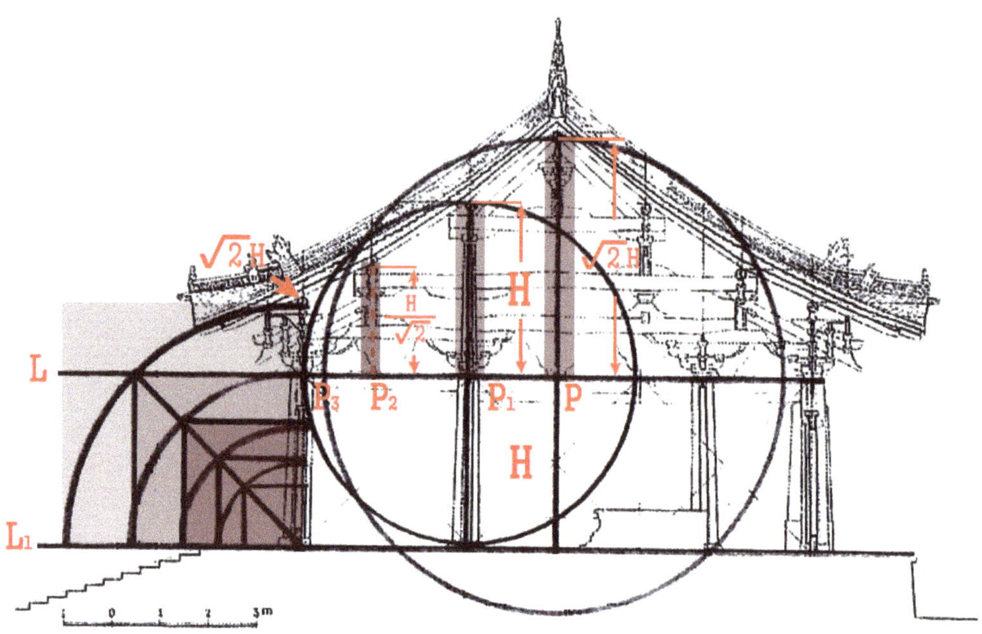

Figure 18. Main Hall of Chuzu'an Hermitage at Shaolinsi Monastery. Proportional analysis of the sectional elevation (H = eave column height, Ø = circle diameter), Adapted from Wang et al. (2011, p. 288).

The Bogwangjeon Hall at Simwonsa Temple and Eungjinjeon Hall at Seongbulsa Temple differed somewhat from the above-mentioned buildings. The Eungjinjeon Hall did not have any column purlins except for its internal and external eave purlins. The distance from the top of the eave purlins to the top of the building platform was $\sqrt{2}$ H times the eave column height. The Bogwangjeon Hall at the Simwonsa Temple had no column purlins but had external eave purlins (Figures 13 and 19). Assuming that the height from the top of the stone plinths to the top of the architraves, including the eave columns, measured 1 unit, a length of $\sqrt{2}$ H times the height of the eave columns would not reach the bottom of the eave purlins. Further, the eave column height was consistent with the top of the main crossbeams, which were made thicker to adjust for the height. Planks were also added below the eave purlins to control the height between the eave purlins and main crossbeams. There are several cases with similar proportions among the Tang, Song, and Liao buildings in ancient Chinese dynasties. They embrace the East Hall of the Foguangsi Temple and the Main Halls of the Zhenguosi, Hualinsi, and Kaiyuansi Temples. The height of $\sqrt{2}$ times the eave columns is slightly larger than the top side of the eave purlins' height, by a ratio of approximately 1.5 times. The majority of buildings in China that have descending cantilevers differ from those in Korea that do not have any descending cantilevers (Wang 2011a, pp. 41–42).

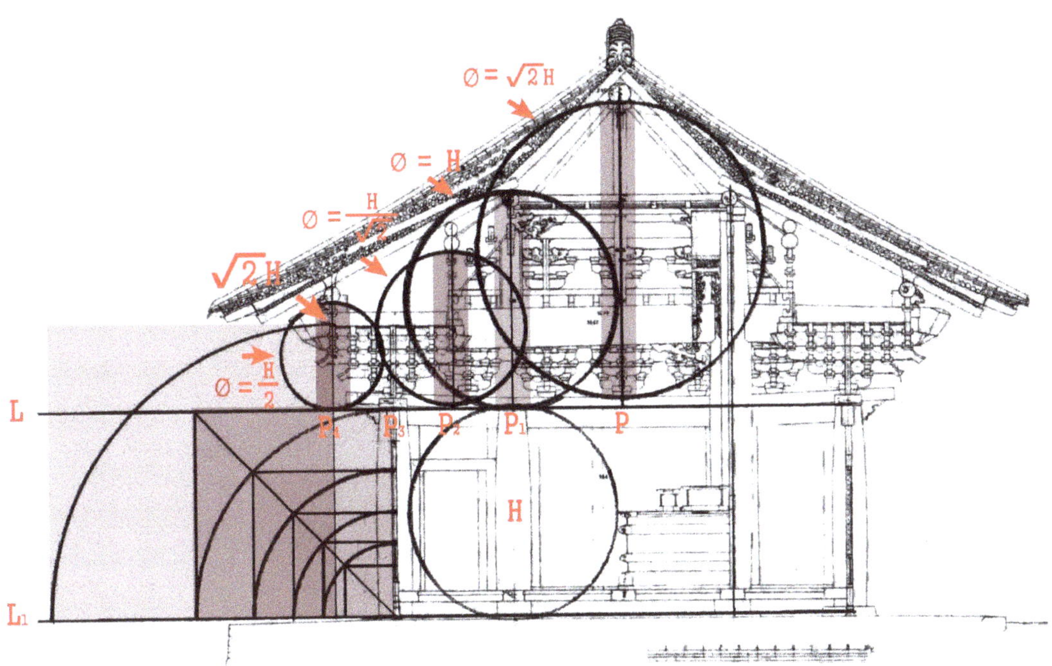

Figure 19. Bogwangjeon Hall of Simwonsa Monastery. Proportional analysis of the sectional elevation (H = eave column height, ∅ = circle diameter), Adapted from CHA (NRICH 1998, p. 49).

For the Geungnakjeon Hall at the Seongbulsa Monastery, there was a difference between (H) and $\sqrt{2}$ H times the height from the top of the plinth stones to the top of the column purlins (Figure 15). The Geungnakjeon Hall's proportional principle complied with a ratio of 1.55H times the eave column height, somewhat higher than $\sqrt{2}$ H times, similar to the proportional rule used for the main halls of the Zhenguosi Monastery (鎮國寺 大殿, 963) and the Hualinsi Monastery (華林寺 大殿, 964) from the Five Dynasties and Ten Kingdoms Period (907–979) (Wang 2011a). However, the two Chinese buildings had relatively larger crossbeams and represented three downward-slanting cantilevers, unlike the structural systems of the Seongbulsa Monastery's Geungnakjeon Hall, which did not use $\sqrt{2}$ H times (H)[11] (Table 2).

Of the nine buildings, seven followed the proportional principle that $\sqrt{2}$ H times the eave column height should meet the top or bottom of the column purlins and the bottom of the eave purlins. The other two buildings (Eungjinjeon at Seongbulsa and Bogwangjeon at Simwonsa Monasteries) did not have column purlins; however, the overall proportional system was the same as for the other seven buildings.

Table 2. The proportional relationship between eave columns' height (H) and $\sqrt{2}$ times.

Building Name	Component Location That Meets $\sqrt{2}$ Times the Eave Column Height (H)	Remarks
Geungnakjeon Hall at Bongjeongsa Monastery	The bottoms of column purlins	Other purlins are positioned at the bottoms
Muryangsujeon Hall at Buseoksa Monastery	The tops of eave purlins	Some purlins are positioned at the bottoms
Daeungjeon Hall at Sudeoksa Monastery	The bottoms of column purlins	Other purlins are positioned at the bottoms
Eungjinjeon Hall at Seongbulsa Monastery	The tops of eave purlins	Other purlins are positioned at the bottoms
Geungnakjeon Hall at Seongbulsa Monastery	A little higher than the tops of column purlins	Other purlins are positioned at the bottoms
Yeongsanjeon Hall at Geojoam Hermitage, Eunhaesa Monastery	The tops of column purlins	Some purlins are positioned at the bottoms
Josadang Hall at Buseoksa Monastery	The tops of column purlins	Some purlins are positioned at the bottoms
Bogwangjeon Hall at Simwonsa Monastery	The bottoms of eave purlins	Other purlins are positioned at the bottom sides
Sammun Gate at Imyeonggwan Guesthouse in Gangneung City	The tops of column purlins	Other purlins are positioned at the tops
The Main Hall at Nanchansi Monastery (Tang, early eighth century)	The tops of eave purlins	
The East Hall at Foguangsi Monastery (Tang, 857)	The tops of eave purlins	
The Main Hall at Hualinsi Monastery (Song, 964)	The tops of eave purlins	
The Main Hall at Geyuansi Monastery (Liao, 966)	The tops of column purlins	
The Shanmen Front Gate at Dulesi Monastery (Liao, 984),	The tops of column purlins	Including plinths height
The Main Hall at Baoguosi Monastery (Song, 1013)	The tops of column purlins	Including plinths height
The Main Hall at Fengguosi Monastery (Liao, 1020)	The tops of column purlins	Including plinths height
The Main Hall at the Chuzu'an Hermitage (Song, 1125)	The tops of column purlins	

4.2. Connections between (H), the Columns' Vertical Positioning, and the Proportional Relationship to $\sqrt{2}$ Times

Pre-Joseon Dynasty documents regarding the Korean Peninsula do not describe a method for determining purlin arrangement to create a roof framework. Further, the *Yeongjo Uigwe* (*Royal Protocols in the Joseon Dynasty*) does not include decisions on building proportions or purlin height. Chinese literature mentions how to arrange purlins, with instructions in the *Kaogongji* (*The Artificers' Record*) (Anonymous n.d.b.) and the *Yingzao Fashi* (*State Building Standards*) of the Song Dynasty, and the *Gongcheng Zuofa* of the Qing Dynasty (Anonymous 1734). Two of these books indicated that in the Ming–Qing Era buildings, the height of each purlin's position had slightly increased, a feature that distinguishes Song-Era buildings from those built during the Ming and Qing Dynasties (Li 1982).

First, in the Jiangren "architect–artisans" section of the *Kaogongji,* in setting out the method for determining roof height, the text of the Jiangren wei gouxu section reads as follows: (For a thatched roof, (the height of the roof from the eaves is) one-third (of the width of the building); for a tiled roof, one-fourth). This not only indicates that the roof height is usually germane to the width of a building, but it also clearly states that the height depends upon which type of roof is under construction, and the ratio between them leads to a progressive increase or decrease in the height of the ridge. Buildings have three modular units ((original annotation) as a falling cone): The (main) crossbeam and what is above it belong to the upper unit; what is above ground (except for the beam structure) belongs to the middle unit; the stairs belong to the lower unit (Wen 2008, p. 126).

Second, the Damiuzuo zhidu 大木作制度 (major carpentry system) in the section of the Yingzao Fashi notices that the construction system of the roof-height curve (juzhe), raising the roof (juwu), and inward-curving roofs (zhewu) correlate with the horizontal length from eave purlins (laoyanfang) to a ridge purlin (jishuan) and the vertical length between them, and the horizontal and vertical lengths have different dimensions depending on round roof tiles (banwa), semi-cylindrical roof tiles (tongwa), multi-story buildings (louge), and hall-type buildings (tingtang), and the construction system of inward-curving roofs, such that the irregular arrangement of the purlins brings about moderate increases or decreases. In all circumstances, the length of the (main) crossbeam governs the height of the ridge purlin, and the ratio between them leads to a progressive increase or decrease in the height of the ridge. Li Jie says that the recent code of the design of the roof-height curve is essentially developed from the code that takes a quarter of the houses to make the height of their roofs. It is in line with the classic text of the *Kaogongji* of the *Zhouli* (Liang and Li 1983).

In contrast, in Korea, anonymous buildings in the pre-Goryeo Era show odd bays at the side façade. In the case of Joseon-Era's buildings constructed since the seventeenth century, most cases subsume an odd row of columns such as one, three, and five bays, and so on. Among them, wooden constructions with three bays at the side façade occupy the majority of cases, and generally, the central bay at the side façade is wider than the outermost bays' widths at the front and back façades.

The roof height from eaves is one-third or one-fourth of the building width mentioned from the *Zhouli* and the *Yingzao Fashi* has the same length as each bay width at the side façade, whereas in ancient buildings in Korea, the bay widths are not equally distributed. These factors show that the ancient Korean buildings are structurally different from the Chinese. Nevertheless, it should be noted that Korean and Chinese buildings shared and preserved universal and standardized design principles for setting purlin locations and bracket-set heights.

The baseline (L) marks the connection between the top of the eave columns in the cross-section from front to back, creating a cross-point intersecting with the baseline (L) and descending from a ridge post to the eave purlins. The eave purlins were generally placed at the lowest point, while the ridge post was at the highest.

The Geungnakjeon Hall at the Bongjeongsa Temple had upper- and lower-middle roof purlins but no middle roof purlins. The lower-middle roof purlins aligned with a row of interior columns in the cross-section. Assuming that the eave column height was H above the baseline (L), a geometric progression with the ratio of $\frac{1}{\sqrt{2}}$ was established as 2H, $\sqrt{2}$ H (=1.414), H, $\frac{H}{\sqrt{2}}$, (=0.7), $\frac{H}{2}$, $\frac{\sqrt{2}H}{4}$, (=0.35), and $\frac{H}{4}$, adjusting the numerical variations (Figure 10). The Bongjeongsa Monastery's Geungnakjeon Hall shows that purlin height dictated the location of the eave purlins, column purlins, lower-middle roof purlins, and ridge purlin as $\frac{\sqrt{2}H}{4}$, $\frac{H}{2}$, $\frac{H}{\sqrt{2}}$, and H times the eave column height, which is a geometric progression. The height of the upper-middle roof purlins was decided as the sum of $\frac{\sqrt{2}H}{4}$ plus $\frac{H}{2}$ times ratio, based on the height of the eave columns and the eave purlins. Determining each purlin height that supported the roof frame of the Geungnakjeon Hall relied on the column height (H).

The purlins of the Buseoksa Monastery's Muryangsujeon Hall were not uniformly distributed but densely placed from the middle roof purlins to the upper-middle roof and ridge purlins. The height proportion of the top of the eave purlins and the bottom of the column purlins above the baseline (L) used the $\frac{H}{2}$ times formula. The height from the column purlins to the upper-middle roof purlins was calculated according to the geometric sequence. That is, if the eave column height was H, the eave purlin height would be $\sqrt{2}$ H, and the middle roof purlin height would be 2H, following a geometric sequence with the $\sqrt{2}$ ratio. The height of the top of the upper-middle roof purlins above the baseline (L) corresponded to the $\sqrt{2}$ H times formula, which had no distinction compared to the ridge purlin height, H plus $\frac{H}{\sqrt{2}}$. The ridge purlin height deviated somewhat from the geometric sequence, as it did not maintain a common difference.

Similarly, the Sudeoksa Monastery's Daeungjeon Hall used a geometric concept with a common ratio of $\sqrt{2}$ from the eave purlins to the ridge post. The heights of the column purlins and the ridge post were decided by their tops, and the other purlins were situated according to their bottoms, revealing a relatively consistent proportional system. The Seongbulsa Monastery's Eungjinjeon Hall had no eave columns, and both the external and internal eave purlins used the height ratio of $\frac{\sqrt{2}\,H}{4}$ and $\frac{H}{2}$ times separately, depending on the column purlin height and geometric progressions. Seongbulsa Monastery's Geungnakjeon Hall added outer bays to the front façade in 1650 (the first year of King Hyojong's reign). Thus, the outer bays were excluded in the present study. The location of the external eave purlins used a somewhat larger value than $\sqrt{2}$ H, while the locations of the other eave purlins used a geometric progression (Table 3).

Vertically downward lines from upper-middle roof purlins, middle roof purlins, and lower-middle roof purlins are considered as P1, P2, P3, P4, and P5, respectively.

The Geojoam Hermitage's Yeongsanjeon Hall at the Eunhaesa Temple employed a geometric progression with the $\sqrt{2}$ ratio from the eave purlins to the middle roof purlins. The ridge purlin height was established based on the sum of $\frac{H}{\sqrt{2}}$ plus $\frac{H}{2}$ times. Its height was higher than (H). The location of the ridge post was relatively high, indicating a proportional system similar to that of the Muryangsujeon Hall at the Buseoksa Monastery (Figures 11 and 20).

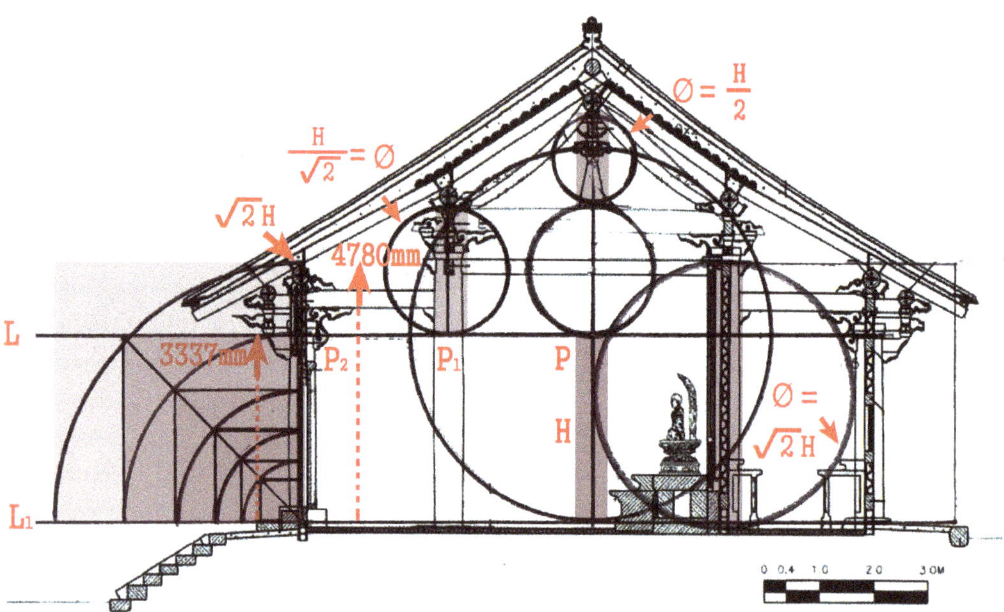

Figure 20. Yeongsanjeon Hall of Geojoam Hermitage, Eunhaesa Monastery. Proportional analysis of the sectional elevation (H = eave column height, Ø = circle diameter), Adapted from CHA (2004, p. 294).

Table 3. Each purlin height in the application of $\sqrt{2}$ times the height of eave columns above the baseline (L).

Building Name		Each Purlin Height Ratio above the Baseline (L) That Meets the Tops of the Front and the Rear Eave Columns					
		Eave (External) Purlins' Height	Column Purlins' Height	Lower-Middle Roof Purlins Height	Middle Roof Purlins' Height	Upper-Middle Roof Purlins' Height	Ridge Purlin Height
Geungnakjeon Hall at Bongjeongsa Monastery		$P_4,\ \frac{\sqrt{2}\,H}{4}$	$P_3,\ \frac{H}{2}$	$P_2,\ \frac{H}{\sqrt{2}}$	None	$P_1,\ \frac{\sqrt{2}\,H}{4}+\frac{H}{2}$	P, H
Muryangsujeon Hall at Buseoksa Monastery		$P_5,\ \frac{\sqrt{2}\,H}{4}$	$P_4,\ \frac{H}{2}$	$P_3,\ \frac{H}{\sqrt{2}}$	P_2, H	$P_1,\ \sqrt{2}\,H$	P, H $+\frac{H}{\sqrt{2}}$
Daeungjeon Hall at Sudeoksa Monastery		$P_5,\ \frac{H}{4}$	$P_4,\ \frac{\sqrt{2}\,H}{4}$	$P_3,\ \frac{H}{2}$	$P_2,\ \frac{H}{\sqrt{2}}$	$P_1,$ H	P, $\sqrt{2}$H
Eungjinjeon Hall at Seongbulsa Monastery	Eave Columns' Height (H)	$P_3,\ \frac{\sqrt{2}H}{4}$	$P_2,\ \frac{H}{2}$ (inner eave purlins)	None	$P_1,\ \frac{H}{\sqrt{2}}$	None	P, H
Geungnakjeon Hall at Seongbulsa Monastery		$P_3,\ \frac{H}{4}$	$P_2,\ \frac{\sqrt{2}H}{4}$	None	$P_1,\ \frac{H}{\sqrt{2}}$	None	P, H
Yeongsanjeon Hall at Geojoam Hermitage, Eunhaesa Monastery		$P_3,\ \frac{H}{4}$	$P_2,\ \frac{\sqrt{2}H}{4}$	None	$P_1,\ \frac{H}{\sqrt{2}}$	None	P, $\frac{H}{\sqrt{2}}+\frac{H}{2}$
Josadang Hall at Buseoksa Monastery		$P_3,\ \frac{H}{4}$	$P_2,\ \frac{\sqrt{2}H}{4}$	None	$P_1,\ \frac{H}{2}$	None	P, $\frac{H}{\sqrt{2}}$
Bogwangjeon Hall at Simwonsa Monastery		$P_4,\ \frac{H}{2}$	None	$P_2,\ \frac{H}{\sqrt{2}}$	None	$P_1,$ H	P, $\sqrt{2}$H
Sammun Gate at Imyeonggwan Guesthouse in Gangneung City		$P_3,\ \frac{H}{4}$	$P_2,\ \frac{\sqrt{2}H}{4}$	None	$P_1,\ \frac{H}{2}$	None	P, $\frac{H}{\sqrt{2}}$
The Main Hall at Nanchansi Monastery (early eighth century)		$P_3,\ \frac{\sqrt{2}H}{4}$	$P_2,\ \frac{H}{2}$	None	$P_1,\ \frac{H}{\sqrt{2}}$	None	P, H
The East Hall at Foguangsi Monastery (857)	Eave Columns' Height (H)	$P_3,\ \frac{\sqrt{2}H}{4}$	$P_2,\ \frac{H}{\sqrt{2}}$	None	P_1, H	None	P, $\sqrt{2}$H
The Main Hall at Hualinsi Monastery (964)		$P_3,\ \frac{\sqrt{2}H}{4}$	$P_2,\ \frac{H}{\sqrt{2}}$	None	P_1, H	None	P, $\sqrt{2}$H
The Main Hall at the Chuzu'an Hermitage (1125)		$P_3,\ \frac{\sqrt{2}H}{4}$	$P_2,\ \frac{H}{\sqrt{2}}$	None	P_1, H	None	P, $\sqrt{2}$H

The Josadang Hall at the Buseoksa Monastery and the Sammun Gate at the Imyeonggwan Guesthouse had three columns in their front façades. They complied with the ratio of $\frac{H}{4}$, $\frac{\sqrt{2}H}{4}$, $\frac{H}{2}$, and $\frac{H}{\sqrt{2}}$ times the eave column height, from the middle roof purlins to the ridge purlin. Whereas the Sammun Gate had the location of each purlin based on the top side of all roof purlins, the Josadang Hall had the placement of each purlin based on the tops of the eave purlins and column purlins, and the bottoms of the middle roof purlins and ridge purlin (Figure 14). The Sammun Gate used the standard height from a stone platform to the top of all roof purlins, and the Josadang Hall used the standard height from a stone platform to the top of the eave and column purlins. The heights of the middle roof purlins and ridge purlin were based on the bottoms of the middle roof purlins and ridge purlin. The Bogwangjeon Hall at the Simwonsa Monastery and the Eungjinjeon Hall at the Seongbulsa Monastery had multilayer bracket systems and showed unique proportional features. Their design principle depended on the eave column height. The Bogwangjeon Hall had a few differences from the Eungjinjeon Hall; for example, the eave purlin height did not correspond to the $\sqrt{2}$ times ratio, although the rest were matched by the ratio of $\sqrt{2}H$ times the eave column height, a standard for composing a complete proportional framework. Based on the eave column height, the location of each purlin, from the eave purlins to the ridge purlin, showed a geometric progression of $\frac{H}{2}$, $\frac{H}{\sqrt{2}}$, H, and $\sqrt{2}H$ times as a proportional system, although there were no column purlins.

Thus, the design characteristics of these nine buildings provide evidence that the builders used (H) as a reference point connected with a geometric series ratio of $\sqrt{2}H$ times to set the height of each purlin.

4.3. The Proportional Relationship between the Outer Bay width and the Eave Column Height (H)

Most buildings in the Goryeo Era had rows of columns at the side façade arranged on the same vertical location as the lower-middle and middle roof purlins. These buildings were evidently distinct from the wood-frame buildings with side columns and roof purlins that have been organized differently since the seventeenth century. Few Goryeo buildings had an even row of columns at the side façade. The Bongjeongsa Monastery's Geungnakjeon Hall and the Sudeoksa Monastery's Daeungjeon Hall had four bays at their side façades. The Seongbulsa Monastery's Eungjinjeon Hall had two bays at the side façade, exclusive of the front outer bays. In comparison to the nine buildings in this study, the Geungnakjeon Hall at the Bongjeongsa Temple had the smallest width in the outer bays at the side façade. The Sudeoksa Temple's Daeungjeon Hall had an almost equivalent lateral width in each bay at the side façade. The Seongbulsa Monastery's Geungnakjeon Hall had high columns that did not meet a ridge post, similar to Sudeoksa Monastery's Daeungjeon Hall, where the columns were sheltered into the bottom of the ridge post. In Bongjeongsa's Geukrakjeon Hall, Seongbulsa's Eungjinjeon Hall, and Simwonsa's Bogwangjeon Hall, the width of the outer bays at the side façade showed a ratio of $\frac{H}{2}$ plus $\frac{\sqrt{2}H}{4}$ times, which was shorter than (H)—0.6H times larger than half the eave column height.

For the Muryangsujeon and Josadang Halls at the Buseoksa Monastery and the Yeongsanjeon Hall at the Geojoam Hermitage in the Eunhaesa Monastery, the outer bay width at the side façade was approximately $\frac{H}{2}$ plus $\frac{\sqrt{2}H}{4}$ (0.85H) times the eave column height. For the Sammun Gate at the Imyeonggwan Guesthouse and the Daeungjeon Hall at the Sudeoksa Monastery, the outer bay width at the side façades was $\frac{H}{\sqrt{2}}$ times (0.7H) the eave column height. Likewise, the Seongbulsa Monastery's Geungnakjeon Hall had high columns at the center of the side façades, divided into two bays from front to back. The outer bay width on the side façades was equal to the eave purlin height (Table 4). This is similar to the proportional rule used in Chuzu'an Hermitage's Main Hall (1125) of the Shaolinsi Monastery (Mount Song, Dengfeng, Henan) and Yuhuagong (1008) of the Yongshou Monastery (Yuci, Shanxi). Nanchansi Monastery's Main Hall (early eighth century) in Wutaixian, Shaanxi, was the earliest wooden building built with three square bays and no inner columns. The outer bay width at the side façade was 0.7H times (H). As was established in earlier sections, the nine buildings used a common proportional rule between (H) and the side façade width as a significant design principle, which was also likely used for other wooden buildings in ancient Korea and China (Figures 17, 18, 21 and 22). The outer bay width at the side façade varied by 0.6H, 0.7H, and 0.85H times the eave column height, depending on (H), supporting the proposition that the builders used the $\sqrt{2}H$ ratio to increase or decrease (H). For example, at Sudeoksa Monastery's Daeungjeon Hall and Imgyeonggwan's Sammun Gate, the (H) width matched the increase or decrease in the ratio of the side bays based on an arithmetic concept. The other buildings had a wood-frame structure incorporating $\sqrt{2}H$ times the height ratio, increasing or decreasing according to arithmetic and geometric concepts.

Table 4. Proportional relationship between outermost bay width and eave column height.

Building Name		The Ratio of Eave Columns' Height (H) to Outermost Bays Width at the Side Façade	Remarks
Geungnakjeon Hall at Bongjeongsa Monastery		$\frac{H}{4} + \frac{\sqrt{2}H}{4}$	
Muryangsujeon Hall at Buseoksa Monastery		$\frac{H}{2} + \frac{\sqrt{2}H}{4}$	
Daeungjeon Hall at Sudeoksa Monastery		$\frac{H}{\sqrt{2}}$	
Eungjinjeon Hall at Seongbulsa Monastery		$\frac{H}{4} + \frac{\sqrt{2}H}{4}$	
Geungnakjeon Hall at Seongbulsa Monastery	Eave Columns' Height (*H*)	H	
Yeongsanjeon Hall at Geojoam Hermitage, Eunhaesa Monastery		$\frac{H}{2} + \frac{\sqrt{2}H}{4}$	
Josadang Hall at Buseoksa Monastery		$\frac{H}{2} + \frac{\sqrt{2}H}{4}$	
Bogwangjeon Hall at Simwonsa Monastery		$\frac{H}{4} + \frac{\sqrt{2}H}{4}$	
Sammun Gate at Imyeonggwan Guesthouse in Gangneung City		$\frac{H}{\sqrt{2}}$	
The Main Hall at Nanchansi Monastery (early eighth century)		$\frac{H}{\sqrt{2}}$	
The East Hall at Foguangsi Monastery (857)		$\frac{H}{2} + \frac{\sqrt{2}H}{4}$	
Yuhuagong Hall at Yongshou Monastery (1008)		H	
The Main Hall at the Chuzu'an Hermitage (1125)		H	

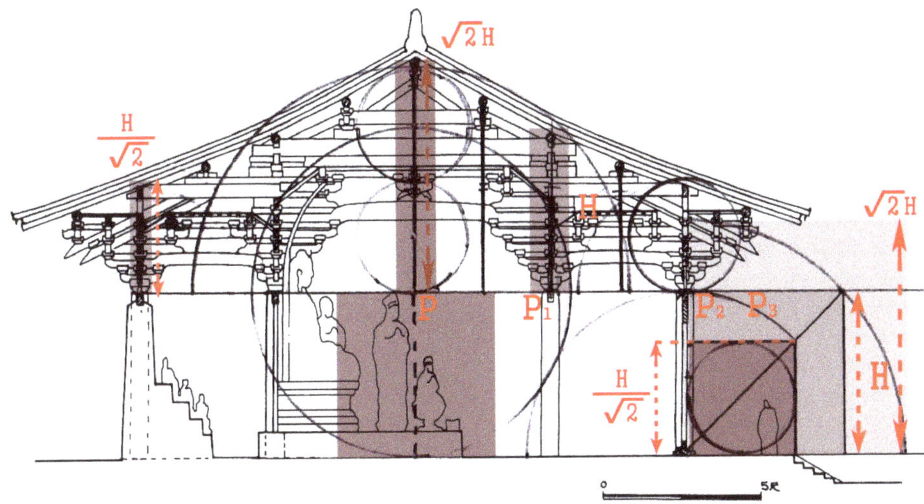

Figure 21. East Hall of Monastery Foguangsi. Proportional analysis of the sectional elevation (H = eave column height, Ø = circle diameter), Adapted from Fu (2001, p. 114).

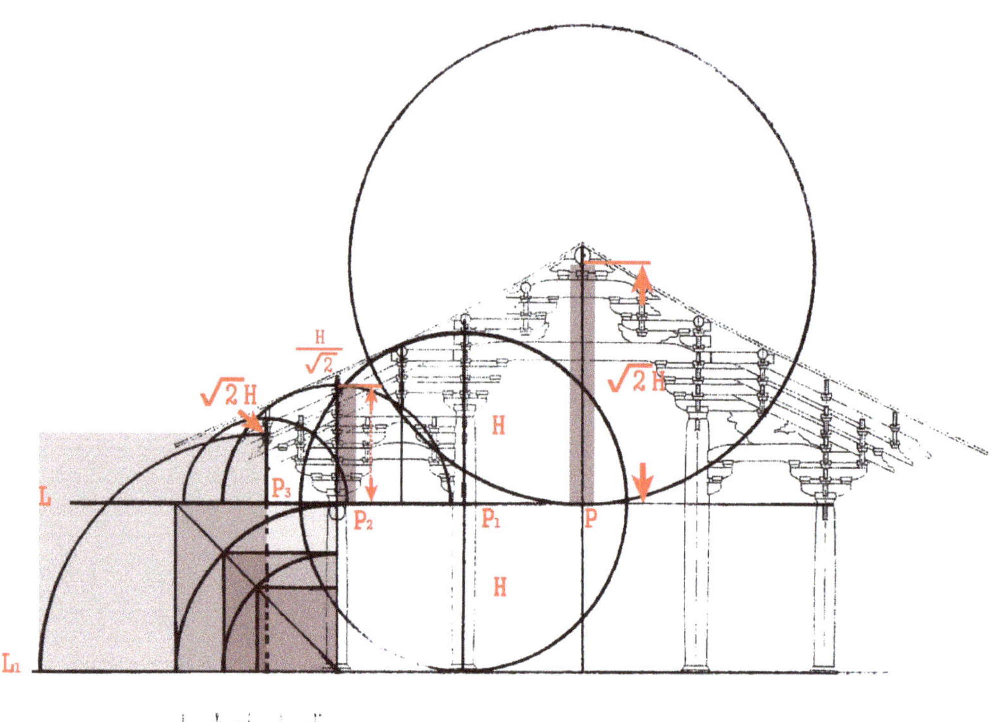

Figure 22. Main Hall of Monastery Hualinsi. Proportional analysis of the sectional elevation (H = eave column height, Ø = circle diameter), Adapted from Wang et al. (2011, p. 143).

4.4. Relationship between the Outer Bay Width and the Vertical Height of the Bracket Sets

Above the eave columns, ancient wooden buildings had bracket set systems to support the column purlins from the tops of eave columns. The height of the bracket sets was calculated by adjusting their alignment using $\sqrt{2}H \times$ eave columns' height (H).

The height ratio of the bracket sets in the nine buildings can be classified into two types. For five buildings (i.e., Bongjeongsa Monastery's Geungnakjeon Hall, Buseoksa Monastery's Muryangsujeon and Josadang Halls, Seongbulsa Monastery's Eungjinjeon Hall, and Bogwangjeon Hall at Simwonsa Monastery), the bracket set height was determined using a ratio of $\frac{H}{2}$ (=0.5H) times complying with half of the eave columns' height (H). However, the other four buildings (i.e., Sudeoksa Monastery's Daeungjeon Hall, Seongbulsa Monastery's Geungnakjeon Hall, Eunhaesa Geojoam Hermitage's Yeongsanjeon Hall, and Imyeonggwan Guesthouse's Sammun Gate) used a ratio of $\frac{\sqrt{2}H}{4}$ (=0.35H) $\times$ eave columns' height (H). The bracket sets in these four buildings were smaller than those in the other five, with a ratio of $\frac{H}{2}$ times the eave columns' height.

Buildings using both classifications were evenly distributed across the Korean Peninsula. This suggests that regional context did not dictate the differences in the height ratio of the bracket-set systems. The universal principle consistently advocated in this study has also been applied to the height ratio of the bracket set systems. This shows that there was a common proportion for constructing buildings across Chinese and Korean territories (Kim 2011). In particular, in the Goryeo buildings, a regular proportion was applied regardless of the type of two bracket-set classifications: *dapo* and *jusimpo*. A major implication is provided on account of the geographical conditions of the isolated peninsula, the existence of the Unified Silla, and of the Goryeo having maintained their power for a long time, as a combined result of which ancient Korean buildings already employed an established and indigenous architectural structure method. The Seongbulsa Monastery's Eungjinjeon

Hall and the Bogwangjeon Hall at the Simwonsa Monastery had intercolumnar bracket systems using the formula $\frac{\sqrt{2}H}{4}$ (=0.35) times eave columns' height (H), which differed from the other buildings with column-top bracket systems. However, this was limited to two cases. As the $\frac{H}{2}$ (=0.5H) and $\frac{\sqrt{2}H}{4}$ (=0.35H) ratios were used in buildings with column-top bracket systems.

In contrast, regarding the proportional association between the bracket set height (H_1) and the outer bay width, most buildings showed a geometric sequence, such as $\sqrt{2}H_1$, H_1, and $2H_1$ times, depending on (H). Each successive term could be obtained by multiplying the previous term by $\sqrt{2}H$ times. Bongjeongsa Monastery's Geungnakjeon Hall and Seongbulsa Monastery's Eungjinjeon Hall were exceptions because they did not maintain the $\sqrt{2}H$ ratio formula (Table 5).

Table 5. The proportional relationship between bracket set height and outermost bay width.

Building Names	Bracket Sets Height (H_1)	Ratio of Bracket-Set Height (H_1) to Outermost Bays Width at Side Façades	Remarks
Geungnakjeon Hall at Bongjeongsa Monastery	$\frac{H}{2} = H_1$		The bracket set height depends on the column purlin height
Muryangsujeon Hall at Buseoksa Monastery	$\frac{H}{2} = H_1$	$\sqrt{2}H_1$	The bracket set height depends on the column purlin height
Daeungjeon Hall at Sudeoksa Monastery	$\frac{\sqrt{2}H}{4} = H_1$	$2H_1$	The bracket set height depends on the column purlin height
Eungjinjeon Hall at Seongbulsa Monastery	$\frac{H}{2} = H_1$		Due to no column purlins, the bracket set height depends on the eave purlin height
Geungnakjeon Hall at Seongbulsa Monastery	$\frac{\sqrt{2}H}{4} = H_1$	$\sqrt{2}H_1$	The bracket set height depends on the column purlin height
Yeongsanjeon Hall at Geojoam Hermitage, Eunhaesa Monastery	$\frac{\sqrt{2}H}{4} = H_1$	$2\sqrt{2}H_1$	The bracket set height depends on the column purlin height
Josadang Hall at Buseoksa Monastery	$\frac{H}{2} = H_1$	$\sqrt{2}H_1$	The bracket set height depends on the column purlin height
Bogwangjeon Hall at Simwonsa Monastery	$\frac{H}{2} = H_1$	H_1	Due to no column purlins, the bracket set height depends on the eave purlin height
Sammun Gate at Imyeonggwan Guesthouse in Gangneung City	$\frac{\sqrt{2}H}{4} = H_1$	$2\sqrt{2}H_1$	The bracket set height depends on the column purlin height
The Main Hall at Nanchansi Monastery (early eighth century)	$\frac{H}{2} = H_1$	$\sqrt{2}H_1$	The bracket set height depends on the eave purlin height
The East Hall at Foguangsi Monastery (857)	$\frac{\sqrt{2}H}{4} = H_1$	$\sqrt{2}H_1$	The bracket set height depends on the eave purlin height
The Main Hall at Hualinsi Monastery (964)	$\frac{\sqrt{2}H}{4} = H_1$	$\sqrt{2}H_1$	The bracket set height depends on the eave purlin height
The Main Hall at the Chuzu'an Hermitage (1125)	$\frac{\sqrt{2}H}{4} = H_1$	$\sqrt{2}H_1$	The bracket set height depends on the column purlin height

The "Bracket Sets Height (H_1)" column spans all rows with: Eave Columns' Height (H)

5. Composition of the Internal Wood-Frame Structures between the $\sqrt{2}$ Times Proportional System and the Eave Column Height (H) in the Buildings of Korea and China

5.1. Common Characteristics in the Proportional Principle of Internal Framed Structures

The East Hall of the Foguangsi Monastery (857 CE) maintained $\sqrt{2}H$ times the eave column height to determine eave purlin height (Figure 21). A baseline (L) marked along the top of the eave columns from the front to the rear façades created a vertical line intersecting with (L) and descending from a ridge post to the eave purlins. A contact point (P) was established where they joined. The distance from the ridge purlin to (P) was $\sqrt{2}$H times the eave column height. The distance of the intersection (P_1), which was perpendicular to the middle roof purlin, was H. With the vertical line below the column purlins marked, the distance from the intersection (P_1) was defined as $\frac{H}{\sqrt{2}}$. Thus, the height of each purlin followed the geometric concept. The eave column height (H) multiplied by $\sqrt{2}$ times created a geometric sequence, such as $\sqrt{2}$H, H, $\frac{H}{\sqrt{2}}$, $\frac{H}{2}$, and $\frac{\sqrt{2}H}{4}$. The outer bay width at the side façade was calculated as $\frac{H}{2}$ plus $\frac{\sqrt{2}H}{4}$. This proportional composition describes the Muryangsujeon and Josadang Halls at the Buseoksa Monastery and the Yeongsanjeon Hall at the Geojoam Hermitage at the Eunhaesa Monastery. The value of $\sqrt{2}$H times the height of the bracket sets below the eave purlins was equivalent to the outer bay width at the side façade (Figures 11, 14 and 20).

When the eave columns' height (H) and the above-mentioned method were applied to Chuzu'an Hermitage's Main Hall at the Shaolinsi Temple, the location of the eave purlins, column purlins, middle roof purlins, and ridge post were found to be in accordance with a proportional composition, and vertical height was maintained with ratios of $\frac{\sqrt{2}H}{4}$, $\frac{H}{2}$, H, and $\sqrt{2}$H, respectively. The ratio of (H_1) to the side bay width was H_1:$2H_1$ (Figure 18). This shows that (H) maintained a proportional system with a $\sqrt{2}$H ratio. This design principle is similar to that used for Sudeoksa's Daeungjeon Hall, revealing a geometric concept with a common $\sqrt{2}$ ratio. The same method was used for Hualinsi Monastery's Main Hall (964) (Fu 2017, p. 277). In the ancient buildings of South and North China, basing increased or decreased adjustments to (H) on the $\sqrt{2}$H ratio for internal timber-framed setting displays was a typical proportional composition (Figure 22). Hence, the geometric concept of the $\sqrt{2}$H ratio has been used with reference to (H) for internal frame structures and purlin placement in ancient wooden buildings in many parts of East Asia.

5.2. Differences between Proportional Principles for Internal Framed Structures in Korea and China

The eave columns' height and setting of internal framed structures in ancient Chinese wooden architecture differed slightly from ancient Korean wooden buildings. First, the main halls at the Nanchansi and Foguangsi Monasteries in the Tang Era, the Guanyinge Pavilion and Shanmen Front Gate at the Dulesi Monastery in the Liao Era, the Hermitage Chuzu'an's Main Hall at the Shaolinsi Monastery, and the Monidian Hall at the Longxingsi Monastery in the Song Era were Buddhist monuments that used the heights of the eave columns and eave purlins as reference points. Similarly, regarding the proportional correlation between eave purlins and eave columns, $\sqrt{2}$ times the height of the eave columns was equal to the height of the eave purlins. For wooden buildings in the Goryeo Era, $\sqrt{2} \times$ (H) did not correspond to the eave purlin height but rather to the heights of the eave columns and column purlins. The Song Dynasty's *Yingzao Fashi* did not mention column purlins for most wooden buildings, owing to the existence of descending cantilevers joined with eave purlins. Even if column purlins existed, their cross-sectional size was smaller than the other purlins. Unlike Chinese buildings, wooden-framed buildings in Korea during the Goryeo Era did not have descending cantilevers, and the cross-sectional sizes of the eave purlins and column purlins were identical. The Eungjinjeon Hall at the Seongbulsa Monastery and the Bogwangjeon Hall at the Simwonsa Monastery did not have column purlins, although the opposite was true for most Goryeo-Era buildings; the two buildings are rare examples. Therefore, how purlin placement was determined using the $\sqrt{2}$H ratio highlights a remarkable difference between ancient wood-frame buildings in China and Korea. Specifically, the Chinese buildings referenced the eave purlins, while the Korean buildings referenced the column purlins.

Second, for Chinese wooden buildings, the measurement differences among the wooden components were determined according to the choice of baseline, such as the bottom of the eave columns or the ground floor (Wang 2011a). Eave column height varied according to whether the lower limit line was defined as the bottom of the foundation stones or the columns. The following buildings applied $\sqrt{2}H \times (H)$ with the column purlin height, including the plinth height: Dulesi Monastery's Shanmen Front Gate (Liao, 984), Baoguosi Monastery's Main Hall (Song, 1013), and Fengguosi Monastery's Main Hall (Liao, 1020). In contrast, the following buildings applied $\sqrt{2}H \times (H)$ with the eave purlin height: Nanchansi Monastery's Main Hall (Tang, 782), Hualinsi Monastery's Main Hall (Wu, 964), and Geyuansi Monastery's Main Hall (Liao, 966). Similar to buildings in the Goryeo Era, eave purlin height was decided according to $\sqrt{2}H$ times the eave column height. This demonstrates the utilization of geometric concepts for internal timber-framed structures in the architectural culture of East Asia (Table 2).

Third, applying $\sqrt{2}H \times (H)$, whether as a fractional or integer ratio, positioned each purlin according to the proportional principle. For the Chinese wooden buildings, each purlin height was determined by using the top and bottom of all purlins interchangeably. In contrast, wooden constructions built during the Goryeo Era had a different appearance. For Bongjeongsa Monastery's Geungnakjeon Hall, Sudeoksa Monastery's Daeungjeon Hall, Seongbulsa Monastery's Geungnakjeon Hall, Buseoksa Monastery's Muryangsujeon Hall, and Simwonsa Monastery's Bogwangjeon Hall, the bottoms of all purlins (which backed up the eaves, columns, and roof framework) were used as standard points. For Imyeonggwan Guesthouse's Sammun Gate and Seongbulsa Monastery's Eungjinjeon, $\sqrt{2}H \times (H)$ referenced the tops of all purlins. Further, Buseoksa Monastery's Josadang Hall and Hermitage Geojoam's Yeongsanjeon Hall of the Eunhaesa Monastery used the tops of eave and column purlins and the bottoms of the middle roof and ridge purlins as standard points. This shows that purlin height was increased or decreased based on each purlin's diameter, which might have reflected the carpenters' personal preferences.

6. Conclusions

This study examined nine existing buildings constructed at the height of the Goryeo Dynasty regarding their use of common mathematical rules as design doctrines related to the column purlin height above the bracket sets. The authors compared these buildings to timber-framed buildings from ancient China and found that they commonly used the ratio of $\sqrt{2}H \times$ eave column height (H) to construct a wooden-frame work. During the Goryeo Era, some timber-framed monuments on the Korean Peninsula were built while a stable imperial power ruled. Local builders embraced mathematical design principles and determined the positioning of roof purlins by applying the $\sqrt{2}H \times (H)$ ratio to calculate outer bay widths and bracket set heights, consciously deciding which ratio to employ on a given project. The stipulations were not always imposed on the provinces; rather, they were combined with the universalized principles based on localized ideals.

The Goryeo-Era buildings embraced the principles of designing ancient Chinese architecture and created new timber-framed structures while adapting to the rules already existing in local buildings (Kim 2011). Then, through the buildings examined in this study, the attitude of building construction in the Goryeo Dynasty proves that East Asian universal and regional design principles were consistently shared in the Korean Peninsula.

The preface of the *Yingzao Fashi* describes the basic principles of ancient mathematics, including the geometric axioms of all things and the numerical values of diagonal length and circumference with reference to rectangles, circles, hexagons, octagons, and other shapes mentioned in the *Jiuzhang Suanshu* and *Kaogongji*. The principles can be inferred from the timber-framed buildings constructed in the Goryeo Era. The present study established that the nine examined buildings used the $\sqrt{2}H \times (H)$ ratio to determine the positions of all purlins, outer bay widths, and bracket set heights to create an internal timber-framed structure. The following conclusions can be drawn from this study.

First, this research makes it possible to identify the components that rely heavily on $\sqrt{2}H \times (H)$ in internal timber-framed structures by implementing the notion of geometric sequences, such as $\frac{H}{4}$, $\frac{\sqrt{2}}{4}$, $\frac{H}{2}$, $\frac{H}{\sqrt{2}}$, H, and $\sqrt{2}H$, depending on advanced mathematical formulas that use L-square rulers and compasses to draw rectangles and circles. Embodied by buildings constructed during the Song and Liao Eras, such design principles have long been used in the wooden structures of East Asia.

Second, for five of the nine studied buildings, the $\sqrt{2}H \times (H)$ ratio was equivalent to the column purlin height. Most buildings in the Tang, Song, and Liao Eras referenced the eave purlins height. However, there were exceptions among the nine: Buseoksa Monastery's Muryangsujeon Hall, Seongbulsa Monastery's Eungjinjeon Hall, and Simwonsa Monastery's Bogwangjeon Hall. For these three buildings, $\sqrt{2}H \times (H)$ was equivalent to the eave purlin height. The Eungjinjeon and Bogwangjeon Halls did not have column purlins, but the opposite was true for Muryangsujeon Hall. A notable finding is that Muryangsujeon Hall embraced the Chinese principle that $\sqrt{2}H \times (H)$ should be equivalent to the eave purlin height. All nine buildings applied the design principles of the Chinese examples to some degree.

Third, baseline (L) is defined as a horizontal line from the top side of the front and rear eave columns. The point at which each purlin met the baseline (L) was defined as P_1, P_2, P_3, and P_4, drawing a vertical line perpendicular to each purlin. Each purlin height (i.e., $\frac{H}{4}$, $\frac{\sqrt{2}H}{4}$, $\frac{H}{2}$, $\frac{H}{\sqrt{2}}$, H, and $\sqrt{2}H$, from the eave purlins to the ridge post) was determined with a ratio of $\sqrt{2}H$ times the baseline (L) to each purlin. This represented an increased or decreased value using $\sqrt{2}H \times (H)$. The locations of the eave purlins, column purlins, and roof purlins were determined using $\sqrt{2}H \times (H)$. Notably, for Bongjeongsa's Geungnakjeon Hall (P, H), Sudeoksa's Daeungjeon Hall (P, $\sqrt{2}H$), and Seongbulsa's Eungjinjeon (P, H) and Daeungjeon (P, H) Halls, the ridge purlin location was determined by H or $\sqrt{2}H$ times the height of the front and rear eave columns. This is similar to Nanchansi's Main Hall (P, H), Foguangsi Monastery's East Hall (P, $\sqrt{2}H$), Hualinsi's Main Hall (P, $\sqrt{2}H$), and Chuzu'an Hermitage's Main Hall (P, $\sqrt{2}H$). Buseoksa Monastery's Muryangsujeon Hall was an exception in that the height of the eave columns reached that of the upper-middle roof purlins (P_1, $\sqrt{2}H$). In addition, for Buseoksa Monastery's Muryangsujeon (P, H $+ \frac{H}{\sqrt{2}}$) and Josadang Halls (P, $\frac{H}{\sqrt{2}}$) and Geojoam Hermitage's Yeongsanjeon Hall (P, $\frac{H}{\sqrt{2}}+\frac{H}{2}$), the concept of an arithmetic sequence was applied to determine the ridge purlin location, showing distinctly Korean characteristics in designing wooden-framed buildings (Table 3).

Fourth, considering the $\sqrt{2}H$ times' proportional relationship between the eave column height (H) and the outermost bay widths at the side façade, the Daeungjeon Hall of the Sudeoksa Monastery, and the Sammun Front Gate of the Imyeonggwan Guesthouse brought the $\frac{H}{\sqrt{2}}$ times the eave column height into contact with the notion of geometric series. Seongbulsa Monastery's Geungnakjeon Hall indicated H times, as well as Yongshou Monastery's Yuhuagong Hall and Chuzu'an Hermitage's Main Hall. The other buildings had a proportional system of $\frac{H}{4} + \frac{\sqrt{2}H}{4}$ (Seongbulsa's Eungjinjon Hall and Simwonsa's Bogwangjeon Hall) or $\frac{H}{2} + \frac{\sqrt{2}H}{4}$ (Buseoksa's Muryangsujeon and Josadang Halls and Geojoam Hermitage's Yeongsanjeon Hall, similar to Foguangsi Monastery's East Hall). This shows that the concept of arithmetic series was used together with the geometric concept in the East Asian sphere (Table 4).

Fifth, regarding the correlation between the eave column height (H) and bracket set height, the bracket set height in all the buildings was proportional to $\frac{H}{2}$ or $\frac{\sqrt{2}H}{4}$ times the eave column height (H) through a geometric progression with a common ratio ($\frac{H}{2} = H_1$) is affiliated with Buseoksa's Muryangsujeon, Seongbulsa's Eungjinjeon, and Simwonsa's Bogwangjeon Halls, which are comparable to the Main Hall of Nanchansi. ($\frac{\sqrt{2}H}{4} = H_1$) is affiliated with Imyeonggwan's Sammun Gate, Bongjeongsa's Geungnakjeon, Sudeoksa's Daeungjeon, Seongbulsa's Geungnakjeon, Geojosam's Yeongsanjeon, and Buseoksa's Josadang Halls, which followed the same ratio rule ($\frac{\sqrt{2}H}{4} = H_1$) as the Main Halls of Foguangsi,

Hualinsi, and Chuzu'an Monasteries in the Tang, Song, and Liao Dynasties (Table 5). The characteristic features of fourteenth-century architecture in the Goryeo Era were the gradual increase of multi-layered bracket-sets and progressive change to inter-columnar bracket-sets systems (*dapo*). However, despite these changes, the purlin placement height and bracket-set height were determined by applying the universal design principle (with a ratio of $\sqrt{2}$H times the eave column height) using the same proportional system, regardless of the format of the inter-columnar bracket set and column top bracket set (*jusimpo*).

Finally, the ratio of the bracket set height (H_1) to the outer bay width at the side façades indicated three types, $2H_1$: H, H_1: H, and $\sqrt{2}H_1$: H, which permitted $\sqrt{2}$H or 2H times with a common ratio. The first type ($2H_1$: H) applied to Sudeoksa's Daeungjeon Hall; the second type (H_1: H) applied to Simwonsa's Bogwangjeon Hall; and the third type ($\sqrt{2}H_1$: H) applied to Seongbulsa's Geungnakjeon, and Buseoksa's Muryangsujeon and Josadang Halls, which are the same as the Main Halls of Nanchansi, Hualinsi, Foguangsi, and Chuzu'an Monasteries. The first and second types represent original building characteristics that used appropriate integer proportions ($2H_1$: H and H_1: H) on the Korean Peninsula, while the third type ($\sqrt{2}H_1$: H) maintained a universal pattern with common languages in East Asia. The comparison in the present study of bracket set and eave column heights reliant on outer bay widths suggests that the closest analogy to the proportionate attributes of existing wood-frame buildings from the Tang–Song Era is the Yeongsanjeon Hall at the Geojoam Hermitage, Eunhaesa Monastery (Table 5). The use of the ratio rule regulated the proportional value of the side façade width, particularly the outer bay width, which controlled the eave column and bracket set heights, as grounded in the proportional rules.

This study shows that the internal arrangement of timber-framed structures was constructed using the geometric series concept based on the increase or decrease in $\sqrt{2}$H × eave column height (H). These design rules were applied to all nine of the studied buildings from the Goryeo Era and to buildings from China's Tang, Song, and Liao Dynasties. This study provides evidence that the wooden architecture of the Goryeo Era evolved in an original way while also accepting architectural principles from ancient China. It further demonstrated that the architecture of the Goryeo Era inherited design principles of ancient China by comparing them with similar buildings designed contemporaneously in Ancient China. The similarities and differences seen in these buildings show that the universal architectural attitude is being integrated with the previous technology of local architecture and adapting to a new architectural form. This research can contribute to new knowledge on East Asian architecture and provide useful information for future building restorations.

Further, the universality of an architectural culture cannot be understood in isolation from its diversity or particularity because the universality of culture is derived primarily from the assembly and harmonization of diversities. On the other hand, the particularity of a culture does not mean an independent culture of some indigenous form. This study indicates that the universality and particularity of culture remain in a state of co-existence.

This study followed a basic review approach to confirm the similarities and differences in the design principles of ancient architecture in Korea and China. In the future, through comparative analysis with the ancient wooden architectures of China and Japan, which were not reviewed in this study, the characteristics of ancient Korean architecture will be understood more clearly. A more fundamental characteristic of the timber-framed proportion system of East Asian ancient wooden architecture will be grasped, and more elaborated findings on universal and regional design principles will be progressed in future studies.

Author Contributions: Writing—original draft, J.-H.C. and Y.-J.K. All authors have read and agreed to the published version of the manuscript.

Funding: This research was supported by Basic Science Research Program through the National Research Foundation of Korea (NRF) funded by the Ministry of Education (No.2021R1I1A3059811).

Conflicts of Interest: The authors declare no conflict of interest.

Notes

[1]   Whereas the *Zhoubi* Suanjing is generally considered the oldest of the mathematical classics completed during the first century BCE, the *Jiuzhang Suanshu* represents a much more advanced state of mathematical knowledge than the *Zhoubi Suanjing* (Cha and Kim 2019, p. 5; Cullen 2007). The *Jiuzhang Suanshu* mentions significant notions through *fa*, *gui*, and *ju*. The methods or rule (*fa*) have been handed down from generation to generation which is used throughout the *Nine Chapters*, just like the compass (*gui*) for drawing circles and gnomon (*ju*) for constructing squares or for surveying in measurement, with which we draw figures (Shen et al. 1999, pp. 53–57).

[2]   The *Samguk sagi*, Book 38, Chapter 7, "Silla Government Offices," records that the *Jiuzhang Suanshu* was utilized as a regular textbook of the Silla Kukhak (the seventh through the eighth century). The earliest written records on the Korean peninsula date from the Three Kingdom period (c. 57 BCE–668 CE) (Cha and Kim 2019, p. 5).

[3]   According to the "Xumu" (序目, the preface and table of contents of a book) in the *Yingzao Fashi*, "the reason that *jian* (橌, flying cantilever), *lu* (櫨, cap-block), *ji* (枅, crossbeam), and *zhu* (柱, column) hold and support each other well is that carpenters built constructions earlier, ruling architectural implements such as *gui* (規, compass), *ju* (矩, L-square ruler), *zhun* (準, water-level instrument), *sheng* (繩, line maker). . . . 橌櫨枅柱之相枝, 規矩準繩之先治. . .]." To construct timber-framed buildings for a long time ago, the ancient carpenters made a circular form with compasses, a straight line with an L-square ruler, a horizontal element even with a water-level instrument, and a straight line with a line maker. The text points denote the five principles of artisanship. Whereas these basic principles of craftsmanship show that the L-square ruler is used for illustrating a squectangle, the compass is for drawing circles as the main architectural tools.

[4]   Fu says that the eave columns' height becomes an important reference of standard proportion for the constructions, mentioning "as the front of the house (房屋, *fangwu*) becomes a multiple of the height of the eave columns, so the height of the eave columns is the basic module (模數, *moshu*) in cross-section" (Fu 2001, p. 5).

[5]   The Goryeo (918–1392 CE) founded by King Taejo (Wang Geon, r. 918–943) ruled all of the peninsula for the better part of five centuries from 935 to 1392. Wang Geon named the state Goryeo after Goguryeo (37 BCE–668 CE), the ancient kingdom that occupied the northern part of the peninsula and parts of Manchuria. The fourth king, Gwangjong (r. 949–975), took measures to consolidate monarchical power, declaring himself *hwangje* (emperor), and renamed Gaeseong the Imperial Capital (*Hwangdo*) (Seth 2019, pp. 79–81).

[6]   This study mainly deals with the affinity between the eave columns and the $\sqrt{2}$ times, and the study on the ground plan and the elevation. However, there is no information on the positioning of the purlins and the proportional system of columns and outermost bays among the internal timber-framed structure; thus, there was no mention about the positioning of each purlin in the wooden building.

[7]   Yoneda Miyoji states that the $\sqrt{2}$ proportional system is applied to the ground plan, elevation, and arch structure of the Seokguram Grotto.

[8]   The "*nuzhi*" (women who weave textiles) in the *Suanshushu* (算數書, Book on Mathematical Procedures) and the *Sunzi Suanjing* (孫子算經, Sun Zi's Mathematical Manual) deal with the same problem. The third question of *Zhang Qiujian Suanjing* (*The Mathematical Classic of Zhang Qiujian*) is also a problem related to geometric series. It proves that the ancient mathematical books of China treated a notion of geometric sequence.

[9]   The original text is as follows: 今有女子善, 日自倍, 五日五尺. 日何? 答曰, 初日一寸, 三十一分寸之九十, 次日三寸, 三十一分寸之七, 次日六寸, 三十一分寸之十四, 次日一尺二寸, 三十一分寸之二十八, 次日二尺五寸, 三十一分寸之二十五. 曰, 直一, 二, 四, 八, 十六列衰, 副法, 以五尺乘未者, 各自, 如法得一尺 . Now given a skilful weaver, who doubles her product every day. In five days, she produces a cloth of five chi. How much does she weave in each successive day? As a response on the question, on the first day, she weaves 1 cun plus 19/31 cun; on the second day, 3 cun plus 7/31 cun; on the third day, 6 cun plus 14/31 cun; on the fourth day, 12 cun plus 28/31 cun; and on the fifth day, 25 cun plus 25/31 cun. As a solution, arrange the rates for distribution: 1, 2, 4, 8 and 16. Take their sum as divisor. Multiply 5 chi by each rate as dividend. Divide by giving the number of chi. This problem is the same as the questions mentioned in the *Suanshushu* and the *Sunzi Suanjing*.

[10]   The $\sqrt{2}$ times are contingent on the geometric progression law, using the numerical representation of the increase and the decrease.

[11]   The exact reason is not known, but it is possible that the building has been modified after the repair work. In the sixteenth century, when the outermost bay was added to the front of this building, the timber-framed structure may have been entirely repaired.

References

Anonymous. n.d.a. *Jiuzhang Suanshu* 九章算術 [*Nine Chapters on the Mathematical Art*].

Anonymous. n.d.b. *Kaogongji* 考工記 [*The Artificers' Record*].

Anonymous. n.d.c. *Zhoubi Suanjing* 周髀算經 [*Astronomy and Mathematics in Ancient China*].

Anonymous. 1734. *Gongbu Gongcheng Zuofa Zeli* 工部工程做法則例 [*Regulations and Precedents Concerning Technical Instructions for the Building Crafts by the Ministry of Public Works*].

Bak, Chan, and Jongman Mun. 1992. Yeongjo beopsik daemokjakgwa buseoksa muryangsujeongwaui bigyoyeon-gu pyeong-myeon mit dan-myeon-jungguk yeongjo beopsikgwa hanguk jeontong geonchukgwaui bigyogochal [A comparative study on the large woodworks at the Ying Tsao Fa shih and Murangsujeon]. *Daehangeonchukakoe Nonmunjip [Journal of the Architectural Institute of Korea]* 8: 79–87.

Cha, Juhwan, and Young Jae Kim. 2019. Reconsidering a proportional system of timber-frame structures through ancient mathematics books: A case study on the Muryangsujŏn Hall at Pusŏksa Buddhist Monastery. *Journal of Asian Architecture and Building Engineering* 18: 457–71. [CrossRef]

CHA. 2002. *Measurement Report of Muryangsujeon at Buseoksa Monastery*. Daejeon: CHA.

CHA. 2003. *Repair-Measurement Report of Geuknakjeon at Bongjeongsa Monastery*. Daejeon: CHA (Cultural Heritage of Administration).

CHA. 2004. *Repair Report of Gangneung Gaeksamun (Imyeonggwan Sammun)*. Daejeon: CHA.

CHA. 2005a. *Measurement Report of Daeungjeon at Sudeoksa Monastery*. Daejeon: CHA.

CHA. 2005b. *Repair-Measurement Report of Josadang at Buseoksa Monastery*. Daejeon: CHA.

Chen, Mingda. 1981. *Yingzao Fashi Damuzuo Yanjiu [Research on Great Carpentry in the Yingzao Fashion]*. Beijing: Wenwu Chubanshe.

Cullen, Christopher. 2007. *Astronomy and Mathematics in Ancient China: The Zhoubi Suanjing*. London: Cambridge University Press.

Fu, Xiannian. 2001. *Zhongguo Gudaichengshiguihua, Jianzhuqunbujuji Jianzhushejifangfa Yanjiu [Research on the Urban Planning, the Arrangement of Architectural Complexes, and the Method of Design in Ancient China]*. Beijing: Zhongguo Jianzhu Gongye Chubanshe.

Fu, Xiannian. 2017. *Traditional Chinese Architecture: Twelve Essays*. Edited by Nancy Steinhardt. Translated by A. Harrer. Princeton: Princeton University Press.

Guo, Shuchun. 2009. *Jiuzhang Suanshu Yizhu [The Nine Chapters on the Mathematical Art]*. Shanghai: Shanghai Guji Chubanshe.

Kim, Chanyeong. 2003. Joseon Hugi Dapo Buljeonui Biryechegyee Gwanhan Yeongu [The Proportion System of Main Worship Hall of Buddhist Temple Buildings with a Multi-Cluster Bracket Style in The Late Choson Period]. Ph.D. dissertation, Yeongnam University, Gyeongsan, Korea.

Kim, Do-Kyoung. 2008. Geonchukseolgye Cheungmyeoneseo bon Sudeoksa Daeungjeonui pyeongmyeonjeokgwa gagu teukseonge gwanhan yeongu [A study on the architectural design characteristics of the plan and the structure in Sudeok Temple's Daeungjeon]. *Geonchugyeoksa Yeongu [Journal of Achitectural History]* 17: 97–112.

Kim, Do-Kyoung. 2014. Suchibunseogeul tonghan buseoksa muryangsujeonui pyeongmyeongwa dan-myeon teukseonge gwanhan yeongu [A Study on the Proportional Characteristics of the Floor Plan and Section in Muryangsujeon of Buseok Temple]. *Daehangeonchukakoe Nonmunjip [Journal of the Architectural Institute of Korea]* 30: 143–52.

Kim, Young Jae. 2011. Architectural Representation of the Pure Land. Ph.D. dissertation, University of Pennsylvania, Philadelphia, PA, USA.

Kim, Young Jae. 2016. Discussing architecture and the city as a metaphor for the human body. *Architectural Research* 18: 1–12. [CrossRef]

Kim, Young Jae, and Sungjin Park. 2017. Tectonic traditions in ancient Chinese architecture, and their development. *Journal of Asian Architecture and Building Engineering* 16: 31–38. [CrossRef]

Li, Jie. 1100. *Yingzao Fashi* 營造法式 *[State Building Standards]*.

Li, Yunxuan. 1982. *Hua Xia Yi Jiang: Zhongguo Gudian Jianzhu She Ji Yuanli Fenxi*. Hong Kong: Guang Jiao Jing Chunbanshe.

Liang, Sicheng, and Jie Li. 1983. *Yingzao Fashi Zhushi*. Beijing: Zhongguo Jianzhu Gongye Chunbanshe.

NRICH. 1998. *Bukanmunhwajaehaeseoljip [Explanation of Cultural Heritages in North Korea] Vol. 2. Sachal Geonchuk pyeon [Temple Architecture Edition]*. Daejeon: NRICH (National Research Institute of Cultural Heritage).

Pan, Guxi, and Jianzhong He. 2005. *Yingzao Fashi Jiedu [Interpreting Yingzao Fashi]*. Nanjing: Dongnandaxue Chubanshe.

Seth, Michael. 2019. *A Concise History of Korea: From Antiquity to the Present*. Lanham: Rowman & Littlefield Publishers.

Shen, Kangshen, John N. Crossley, Anthony Wah-Cheung Lun, and Hui Liu. 1999. *The Nine Chapters on the Mathematical Art: Companion and Commentary*. Oxford: Oxford University Press.

Sim, Daeseop, and Jeongsu Kim. 1983. Hansigmokjogujoui Gongpogujowa Biryeguseong [Structure and Proportion of Gong-Pou in Traditional Korean Wooden Construction]. *Daehangeonchukakoe Chugyehaksulbalpyo Nonmunjip* 3: 31–34.

Stevens, Garry. 1990. *The Reasoning Architect: Mathematics and Science in Design*. New York: McGraw-Hill College.

Vitruvius Pollio, Marcus, and Morris Hicky Morgan. 1960. *Vitruvius: The Ten Books on Architecture*. Translated by Morris Hicky Morgan. New York: Dover Publications.

Wang, Guixiang. 2011a. $\sqrt{2}$ Yu Tang Song Jianzhu Zhuyan Guanxi [Relationship between $\sqrt{2}$ and Eaves Columns in the Tang and Song Architecture]. In *Zhongguo gudai mugou jianzhu bili yu chidu yanjiu [Research on the Proportion and Scale of Ancient Chinese Wooden Buildings]*. Beijing: Zhongguo jianzhu gongye chubanshe, pp. 40–47.

Wang, Guixiang. 2011b. Tang Song Danyan Mugou Jianzhu Pingmian Yu Limian Biliguilu De Tantao [Discussion on Regular Proportion between Plans and Façades of Tang and Song Wooden Architecture with Single Eaves]. In *Zhongguo gudai mugou jianzhu bili yu chidu yanjiu*. Beijing: Zhongguo jianzhu gongye chubanshe, pp. 48–63.

Wang, Guixiang, Chang Liu, and Zhijun Duan. 2011. *Zhongguo gudai mugou jianzhu bili yu chidu yanjiu*. Beijing: Zhongguo jianzhu gongye chubanshe.

Wang, Nan. 2017a. Guijufangyuan Futuwanqian [Rules of Square and Circle, Great Diversity in Buddhist pagodas]. *Zhongguo Jianzhushi Lunhuikan [Journal of Chinese Architecture History]* 2: 216–56.

Wang, Nan. 2017b. Guijufangyuan Fuzhijusuo [Circles, Squares, and the Place Where Buddha Lives]. *Jianzhu Xuebao [Architectural Journal]* 6: 29–36.

Wang, Nan. 2017c. Xiangtianfadi, Guijufangyuan [The Principle of Modelling Heaven and Earth, the Rules of Square and circle]. In *Jianzhushi [Architectural History]*. Beijing: Tsinghua University Press, vol. 2, pp. 77–125.

Wang, Nan. 2018a. Guijufangyuan Duxianggouwu [Rules of Square and Circle, Sculpture Sizes in Architectural compositions]. In *Jianzhushi [Architectural History]*. Beijing: Tsinghua University Press, vol. 1, pp. 103–25.

Wang, Nan. 2018b. Jinchenggongque Taiziyuanfang [The Forbidden City and Square-Circle composition]. In *Jianzhushi [Architectural History]*. Beijing: Tsinghua University Press, vol. 2, pp. 93–128.

Wang, Nan. 2019. Guijufangyuan Tiandizhihe [Rules of Square and Circle, Central Axis of Heaven and Earth]. *Beijing Guihua Jianshe [Beijing Planning Review]* 1: 138–53.

Wang, Nan, Zhuonan Wang, and Hongyu Zheng. 2021a. Tiandìyuanfang taxiangheyi [Round Heaven and Square Earth, Unity of Pagoda and Statuary]. *Jianzhushixuekan [Journal of Architecture History]* 2: 71–95.

Wang, Shushegn, Xu Chen, Yuqian Xu, Kai Wang, Jialong Lai, and Lu Shi. 2021b. Morphological Study on Multi-Storied Brick Pagodas of the Tang Dynasty: An Analysis Method Based on Historical Patterns and Mathematical Models. *International Journal of Architectural Heritage* 15: 1655–70. [CrossRef]

Wen, Renjun. 2008. *Kaogongji Yizhu*. Shanghai: Shanghai Guji Chubanshe.

Yoneda, Miyoji. 1944. *Chosen Jodai Kenchiku No Kenkyu [Research on Ancient Korean Architecture]*. Osaka: Akitaya.

Yoneda, Miyoji, and Yeonghun Sin. 1976. *Hangugui Sangdaegeonchug Yeongu [Study of the Ancient Korean Architecture]*. Seoul: Hangung-munhwasa.

Yun, Jang-Seop. 1975. Hangugui Yeongjocheokdo [The Measuring Units of Scale in Korean Architecture]. *Geonchuk (Architecture)* 19: 2–9.

religions

MDPI

Article

The Remaining Buddhist Architecture in Fu'an, the Core Hinterland of the Changxi River Basin

Jie Liu [1,*], Yincheng Jiang [2] and Chen Cao [2]

[1] Department of Architecture, School of Design, Shanghai Jiao Tong University, Shanghai 200030, China
[2] Design and Research Institute, Shanghai Jiao Tong University, Shanghai 200030, China; yincheng.jiang@swedishwood.cn (Y.J.); chen.cao_cc@outlook.com (C.C.)
* Correspondence: jackliu@sjtu.edu.cn

Abstract: The Changxi River Basin is a small root-like watershed, surrounded by mountains on three sides and facing the sea to the southeast. It is located on the border between Fujian and Zhejiang on the southeast coast of China. The area gave rise to the Changxi Culture that began in the Sui and Tang Dynasties and flourished in the Song Dynasty. Buddhism in the Changxi Basin was introduced no later than the 9th century. As the core hinterland of the Changxi Basin, Fu'an has always been an important center for Buddhism in Eastern Fujian. It reached its peak in the 10th to 13th centuries during the Song Dynasty. This article conducts a comprehensive investigation and study of the existing Buddhist temple sites and relics in Fu'an. It highlights these structures' single-bay pattern of construction, based on rectangular plans in which the longitudinal axis extends along the plan's direction of depth. This is a pattern rarely seen in the history of Chinese Buddhist architecture. The paper also summarizes a common element in these temples, their petal-shaped corrugated stone pillars which are divided into eight segments. Lastly, it illustrates the evolution of the temples in the Changxi River Basin from single-bay layouts to those with widths of multiple bays and indicates the unique status and associated values of single-bay Buddhist temples in the history of southern Buddhist architecture. The study examines new local findings and ideas for the study of Chinese Buddhist architectural history, providing academic support for the protection and research of Buddhist architectural heritage in Southeast China.

Keywords: Changxi River Basin; Buddhist architectures; Fu'an; relics; evolution

Citation: Liu, Jie, Yincheng Jiang, and Chen Cao. 2021. The Remaining Buddhist Architecture in Fu'an, the Core Hinterland of the Changxi River Basin. *Religions* 12: 1054. https://doi.org/10.3390/rel12121054

Academic Editors: Shuishan Yu and Aibin Yan

Received: 23 September 2021
Accepted: 1 November 2021
Published: 26 November 2021

Publisher's Note: MDPI stays neutral with regard to jurisdictional claims in published maps and institutional affiliations.

1. Introduction

The Changxi River Basin in the areas bordering Fujian and Zhejiang provinces includes most of the areas of Fu'an 福安, Jiaocheng 蕉城, Pingnan 屏南, Zhouning 周宁, Shouning 寿宁, Zherong 柘荣, Fuding 福鼎, and Xiapu 霞浦 counties in Fujian 福建 province and part of the areas of Qingyuan 庆元 and Taishun 泰顺 counties in Zhejiang 浙江 province (Figures 1 and 2). The area is typical of the southeast coast of China that is covered with many such small river basins. The whole river basin unfolds in a complete tree root and fan-shaped distribution. Only the south side is open to the East China Sea. The other sides are surrounded by mountains. The social economy and cultural development of the Changxi River Basin was thus partly closed and partly open. The early production technology and civilization in this area was basically imported from the outside through its estuaries. Through these estuaries also flowed the mineral resources, timber, and tea produced from this area to inland China and even to many parts of the world.

From the south of the Qiantang 钱塘 River to the north of the Zhujiang 珠江 River, most of the areas are vertical and crosscut the Zhongshan landform, surrounded by mountains. The main courses of these mountain-enclosed rivers basically do not penetrate beyond the province. In other words, this area is defined instead by the distribution of small river basins. The direction of the main stream is basically northwest–southeast, flowing east or south into the East China Sea. In such a particular geographical environment, these areas are surrounded by mountains on the west, north and east sides. The vast sea serves as a barrier on the south or southeast side. In contexts where there were not domestic disturbances or foreign invasions, the imperial central government ordered the development of these small river basins, doing so much later than similar plans enacted throughout large river basins and across central China. At the beginning of such development, the ancient counties, provinces and cities set up by the central governments here were all near estuaries. It is clear that the central governments' administration and control of the small river basins in these southeast coastal areas started from the estuaries. Hence, this "central-to-local pipeline" was first directed from the sea. The sea routes backtracked along the main river channels and the tributaries to the hinterlands of the river basins and then arrived at the river basin's marginal areas. Therefore, the production, economic relations and trade in these small river basins were relatively simpler than that of other large river basins. The formation, development and evolution of architectural technology here was also relatively clear and easy for us to investigate and study.

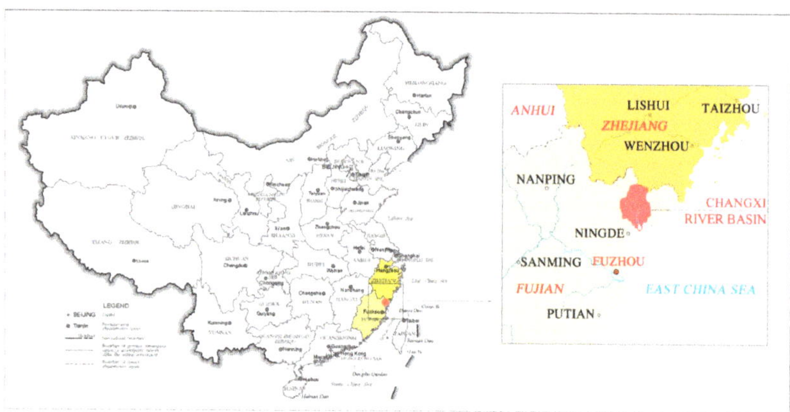

Figure 1. Map of the location of the Changxi River Basin in mainland China. Diagram by authors.

Since the late 19th century, foreign scholars were the first to start systematic investigations and research on ancient Chinese architecture, among which the more representative ones are German scholars Ernst Boerschmann, Japanese scholars Sekino Tadashi, Ito Chutaand Tokiwa Osada. Although these scholars had different goals and methods of investigation, their research had all covered several provinces of China. Among them, architecture scholar Ito Chuta conducted two months-long expeditions along the Yangtze River Basin in China. The religious scholar Tokiwa Osada conducted several thematic expeditions in the 1920s on the origins of Buddhism in China and Chan temples. He arrived in Fujian in 1929 and made an in-depth investigation of three Chan temples, including Xuefeng 雪峰 Chongsheng 崇圣 temple in Minhou 闽侯, Gushan 鼓山 Yongquan 涌泉 temple in Fuzhou 福州, and Huangboshan 黄檗山 Wanfo 万佛 temple in Fuqing 福清 (Osada and Tadashi 1939, p. VI-69-120; He 2013, pp. 91–112). Unfortunately, that expedition did not include the northeastern part of Fujian, which is not far from Fuzhou. There have been studies on small river basins in the mountainous areas of the southeast coast of China, as well architectural academics' work on the architecture of certain river basins. For the former, Wu Songdi at the Center of Historical Geographical Studies of Fudan University systematically discussed the relationship between the economic development, industrial characteristics and geographical environment of the southeastern coastal areas from the south of Ningbo 宁波 to the north of Chaoshan 潮汕 (Wu 1990). Ma Xueqiang's team from the Shanghai Academy of Social Sciences systematically studied and investigated the Oujiang 瓯江 river basin, the largest river in the south of Zhejiang Province (Ma 2016). In the 1980s and 1990s, Chen Zhihua of Tsinghua University used sociological and architectural methods to conduct research on vernacular architecture distributed in the Nanxi 楠 溪 river basin that is a tributary of the Oujiang River Basin, mainly on the planning of the villages and the type of vernacular architecture, such as residential houses, ancestral halls, and temples. The characteristics of the vernacular architecture, craftsmanship and village planning principles in this river basin are discussed clearly in this book (Chen 1992, 2004). In the past two decades, the author's team has conducted in-depth research on the Changxi River Basin's timber covered bridges and vernacular architecture. They have organized and summarized a large amount of comprehensive information on local architecture styles, structural types, and construction techniques in several books (Liu 2017; Liu and Chen 2016; Liu and Hu 2011; Liu and Shen 2005).

Figure 2. Topographic Map of the Changxi River Basin. Diagram by authors.

For a while, there has been a dearth of research into the remains of Buddhist architecture in the entire Changxi Basin from the perspective of the river basin. Currently, most of the information that can be found is based on the statistics of the traditional administrative divisions of provinces, counties, townships, and other units. From the perspective of the river basin's economy, social development is accompanied by the successive development of economic activities along estuaries, main streams and tributaries chronologically. Therefore, this study focuses on Fu'an, located on the main stream area of the basin in order to grasp the development of the early remains of Buddhist architecture in the Changxi basin.

Since 2015, the author's team has carried out interdisciplinary investigations and research with religious scholars on the Buddhist relics of Fu'an. At present, the remains of ancient buildings in Fu'an mainly include temples, palaces, temples, residential buildings, and bridges. Among them, the Buddhist temples are comparatively older and better preserved.

2. The Core of the Changxi River Basin—The Remains of Ancient Buddhist Buildings in Fu'an

2.1. The Formation of Fu'an Buddhist Temples

Buddhism was introduced to China through its Western Regions as early as the 1st century C.E., then becoming a revered faith in the Central Plains. Fujian, located in the southeast corner of the Chinese mainland, has a closed geographical environment surrounded by mountains on three sides and the sea on the other, delaying its development. Controversy surrounds the questions of when and where Buddhism arrived in Fujian. Most scholars believe that it was transmitted by land from the Central Plains to the south in the late 2nd century C.E. (Wang 1997, pp. 1–4). Some scholars also put forward the theory that "Buddhism spread into China by sea" (Wu and Zheng 1995). Regardless of the path of Buddhism's spread in China, Buddhism, as a foreign religion, has experienced conflicts, integration and development with traditional Chinese culture, and gradually formed the system of Chinese Buddhist denominations in the Sui 隋 and Tang 唐 Dynasties, which mainly included eight schools: Tiantai Zong 天台宗, Sanlun Zong 三论宗, Faxiang Zong 法相宗, Lv Zong 律宗, Huayan Zong 华严宗, Jingtu Zong 净土宗, Mi Zong 密宗, and Chan Zong 禅宗, with the Tiantai Zong 天台宗, Huayan Zong 华严宗, and Chan Zong 禅宗 schools having the most Chinese characteristics (Yang 1995, p. 67). It is generally believed that Chan Zong was founded by Dharma 达摩 and inherited by Huike 慧可, Sengcan 僧璨, Daoxin 道信 and Hongren 弘忍. After the death of the fifth ancestor Hongren, Chan Zong was divided into two schools by his disciples: the Southern School of Chan, represented by Huineng 慧能 (639–713), and the Northern School of Chan, represented by Shenxiu 神秀 (606–706). The northern school of Chan was prevalent in the northern region, centered on the capitals of Chang'an 长安 and Luoyang 洛阳 in the Tang dynasty, and was worshipped by the ruling class. Therefore, Shenxiu and his disciple Puji 普寂 (651–739) once gained a high Buddhist status. Meanwhile, the southern school of Chan was reformed and developed by Huineng in the south of five ridges (Lingnan 岭南), and its influence was increasing and spread throughout the south area. Later, his disciple Shenhui 神会 (684–758) went to the north area and had a long-term debate with the northern school of Chan. Finally, the Southern school of Chan won the debate and replaced the northern school of Chan as the mainstream of Chan Zong in the late Tang dynasty (Yang 1995, pp. 157–77). Huineng's disciples inherited and developed his theory of Chan, and created many new schools of Chan. In addition to Shenhui who created the Heze Zong 菏泽宗, there are other important inheritors, such as Nanyue Huairang 南岳怀让 (677–744) and Qingyuan Xingsi 青原行思 (671–740), who preached in Hunan 湖南 and Jiangxi 江西 and have significant influences on the later generations (Yang 1995, pp. 195–96).

In the Sui and Tang Dynasties, various schools of Buddhism were popular to varying degrees in some places. In particular, with the strong rise and rapid development of Chan Zong, from around from the late of 8th century to the late of 9th century, the southern school of Chan formed many Dharma centers in China, many of which became the origin of various tradition of Chan Zong (Yang 1995, p. 251). Among them, the most influential tradition are Hongzhou Zong 洪州宗 founded by Mazu Daoyi 马祖道一 (709–788), the disciple of Nanyue Huairang, and Shitou Zong 石头宗 founded by Shitou Xiqian 石头希迁 (700–790), the disciple of Qingyuan Xingsi. Mazu Daoyi once went to Jianyang 建阳 (now Jianyang District, Nanping City) to preach for a short time in the early Tianbao 天宝 period of the Tang dynasty (around 742 C.E.) and received several disciples from Fujian. After that, Mazu Daoyi left Fujian and went to Jiangxi, where he taught disciples in Linchuan 临川 (now Linchuan County, Jiangxi Province) and Hongzhou 洪州 (now Nanchang City, Jiangxi Province). According to historical records, Mazu Daoyi has more than 139 disciples, including some from Fujian, such as Baizhang Huaihai 百丈怀海 (750–814, from Changle

长乐) and Dazhu Huihai 大珠慧海 (from Jianzhou 建州), and the regional scope of his disciples' missionary work covered the vast region of China (Yang 1995, p. 266). Although Daoyi and his disciples from Fujian did not preach Dharma in Fujian for a long time, the Chan Zong still has a lasting impact on Fujian Buddhism. There were many Fujian monks go to Jiangxi and Hunan to worship the Dharma and return to preach (Wang 1997, p. 85). For example, Huangbo Xiyun 黄檗希运 (?–855), a disciple of Baizhang Huaihai, went to Hongzhou to worship the Dharma after becoming a monk in Huangbo Moutain 黄檗山, which was in his hometown, Fuzhou 福州. Later, Xuefeng Yicun 雪峰义存 (822–908), the fifth-generation inheritor of Shitou Zong 石头宗, was from Nan'an 南安, Quanzhou泉州. He was respected by the local governor at that time, and established Xuefeng temple on Xianggu Moutain of Fuzhou in 870 C.E. His disciples also successively created Yunmen Zong 云门宗 and Fayan Zong 法眼宗 in the Chan School, and the temple later became an important Chan temple in the south of the Yangtze River (Yang 1995, pp. 320–21). The emergence of local representative Chan masters and the establishment of Chan temples in Fujian indicated the popularity and development of the Chan School in Fujian at that time. Due to Buddhism being prosperous among the society and his great influence, Yicun was known as the famous figure of the Chan School in the late Tang dynasty. He had 56 disciples, 20 of whom preached Dharma in Fujian, and the footprints of other disciples also spread throughout most parts of China (Wang 1997, p. 129). In this background, a large number of Chan temples were erected in the 9th century, although it is difficult to find the initial temples they erected today. However, most of the Chan temples were founded under their influence and within this context.

According to the Sanshanzhi Records 三山志, Qishan Yuan 栖善院 Temple, founded in 849 C.E., is one of the earliest Buddhist temples in Fu'an with a known date. In the following hundred years of development, up to the Five Dynasties period (907–960), there were more than 30 temples erected in Fu'an alone. From the perspective of the spread and development of Chan Buddhism in this area, most Buddhist temples were Chan temples, most of which were built by local clans who "donated their houses as temples." For example, the genealogy of the Ruan 阮 Clan in Chenliu 陈留, Shuyang 枢洋 Village, and Tantou 潭头 Town, Fu'an recorded that the ancestors of the Ruan family laid a foundation in Longyan 龙岩 in 869 C.E. After serving as a resident for a few years, tigers caused calamities and alarm. The clan ancestor suspected that this showed the existence of the Buddha's spirit, so he returned to his residence, and transformed the building into a Buddhist temple. This is a relatively early record of the formation of Buddhist temples in the Changxi River Basin (Lan 2021, p. 97).

2.2. Remains of Fu'an Buddhist Temple

Early Chan Buddhism advocated only building Dharma halls rather than grand Buddhist halls. The Dharma hall was the most important building in an early temple. It had the functions of meditation and lectures and was also called the Chan hall. Unfortunately, investigation has not revealed any Buddhist temples with Chan halls left from the Tang Dynasty in the Changxi River Basin. According to the statistics in the Sanshanzhi Records quoted by Bucolic Fu'an, it was found that there were 138 temples in Fu'an alone from the 12th to 13th Century (Liu and Chen 2016, p. 260). Many of them are preserved, such as Shifeng 狮峰 Temple, the Sanbao 三宝 Temple, the Xingyun 兴云 Temple, and more.

Shifeng Temple is located at the foot of Shifeng 狮峰 Peak in Xibing 溪柄 Town, Fu'an. It was founded in 892 C.E. and has been rebuilt many times since. Today, the oldest building in the temple is the main hall, which was rebuilt in 1612 C.E. It has a width of 11.75 m with three bays and a depth of 16.20 m with five bays. The floor plan is a longitudinal rectangle, and the maximum height of the building is 14.24 m. The main hall has a Xieshan 歇山 roof with double eaves. The main load-bearing pillars are stone with a square section except for the gate's pillars that are made of wood. There are 28 pillars in total, including 24 stone pillars and 4 wooden pillars (Figures 3 and 4).

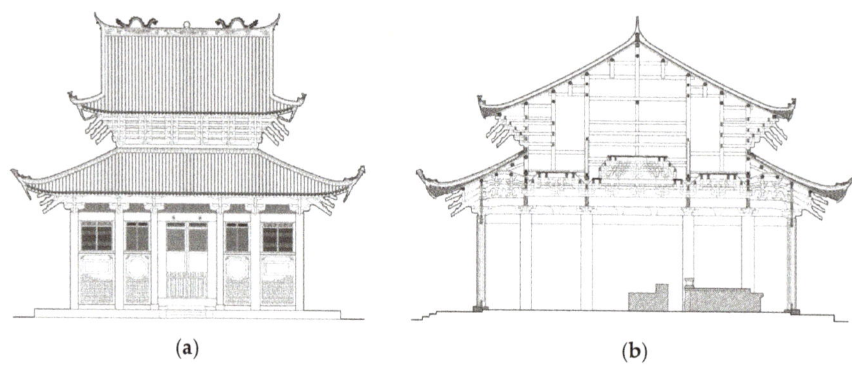

Figure 3. The elevation and section drawings of the Shifeng Temple's main hall (rebuilt in 1612 C.E.): (**a**) Elevation; (**b**) Section. Source: (Liu and Chen 2016, pp. 230–31).

Figure 4. The floor plan drawings and photograph of the Shifeng Temple's main hall (rebuilt in 1612 C.E.): (**a**) Floor plan; (**b**) Photograph. Diagram and photograph by authors.

Sanbao Temple is located in Fu'an County. It was first built in the Song Dynasty as a school and renovated as a Buddhist temple in the Yuan Dynasty. It was rebuilt many times in the Ming and Qing dynasties. The main hall was rebuilt in 1733 C.E. It has a width of three bays measuring 11.21 m and a depth of four bays measuring 15.98 m. The plan is a rectangle with the longitudinal axis in the direction of depth. The main hall has a Xieshan roof with double eaves. The 22 pillars in total are made of stone. The section of the eight internal pillars is square-shaped. The body of the pillars has grooves, and some inscriptions are faintly visible. (Figures 5 and 6).

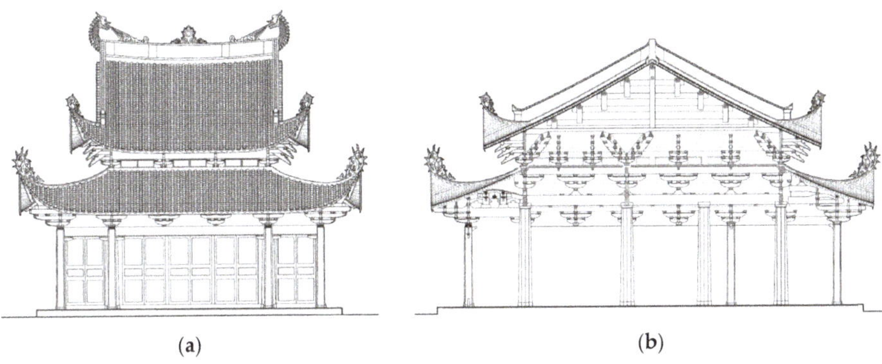

(a)   **(b)**

Figure 5. The elevation and section drawings of the Sanbao Temple's main hall (rebuilt in 1733): (**a**) Elevation; (**b**) Section. Source: (Liu and Chen 2016, pp. 239–40).

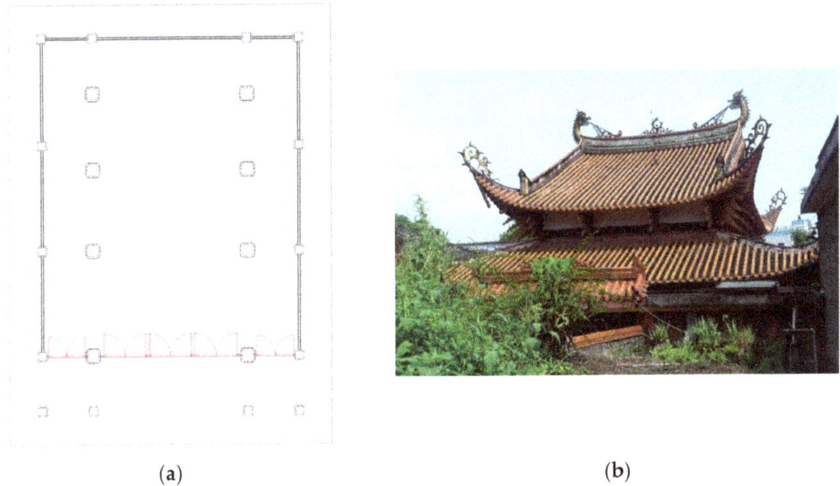

(a)   **(b)**

Figure 6. The floor plan drawings and photograph of the Sanbao Temple's main hall (rebuilt in 1733): (**a**) Floor plan; (**b**) Photograph. Diagram and photograph by authors.

Xingyun Temple, located in Rongtou 榕头 Village, Xibing 溪柄 Town, was founded during 1098~1100 C.E. The main hall of the temple was built in about 1170 C.E. and was originally named Rulai (from the Buddhist term "Tathagata") Hall 如来殿. It was rebuilt in 1776 C.E. and has a width of five bays measuring 14.4 m and has a depth of six bays measuring 13.0 m. The middle bay along its width is 5.54 m. It has a single Xieshan eave and an almost square plane. There are 37 pillars in total in the temple. The four inner pillars in the center are the eight-petal or pumpkin-shaped circular stone pillars, relics from the beginning of the Southern Song Dynasty. The inner pillars in the center are 3.14 m in height and about 50 cm in diameter. Notches are cut in three directions in the capitals of the pillars. The body and the foot of the pillars are an integral whole in stone and there is no plinth. The pillars are clearly tapered at both ends. There are two identical stone pillars which have been found discarded outside the hall. Thus, it can be presumed that, originally, the temple had at least six stone pillars and that only four were used when the hall was rebuilt in the Qing Dynasty (Figures 7–9).

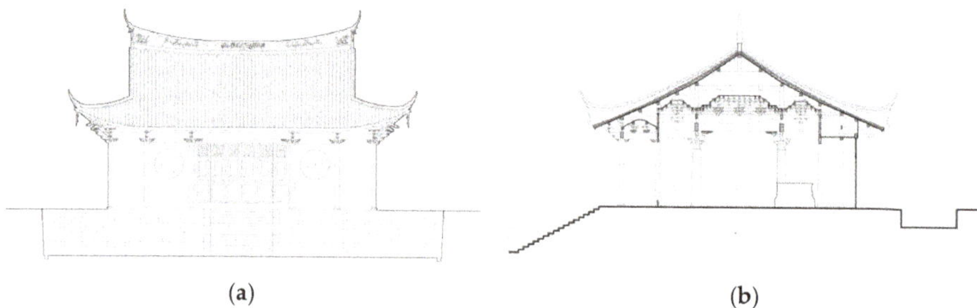

Figure 7. The elevation and section drawings and photograph of the Xingyun Temple's main hall (rebuilt in 1776): (a) Elevation; (b) Section. Source: (Liu and Chen 2016, p. 243).

(a)   (b)

Figure 8. The floor plan drawings and photograph of the Xingyun Temple's main hall (rebuilt in 1776): (a) Floor plan; Source: (Liu and Chen 2016, p. 243.) (b) Photograph., Photograph by authors.

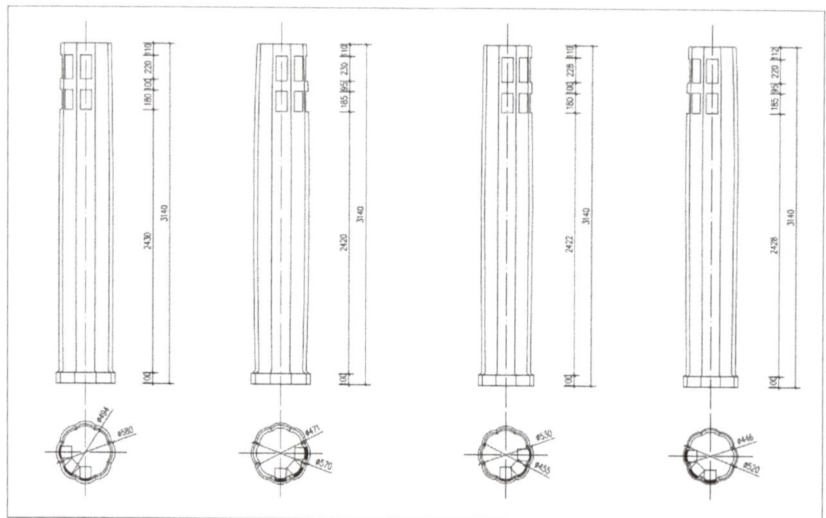

Figure 9. The petal-shaped corrugated stone pillars of four segments used in the main hall of Xingyun Temple. Diagram by authors

In addition to the above Buddhist temples, there are still many Buddhist temples in Fu'an that were built in the Tang and Song Dynasties. However, most of the halls in the temples have been rebuilt in modern times and their historical features have been changed completely. Only early stone pillars, stone grooves, stone foundations, and other architectural parts remain. For example, Bao'en 报恩 Temple, near Xingyun Temple, built from 1098 to 1100 C.E., has at least four round grooved pillars from the Song Dynasty (Figures 10 and 11). The Suoquan 锁泉 Temple in Shouyang 首洋 Village, Xiaoyang 晓阳 Town, founded in 1099 C.E., also has four original pillars from the Song Dynasty in the main hall. Xingqing 兴庆 Temple, founded in 968 C.E. in Panxi 蟠溪 Village, Xitan 溪潭 Town, previously retained its six original Song Dynasty pillars. In the new hall in 2014, regretfully, only four petal-shaped stone pillars were retained. The other two were left for display.

Several remarkable regional characteristics can be summarized on the basis of the existing Buddhist temple halls in Fu'an built from the 9th to 12th centuries:

1. Most Buddhist temples built hundreds or more than a thousand years ago retain their rectangular plans despite reconstruction or expansion in subsequent dynasties. Most of their parts and members have been replaced except for a few stone components and base elements that are not easily damaged or not easily replaced. Hence, very few original components remain that can reflect the original appearance assumed by these structures in the Tang and Song dynasties.

2. Stone pillars are commonly used as vertical load-bearing components. The petal-shaped corrugated eight-segment stone pillars commonly utilized in the 12th century still serve as the inner pillars in the middle of these Buddhist temples. At the same time, square stone pillars or wooden pillars were used by later generations to support extensions of the structures. The petal-shaped corrugated stone pillar (also called "segment pillars" in the Yingzao Fashi 营造法式 (Li 1933, p. 206) can be used to identify the age of the temples. There are many petal-shaped corrugated stone pillar relics in South China, and most of them were made in the Song Dynasty, such as the stone pillars of the main hall of Luohan Yuan罗汉院 in Suzhou and for the Kaiyuan Monastery 开元寺 in Chaozhou 潮州. Wood is not as durable as stone, and consequently, petal-shaped corrugated wooden pillars are not as well preserved. The only remaining case is the four petal-shaped, eight-segment corrugated stone pillars in the main hall of the Baoguo Monastery in Ningbo. The petal-shaped corrugated stone pillars are very common in the Changxi River Basin and can been seen in almost all the Buddhist temples. Most of them are petal-shaped corrugated stone pillars of eight segments, tapered at both ends. The petal-shaped corrugated stone pillars in one Buddhist temple are usually all the same size, about 3~4 m tall with a diameter of about 450~700 mm. The diameter-to-height ratio is about 1/6~1/8, which is less than the ratio of normal wooden pillars, making the stone pillars seem sturdy and solid by comparison.

3. Combining the characteristics of the plan and the structural frame, the evolution of most Buddhist temples from single-bay to three-bay or five-bay structures can be clearly seen. For example, in a three-bay temple, the middle bay is significantly wider than the adjacent bays. The ratio of the width of the middle bay to that of the adjacent bays is usually over two. Furthermore, the maximum middle bay width in these structures can reach 7.1 m, much more than the upper limit of 18 chi 尺 (6 m) that is the maximum width for middle bays of main hall specified in the Song Dynasty Yingzao Fashi. The extremely large middle bay is rare in folk houses of the same period or later. Therefore, the extremely large middle bay is a significant indication that single-bay halls in Buddhist temples initially functioned to enclose an internal space.

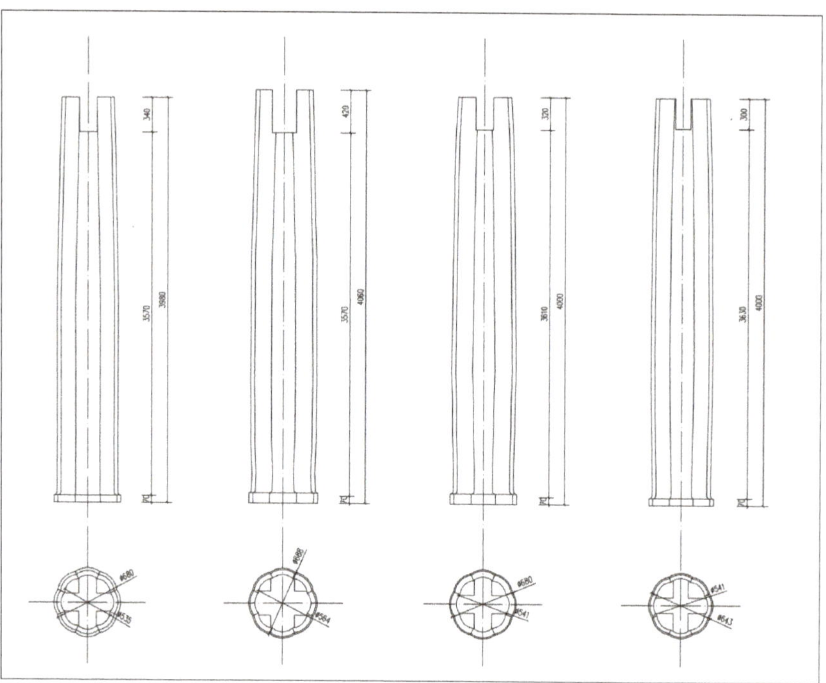

Figure 10. The drawings of four stone pillars used in the main hall of Baoen Temple. Diagram by authors.

Figure 11. The petal-shaped corrugated stone pillars of four segments used in the main hall of Baoen Temple. Photograph by authors.

3. The Restoration of the Early Forms of the Buddhist Halls in the Changxi River Basin

Existing investigations reveal that most of the temples remaining in the Changxi River Basin were rebuilt in the Ming and Qing dynasties. Their planes are nearly square, or they are rectangular, three or five bays wide, and three or five bays in deep. Their roof is often a single or double Xieshan roof, and its structure is based on local practice in Fujian since the Ming Dynasty. The few Buddhist temples that can be traced back to the Tang and Song dynasties can only be verified from remaining components. There are only the petal-shaped corrugated stone pillars of four or eight segments that were left from the Song Dynasty. From the perspective of the traditional Chinese concept of fengshui 风水, the site selection and orientation of ancient Buddhist temples were generally strictly considered by founders at the beginning of construction, and the inheritors of the temple also remained devout to such principles. If there is no disaster during later periods of operation, the previous decision is considered to have been correct. There is a tradition of rebuilding on the original site of construction as well. Clearly, Buddhist temples created by predecessors were destroyed in future generations in the course of natural disasters or man-made disasters. Most of the components have disappeared, and only stone pillars, stone beams, and other stone components survive. This shows how stone components serve as a symbol of longevity in Buddhist temples. However, at the same time, the preservation and use of these finely selected and exquisitely crafted stone pillars donated by predecessors in the reconstruction of the temples by later generations undoubtedly demonstrates respect and the continuation of the history of temple construction. Therefore, in the above-mentioned cases, when the Buddhist temples were rebuilt during the Ming and Qing dynasties, the Song Dynasty stone pillars were used. However, just as a "stratum" becomes disturbed in archaeology, subsequent generations rebuilt the Buddhist temple on top of the relics of the Song Dynasty. The degree of such interventions on the original structure cannot be completely ascertained, so it is difficult to reveal the "original state" of buildings through current research. Although there are not a few Song Dynasty stone pillars in Fu'an Buddhist temples, the original arrangement of stone pillars has been possibly retained only in the Shifeng Temple Hall and Sanbao Temple Hall, based on an examination of the base site only. The remaining stone pillars in other temples are either inadequate or they have been moved. Therefore, it is difficult to use these remains to restore the appearance of the Song Dynasty Buddhist Temple. Here, we only make a speculative restoration based on relevant references.

3.1. The Single-Bay Hall

In a main hall of a temple, if the remaining six pillars in three rows or the eight pillars in four rows are regarded as eave pillars, the hall can be restored to a single-bay structure, extending multiple bays in depth. Although this pattern is not common, there are remains around the Changxi River Basin. For example, the Zushi 祖师 Ancestor Hall of West Mingshanshi 名山室 located on the hillside of Gaogai 高盖 Mountain in Yongtai 永泰 County, Fujian Province, was built at the entrance of a mountainside cave and was founded in 1103 C.E. (Figure 12). The hall is a single bay in width and two bays in depth with a single Xieshan roof. The six eave pillars are all petal-shaped corrugated stone pillars arranged in three rows. Although the structure has been repaired by later generations, it still follows the Song system and is a rare single-bay hall relic of the Song Dynasty. Although the Zushi 祖师 Ancestor Hall is not huge and is limited by the size of the cave, the structure proved that a single-bay width, multi-bay depth hall with a Xieshan roof is completely feasible.

There also are similar single-bay Buddhist temple remains found in archaeological sites, for example, the hall of Guoxing 国兴 Temple built in 1011 C.E. in the Song Dynasty located on Taimu 太姥 Mountain in Fuding 福鼎. (Figures 13 and 14) The main hall of Guoxing Temple is one bay in width and three bays in depth. The eight stone pillars are arranged in four rows. The pillars stand on the original position except for one that is lost. Although Guoxing Temple is located in a mountain col, the area for construction is spacious. The main hall should not have been restricted by the site when it was initially built. Therefore, the plan pattern of plan should have been one of the choices popular at that time (Gao et al. 2021, pp. 7–11).

Figure 12. The Zushi ancestor hall at the entrance of the mountainside cave. Diagram by authors.

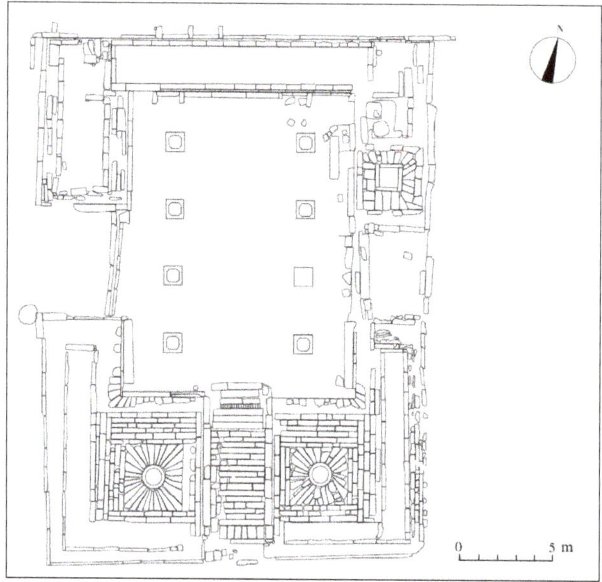

Figure 13. The archaeological site plan of the main hall's relics from Guoxing temples in Fuding. Diagram courtesy of Guoping Pan.

Figure 14. The main hall's relics from Guoxing temples in Fuding. Photograph courtesy of Guoping Pan.

The main hall of Guoxing Temple is the only excavated site built in the Song Dynasty so far. Similar cases of this pattern of single bay in width and multiple bays in depth can be found in early murals and other paintings. For example, the pattern can be found in the Hui Yinguo Scripture, painted in the Nara Period (710–794), collected in the Nara National Museum, Japan. (Figure 15) The Hui Yinguo Scripture 绘因果经 is a picture book of the Past and Present Yinguo Scripture 过去现在因果经 translated by Gunabhadra 求那跋陀罗, who was an Indian monk preaching in the southeastern China during the 5th century. Similarly, a gate tower and a scene of construction of a Buddhist temple, both of single-bay in width and three-bay in depth, can be seen in Mogao 莫高 Grottoes murals in Dunhuang 敦煌 painted in the 8th century, which indicates that the single-bay pattern had already appeared in the Tang and Song Dynasties (Figure 16).

Figure 15. The image of the single-bay pavilion in Hui Yinguo Scripture. Diagram by the authors based on the paintings collected by Nara National Museum.

(a)   (b)

Figure 16. The image of gate tower and a construction scene of a Buddhist temple in Mogao 莫高 Grottoes murals in Dunhuang 敦煌: (**a**) A gate tower with built; (**b**) A Buddhist hall under construction. Photograph courtesy of Liliang Song.

3.2. The Tansformation from Single Bay to Three Bays

From the perspective of roof structures, a single-bay hall with three or four rows of pillars of the same height forming a rectangular plan can have either been a Xuanshan 悬山 or a Xieshan 歇山 roof. When it comes to a Xuanshan roof, there are two possibilities: one is that the roof ridge is set in accordance with the direction of the depth, placing the entrance at the gable side. This can be found in very early times in the wooden structures of South China. It is derived from the "long houses" of matriarchal clans in the Neolithic Age. This practice later spread to Japan, and there are still remnants of such construction today. The other second possibility is that the roof ridge is set in the direction of the width of the hall. As a result, it was likely to be eroded by rain and damaged. Xieshan roofs, on the other hand, have slopes on all four sides, with the four eaves and four original pillar rows of the same height thus best suiting this form of roof. Thus, whether from the perspective of functionality or the requirements of ritual, the Xieshan roof was undoubtedly a better choice.

With the popularity and development of Chan Buddhism, the small single-bay Buddhist Hall developed and spread in the Changxi River Basin, gradually becoming the central building type for Buddhist temples in the 12th century. Obviously, the narrow indoor space formed by a single bay could no longer meet the requirements of Buddhist activities. Therefore, there were two main possibilities for expansion based on the foundation of the original single-bay hall.

Firstly, a line of eave pillars could be added on both sides and the roof extended. This is even common in the folk houses in Zhejiang and Fujian nowadays. The museum in Zhouning 周宁 County, located in the upper reaches of the Changxi River Basin, houses a ceramic barn built in the 12th century with a hall of this type on its roof (Figure 17). This kind of roof extension is relatively easy to build. Only the width of the hall is increased, and the depth stays the same. The whole plan thus transforms to nearly become a square.

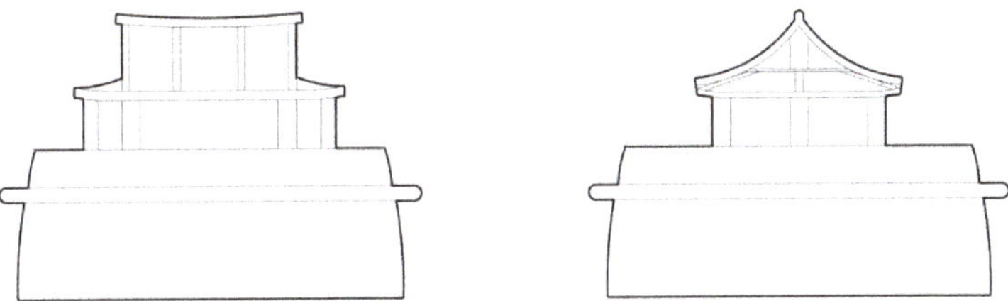

Figure 17. The hall with three bays width on the top of the ceramic barn housed by Zhouning County. Diagram by authors.

The second possibility is to add eave pillars around the Buddhist hall consisting of a single bay in width and covered by a Xieshan roof. This forms a porch around the structure and a total width of three bays. The plan is still a longitudinal rectangle. Typical cases, such as the existing Shifeng Temple and Sanbao Temple, were rebuilt in the style of double-eave Xieshan roofs in the later generations. Another example is the Chen Taiwei 陈太尉 Palace in Luoyuan 罗源 at the southern end of the Changxi River Basin (Figure 18). It was built in 1239 C.E. and rebuilt many times subsequently. Its main hall is now three bays in width and six bays in depth. According to the shape and structure of the components, it is presumed that the original structure of the Song Dynasty consisted of a single bay in the middle forming its width and the first two bays in depth, and the rest was added by later generations. The expansion of the Chen Taiwei Palace in both width and depth caused the double-eave Xieshan roof to take on a somewhat weird shape (Zhang 1999, pp. 67–68).

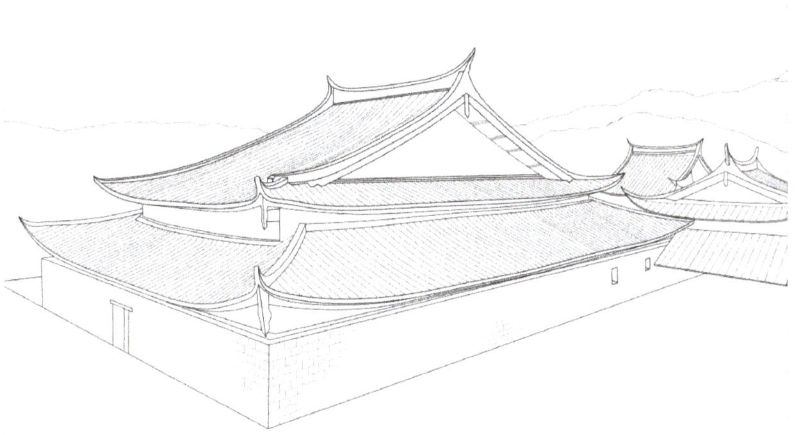

Figure 18. The appearance of Chen Taiwei Palace. Diagram by authors.

The main structure of the single-bay hall with a Xuanshan or Xieshan roof was not changed by the above two types of expansion (Figure 19). However, when the original single bay width is expanded to three bays and the original roof structure is changed to fit three bays, the roof structure is undoubtedly changed. It is found through the investigation into the remains of the temples in the Changxi River Basin that there have been temples of three bays in width with single-eave Xieshan roofs built from the 10th to 14th centuries.

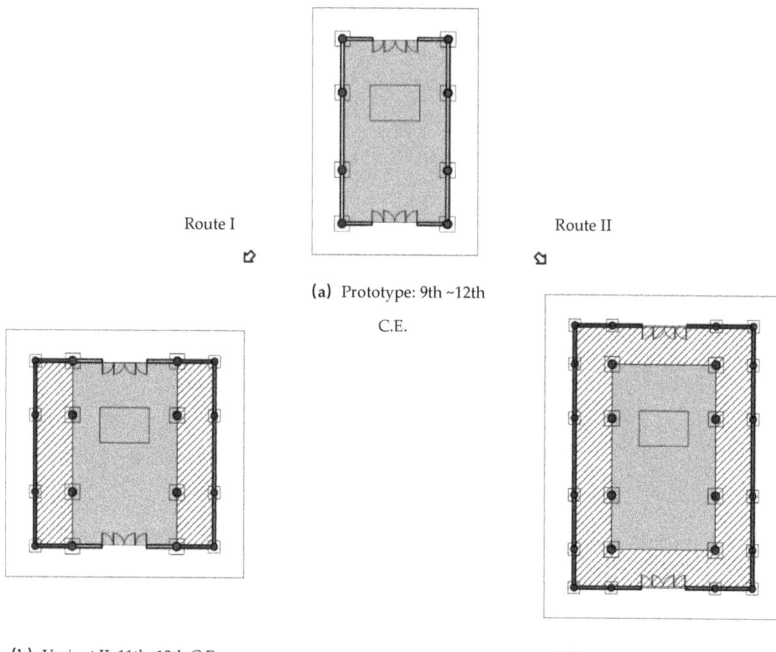

Figure 19. Transformation Paths from single bay to three bays. Diagram by authors.

The transitional roof seems to be a temporary rather than a permanent structure and an intermediate product of the change from the Xuanshan to Xieshan roof. Although there are not many remains in the Changxi River Basin, from the perspective of technological evolution, it is reasonable to infer that the rectangular plan of the Buddhist hall with a single bay in width and three bays in depth with the longitudinal axis in the depth direction changed to a square plan with three bays in both width and depth. Such a path of change seems possible. The rectangular plan has been expanded into a square plan by adding two sides. Later reconstructions are often completed based on the square plan. The roof structure also has been improved and the single-eave Xieshan roof has prevailed. The remains of this type of Buddhist temple can be found in the South of the Yangtze River area, but it is not common in the Changxi River Basin.

4. The Reasons for the Evolution of Buddhist Halls in Changxi River Basin

The development of Buddhist temples in the Changxi River Basin began in the 9th century C.E. Now, after more than 1000 years of developments, there are still dozens of Buddhist temples that were founded during that era. Although most of this Buddhist architecture has been greatly changed due to reconstruction in previous dynasties, the appearance of Buddhist temples in various historical stages in the Changxi River Basin can still be analyzed and appreciated through the study of several sites and relics. As mentioned above, from the 9th to the 12th century C.E., Buddhist temples with widths composed of wide single bays and depths that extended across several bays prevailed in the Changxi River Basin. These plans were composed of longitudinal rectangular planes. Since the 14th century, these single-bay Buddhist halls were often transformed into larger Buddhist temples encompassing three or five bays in width. The direct reason for this evolution was to achieve larger indoor spaces. However, in terms of the larger history of Buddhist architecture, this expansion from small Buddhist halls into medium Buddhist temples was accompanied by the enrichment of spiritual space through developments such as as Chan teachings and rituals.

In the early days of Chan Buddhism, monks lived in seclusion and traveled to learn. Later, they gradually settled in groups and set up fixed places for preaching. At the end of the 8th century, the Chan master Baizhang 百丈 created the rules of Chan Buddhism and proposed regulations for construction that set forth the requirement to build only Dharma halls rather than Buddha halls in temples. These rules and regulations were followed and observed by the inheritors of the tradition of Chan Buddhism. Later, from the 10th to 13th century, Chan Buddhism spread widely and became prosperous in South China. More and more Chan Buddhist temples were erected, and their scale was gradually enlarged. The rules and regulations of Chan Buddhism then gradually changed. The Buddhist halls, as the places for Buddhist faith, became more important than the Dharma hall (Zhang 2002, pp. 72, 73). Thus, it was erected as the main hall of a temple in the 13th century. The Buddhist halls with three or five bays in width thus became mainstream. Under the layout rule of "Galan Seven Halls 伽蓝七堂" that matured and took shape at the end of 13th Century, the Buddhist hall became the core of a temple and has expanded continually in subsequent times.

On the other hand, the doctrine of Chan has also constantly evolved with the passage of time. The original "non-written" oral teachings were subsequently written down, read, and taught. The original meditation ritual has become a part of daily life. The worship of the single Buddha of Sakyamuni 释迦牟尼 has been changed to include the worship of multiple Buddhas. Such changes have brought about a series of changes in the space of the temple, such as the Dharma hall gradually becoming the lecture hall, and the function of the main hall has become an important space for rituals in Chan temples. In short, larger spaces became necessary.

It is worth noting that this change in Buddhist architecture caused by the development of Buddhist doctrines and rituals is reflected not only in the Changxi River Basin but also in the South of the Yangtze River region and other surrounding areas. As mentioned concerning Buddhist halls with square plans and with three bays in width and a single-eave Xieshan roofs in the south of the Yangtze River from the 10th to 13th century, most of such temples have been rebuilt or renovated during later dynasties, such as the main hall of Hualin 华林 Temple and Baoguo 保国 Temple (Zhang 2012, pp. 188, 204), Yanfu 延福 Temple, and Zhenru 真如 Temple. These temples were all expanded with a porch around them based on the original structure, forming five bays in width with a double-eave Xieshan roof (Figure 20). There are several reasons behind these expansions. One is to meet the new requirements of the space in the main halls for later generations. The other is the pursuit of a more magnificent and solemn space for worship, which may have been caused by the increasing worship of Buddha statues in the process of an increasing focus on laity.

Figure 20. The present appearance of the Baoguo Temple's main hall after extended in the Qing Dynasty. Photograph by authors.

5. Conclusions

Buddhist halls with a single bay in width and two or three bays in depth were popular in the Changxi River Basin from the 9th to the 13th centuries C.E. The roofs were usually supported by three or four rows of the petal-shaped corrugated stone pillars. The plan was a longitudinal rectangle, and the roof was a single-eave Xieshan 歇山 roof. After the 13th century C.E., due to practical and symbolic needs, many temples were expanded by adding a porch around the original structures. The double-eave Xieshan roof was thus formed, and a few of them remain today.

Therefore, the following are concluded based on the above analysis:

1. The single-bay Buddhist hall in the Changxi River Basin was either a small-scale Buddhist Hall at the time of the implementation of the rule to build Dharma Halls and not Buddhist Halls, or it was the form of an early era Dharma Hall. The Buddhist hall became important in a temple in the 10th century. At that time, the traditional regulations on the construction of Chan Buddhist temples had been broken or developed. In 13th Century, the development of Chan Buddhism reached its peak, and its rules were popularized. Then, the main hall of temples of three or five bays in width were built by either demolishing the old temples on the original site and rebuilding a new one or by expanding the old temples.

2. The closed geographical environment of the Changxi River Basin has long been far away from the central or regional government center, allowing local Buddhism to develop more spontaneously and be influenced by local clans and believers. Therefore, there was a more conservative development of Buddhist architecture. The changes to the structure of the Chan temples were not as drastic as in the surrounding areas. The more ancient forms still exist in this area. The rectangular plan of the temples where the longitudinal axis is in the depth direction driven by the single bay in width remains to this day.

This study is an initial examination of single-bay Buddhist temples and their evolution in the Changxi River Basin through an investigation into these temple's remains. It provides new references and academic support for the study of ancient Chinese Buddhist architecture and for the protection of the heritage in South China. More in-depth discussions on the techniques and evolution of the single-bay Buddhist temple will be carried out based on this study. This will include work regarding the roof forms, beam frames, and the direction of the Xuanshan roof of single-bay Buddhist temples. Through such research, we can better appreciate the significance of the Buddhist temples built in the Changxi River Basin from the 10th to 13th century.

Author Contributions: Conceptualization, J.L.; methodology, J.L.; software, C.C.; investigation, J.L., C.C. and Y.J.; resources, J.L. and C.C.; writing—original draft preparation, C.C., Y.J.; writing—review and editing, J.L., C.C.; All authors have read and agreed to the published version of the manuscript.

Funding: This research was funded by [National Natural Science Foundation of China] grant number [50308015].

References

Chen, Zhihua. 1992. *Research on the Vernacular Architectures in the Middle Reaches of Nanxi River* 楠溪江中游乡土建筑. Taiwan: Hansheng Magazine.

Chen, Zhihua. 2004. *Research on the Acient Villages of the Upper Reaches of Nanxi River* 楠溪江上游古村落. Shijiazhuang: Hebei Education Press.

Gao, Jianbing, Guoping Pan, and Wenda Chen. 2021. Brief report on archaeological investigation and exploration of the relics of Guoxing Temple in Taimu mountain during 2015–2016 太姥山国兴寺遗址 2015-2016 年度考古调查, 勘探简报. *Fujian Wenbo* 1: 6–14.

He, Meifang. 2013. Re-Examining the Investigation and Image Recording of Chinese Architecture Taken by the Modern Japanese Scholars 解读近代日本学者对中国建筑的考察与图像记录. Doctoral thesis in architecture, Tianjin University, Tianjin, China.

Lan, Jiongxi. 2021. The social foundation of Fu'an Buddhism 福安佛教生成的社会基础. *Journal of Fujian Studies* 2: 93–106.

Li, Jie. 1933. *Yizao Fashi* 营造法式. Shanghai: The Commercial Press, p. 206.

Liu, Jie, and Changdong Chen. 2016. *Bucolic Fu'an* 乡土福安. Beijing: Zhonghua Book Company, pp. 20, 243, 293–318.

Liu, Jie, and Gang Hu. 2011. *Bucolic Qingyuan* 乡土庆元. Hangzhou: Zhejiang Ancient Books Publishing House, pp. 296–318.

Liu, Jie, and Weiping Shen. 2005. *Lounge Bridges in Taishun* 泰顺廊桥. Shanghai: Shanghai People's Fine Arts Publishing House, pp. 6–20.

Liu, Jie. 2017. *The Architectural Artistry of China's Timber Arch Covered Bridges* 中国木拱廊桥建筑艺术. Shanghai: Shanghai People's Fine Arts Publishing House, pp. 262–66.

Ma, Xueqiang. 2016. *Eight Hundred Miles of Oujiang* 八百里瓯江. Beijing: The Commercial Press.

Osada, Tokiwa, and Sekino Tadashi. 1939. *The Historical Sites of Chinese Culture* 支那文化史迹. Kyoto: nihonnoshingaku, vol. 4, p. VI-69-120.

Wang, Rongguo. 1997. *The History of Buddhism in Fujian* 福建佛教史. Xiamen: Xiamen University Press, p. 4.

Wu, Songdi. 1990. Economic Development of the Mountain Areas in the Southeast Coast of the Song Dynasty 宋代东南沿海丘陵地区的经济开发. *Historical Geography* 7: 14–27.

Wu, Tingqiu, and Pengnian Zheng. 1995. Research on the Introduction of Buddhism Spread into China by sea 佛教海上传入中国之研究. *Historical Research* 2: 20–39.

Yang, Cengwen. 1995. *History of Chan Buddhism From Tang Dynasty to Five Dynasties* 唐五代禅宗史. Beijing: China Social Sciences Press.

Zhang, Shiqing. 1999. The Architecture of Chen Taiwei Palace in Luoyuan, Fujian 福建罗源陈太尉宫建筑. *Cultural Relics* 1: 67–75.

Zhang, Shiqing. 2002. *Buddhist Temples of Jiangnan in China* 中国江南禅宗寺院建筑. Wuhan: Hubei Education Press, pp. 72–73.

Zhang, Shiqing. 2012. *The Main Hall of Baoguo Temple in Ningbo: Survey Analysis and Basic Research* 宁波保国寺大殿: 勘测分析与基础研究. Nanjing: Southeast University Press, pp. 188, 204.

Article

Comparative Review of Worship Spaces in Buddhist and Cistercian Monasteries: The Three Temples of Guoqing Si (China) and the Church of the Royal Abbey of Santa Maria de Poblet (Spain)

Weiqiao Wang

Escuela Técnica Superior de Arquitectura de Madrid, Universidad Politécnica de Madrid, 28040 Madrid, Spain; weiqiao.wang@alumnos.upm.es

Abstract: Although the two parallel architectural forms, Han Buddhists and the Cistercian monasteries, seem, on the surface, to be very different—belonging to different religions, different cultural backgrounds, and different ways of construction—they share many similarities in the internal institutional model of monks' lives and the corresponding architectural core values. The worship space plays the most significant role in both monastic life and layout. In this study, the Three Temples of Guoqing Si and the Church of the Royal Abbey of Santa Maria de Poblet are used as examples to elucidate the connotations behind the architectural forms, in order to further explore how worship spaces serve as an intermediary between deities, monks, and pilgrims. Based on field research and experience of monastic life, this comparative study highlights two fundamental similarities between the Three Temples and the Church: First, both worship spaces are derived from imperial prototypes, have a similar priority of construction, occupy the most important place in both sacred venues, and both serve as a reference for the development of monastic layout. Second, both worship spaces are composed of a similar programmed functional layout, including similar space dominators as well as itineraries. Beyond the surface similarities, this article further analyzes the reasons behind the three differences found. Due to their different understanding of deities, both worship spaces show different ways of worship, images of deities, and distances towards them.

Keywords: worship space; the Three Temples; the Church; comparative study; monastic life; spatial layout; Guoqing Si; the Royal Abbey of Santa Maria de Poblet

Citation: Wang, Weiqiao. 2021. Comparative Review of Worship Spaces in Buddhist and Cistercian Monasteries: The Three Temples of Guoqing Si (China) and the Church of the Royal Abbey of Santa Maria de Poblet (Spain). *Religions* 12: 972. https://doi.org/10.3390/rel12110972

Academic Editor: Jaime Vázquez Allegue

Received: 21 July 2021
Accepted: 27 October 2021
Published: 8 November 2021

Publisher's Note: MDPI stays neutral with regard to jurisdictional claims in published maps and institutional affiliations.

1. Introduction

1.1. Study Background and Objectives

"Typical of the plan is first of all the arrangement of its buildings in three groups. The whole layout is arranged around three longitudinal axes. The structures lying along the central one form the group which stands as the official embodiment and expression of the idea on which the existence of the whole layout is based and will contain the chief halls to which the public have access."(Prip-Møller 1937)

Since Chinese scholars first began to investigate Buddhist architectures in China (Liang 1961), the central axis has always been considered as a key issue. From its evolution and formation (Wang and Xu 2011), speculation on the basic spatial sequence (Su 2009), space configuration (Wang and Lu 2000), and its role in the development of the whole monastic layout (Wang 2016), it has emerged as a fundamental unit in Buddhist monasteries. Therefore, it has always received great attention through records (Liang 1932), drawing (Liang and Fairbank 1984; Zhang 2012), and mapping (Wang 2011).

However, the above studies mostly discuss the central axis from a single view of architectural meaning, but rarely from its essence of religious space. As a worship space, what is the internal relationship with deities, monks, and pilgrims? How is it related to the whole layout? How is the range of worship space defined? How is the worship itinerary generated?

At the same time, we have also seen that Western as well as Japanese scholars have been consciously forced to think about the connotations of religion in the study of Chinese Buddhist monasteries, from the daily rituals of monks (Boerschmann 1912), monastic layout (Itō 1931), space functions (Sekino and Tokiwa 1926), and itinerary of worship (Prip-Møller 1937).

However, in the Western context, the worship space refers to a church in a monastery. It is worth considering the fact that the worship space in Buddhist monasteries does share some similarities with the church. Despite their apparent different appearances, relatively little research has been done to thoroughly investigate the similarities and differences in the worship spaces of Buddhist and Christian monasteries. Furthermore, how can we define the range of worship spaces in both monasticisms?

The configuration of the central axis of Buddhist monasteries has gone from being centered on the pagoda to being centered on the main hall (Liu 2007). The main hall, where the central altar of the Buddhist image is located, is regarded as the most important sacred space (Sun 2017) and the climax of the whole layout, no matter the spiritual level or the architectural level (Prip-Møller 1937). In order to make the comparative research more scientific, this article regards the entrance to the main altar in the central axis space as a complete worship space based on the religious and spatial relationship between deities, monks, and pilgrims.

With the evolution of Western monasticism (King 1999), church architecture has also gone through evolution and transformation (Braunfels 1972); however, it has always been considered and constructed in a whole volume. It has maintained the religious function (Walsh 1932) and the direction of approach from the west to the east. Furthermore, the location of the church is always set next to a cloister with a solid façade. These all contribute to the realization of the independence and integrity of the church space as a worship space.

Buddhism and Christianity have always been studied comparatively from historical (Boisvert 1992), spiritual (Henry and Swearer 1989), and religious practice (Moon 1998) aspects, but there is not enough comparative analysis to illustrate their similarities and differences in terms of their architectural aspects. Therefore, the author believes that a comparative analysis of worship spaces in Buddhist and Christian monasteries could fill this historiographic gap and provide materials and reflection for a better comprehension of the similarities and the differences between the two spaces.

Because this is a wide-ranging topic, it is better to focus on detailed cases studies; therefore, two monasteries, a Buddhist monastery, the Guoqing Monastery, and a Cistercian monastery, the Royal Abbey of Santa Maria de Poblet, have been selected to conduct a comparative analysis. The reasons for their selection will be explained in the following sections.

1.2. Research Methods and Scope

Before discussing the worship spaces in the Guoqing Monastery (hereafter Guoqing Si) and the Royal Abbey of Santa Maria de Poblet (hereafter Poblet Monastery), it is fundamental to consider the following: Why have Han Buddhist and Cistercian monastic traditions been selected as the foci of this comparative analysis?

Many religions active in the world today impress upon their followers their respective teachings and methods of practice, but not all emphasize a strict cenobitic approach. Indeed, among the three largest world religions, only Buddhism and Christianity encourage and practice cenobitic life, whilst Islam criticizes and forbids it; the Quran teaches acceptance of life, not rejection or withdrawal, and by dint of this, monasticism and asceticism are not allowed in Islam. In comparison, Buddhism and Christianity share very similar attitudes towards their cenobitic practices. However, within both Buddhism and Christianity, one can see examples of many types of cenobitic life.

Christianity originated in the Middle East, and is the largest religion in the world. It is formed of three main branches; the Catholic Church, Protestantism, and the Orthodox Church. Catholicism can be considered as the most extensive of the Christian traditions, and has had a decisive influence across Western European culture and history, particularly in Italy, France, and Spain. Within this, the Order of Cistercians is the most influential of the monastic traditions. In contrast, Buddhism originated in ancient India, and has a widespread following across Asia, particularly in east Asia, as well as a significant number of devotees in Europe, Africa, and North America. China has the largest number of Buddhist monuments and Buddhists in the world.

With Han Buddhist and Cistercian having developed globally as the main monastic traditions, specific cenobitic traditions have in turn developed within each, with some clear similarities between them. Both monasticisms derived from the hermit tradition, whereby hermits lived in deserts and followed a self-sufficient lifestyle. Adherents to both have similar pursuits for a strict and austere religious life. It is also worth considering, therefore, whether there exists a further similarity in the architectural organization of the cenobitic lifestyle.

It is undeniable that there are many differences between the architecture of Han Buddhist and Cistercian monasteries. The purpose of this research is not to simply compile a catalog of similarities and differences, but rather to shed light on the reason behind common characteristics shared by both monastic spaces. In his preface to *For the Sake of the World: The Spirit of Buddhist and Christian Monasticism*, Patrick G. Henry mentioned a prediction made by Arnold J. Toynbee: *"It seems more meaningful to cheer on the commons than point out the difference between Buddhism and Christianity. More than forty years ago, Arnold J. Toynbee predicted that when historians in the twenty-first century write about our own, they will be more interested in the interaction between Buddhism and Christianity than in the conflict between communist and democratic systems."* (Henry and Swearer 1989). The similarity is the start point of comparison, and comparison is the structure of analysis. In the comparison procedure, one case will intrigue questions for the other that this dynamic is originally built into it. *"If you analyze a single thing, you learn more about it, but it confirms itself in a way there's no dynamic built into the procedure. ... But you reveal sameness and difference and the layering of architectural expression when you compare two works that are from different points of view, answering more or less the same problem."* (Frampton 2015).

First emerging in France in the 11th century, following the rule of Saint Benedict (Benedict of Nursia), Cistercian monasteries quickly expanded in Western Europe in the 12th and 13th centuries. Almost in the same period, from the 11th to 13th centuries, the Song Dynasty of China witnessed the flourishing of Chan Buddhist monasteries. These followed the Imperial Edition of the Pure Rules of Baizhang (Dehui 1335) and became the dominant form of monasticism in Asia.

As these two traditions emerged, it is worth noting that both shared similar preferences in site selection: concealed in valleys, located in wilderness remote from villages, or in selected desolate and uninhabited sites near the sea. Such selections are closely related to similar preferences in the style of religious daily life; self-sufficient and the self-sustained, desiring to maintain their lives through their own works and actions, rather than by relying upon the support of wider society. Furthermore, following from the way in which both types of monastic life developed from a hermit origin, followers must also dedicate all their time to worship, meditation, work, and reading (Tables 1 and 2); among these, worship is the most important.

Table 1. Daily Routine for Buddhist Monastic Life (Guoqing Si).

Time	Office	Location
04:00–06:00	Morning Chanting	Main Hall
06:00–06:15	Breakfast	Refectory
06:15–07:00	Rest	Dormitory
07:00–10:30	Work	Farm, Kitchen, Laundry, etc.
10:30–11:00	Lunch	Refectory
11:00–13:00	Rest	Dormitory
13:00–16:30	Work	Farm, Kitchen, Laundry, etc.
16:30–16:45	Dinner	Refectory
17:00–19:00	Evening Chanting	Main Hall
19:00–19:45	Meditation	Meditation Hall
19:45–21:00	Reading	Dormitory
21:00	Sleep	Dormitory

Table 2. Daily Routine for Cistercian Monastic Life (Poblet Monastery).

Time	Office	Location
05:15–06:00	Vigilias	Church
06:00–07:00	Meditation	Dormitory or Cloister
7:00	Laudes	Church
08:00–08:45	Conventual mass	Church
08:45–09:00	Breakfast	Refectory
09:00–13:00	Work	Kitchen, Laundry, etc.
13:00–13:15	Noon prayer	Church
13:15–13:45	Lunch	Refectory
13:45–18:30	Work	Kitchen, Laundry, etc.
18:30	Visperas	Church
19:00–19:30	Dinner	Refectory
19:30–20:15	Work	Kitchen, Laundry, etc.
20:15–20:30	Reading	Chapter room
20:30–21:00	Completas	Church
21:00	Sleep	Dormitory

The essence of the monastery is the spatial layout and unity of the main buildings. Indeed, the very construction of the main buildings is done in such a manner as to reflect the relationships between deities, monks, and pilgrims. Typical spaces within the monastery include worship spaces, a Dharma room or chapter room, refectories, dormitories, and meditation halls. Among these, those with highest priority are the worship spaces, and thus these are often the most complex of all. It is possible to build a larger worship space as an aspect of cenobite life[1], and this can be seen in both monastic traditions. To strictly follow their monastic rules, both Buddhists and Cistercians require an ideal layout for their monasteries, with Buddhists following the *Illustrated Scripture of Jetavana Vihara of Sravasti in Central India* as drawn in the 7th century (Figure 1), and Cistercians following the 8th century *Plan of St. Gall* (Figure 2). From both ideal layouts, we can find that the axis of worship is emphasized, and a functional partition is set to delineate between deities, monks, and pilgrims (Ho 1995; Horn et al. 1979). The more central an architectural aspect is in the layout of the monastery, the more sacred; the further to the outside, the more secular. At the heart of the monastery are the temple and the church—essentially considered the residences of Buddhas and God, located in the holiest place and constructed with the highest standard. They are the largest buildings in the monastery and within them are held the most significant and ordinary liturgies.

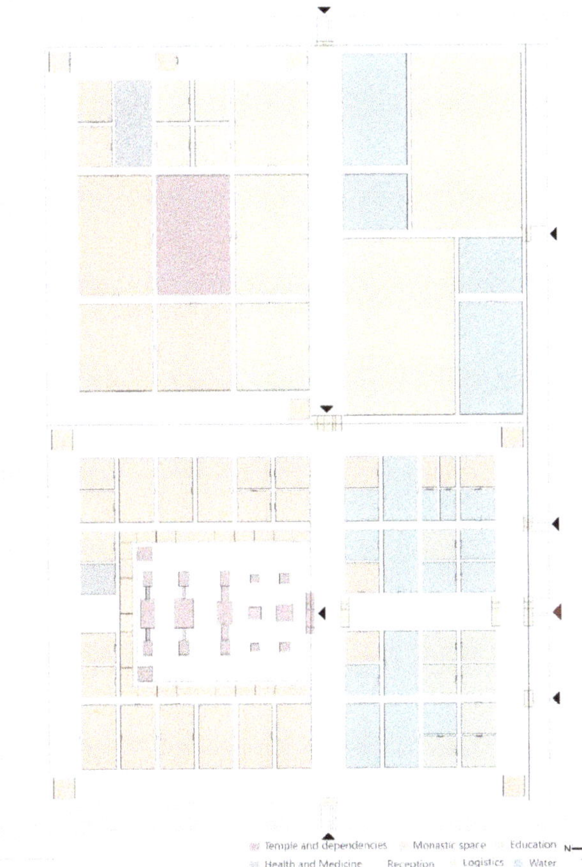

Figure 1. Illustrated Scripture of Jetavana Vihara of Sravasti in Central India (667).

To further explore how the layout of these areas conveys the spirit of the deities and affects the lives of monks and pilgrims, the Three Temples of Guoqing Si[2] and the Church of Poblet Monastery are taken as typical cases based on a preliminary comparative study of 20 Han Buddhist Monasteries in Southern China and 20 Cistercian Monasteries of Western Europe (Wang 2021). This study conducts a parallel analysis on how the Three Temples and the Church function, which can be used as a reference for the overall development of monastic layouts. How does the layout of these monasteries serve as the medium of communication among deities, monks, and pilgrims? What kind of dominators occupy the spaces? How do they materialize into concrete functional layouts, and how are they perceived by monks and pilgrims through worship itineraries?

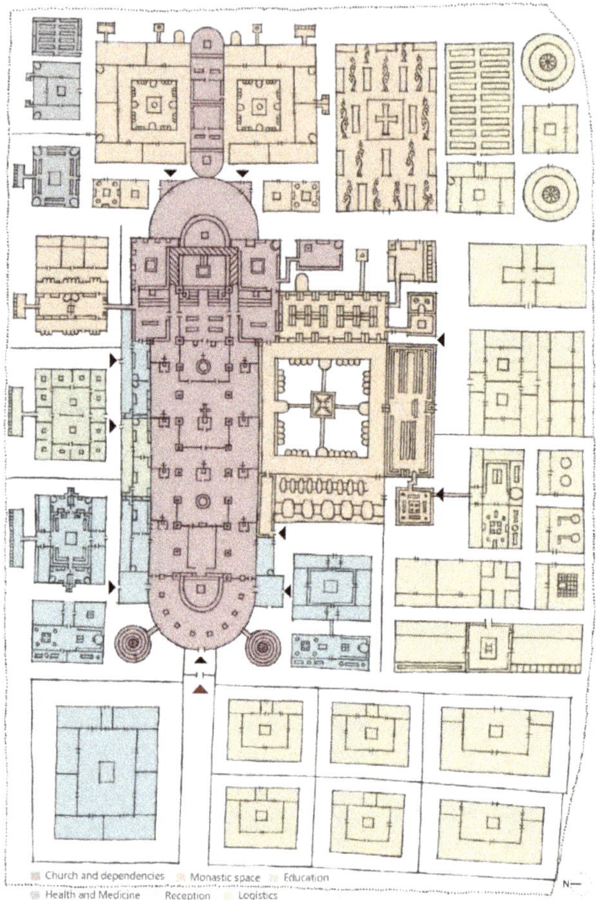

Figure 2. Plan of St. Gall (816).

2. Worship Spaces in Both Monasteries

2.1. The Three Temples in Guoqing Si

Guoqing Si is situated in Mount Tiantai, in Taizhou, Zhejiang Province (Figure 3). It was originally built in 598 during the Sui Dynasty and has witnessed more than 1400 years of history. It is the cradle of the Tiantai Buddhist sect, considered the first to evolve after the spread of Buddhism to China. The monastery is located in the valley among five peaks and is surrounded by two creeks (Zhang 1717) (Figure 4).

Guoqing Si covers an area of 17,501 square meters, with 22 main halls distributed in five axes. These can be separated into three main functions: divine space, monastic space, and pilgrim space (Figure 5). There are more than 60 buildings along these axes, which also perform the auxiliary function of helping to enclose the courtyard. The functional division of the monastery can be understood as the following: the western part is mainly for Buddhist rituals, sutra study, and monks' day-to-day lives. The eastern part is mainly made of the supporting facilities for daily life at the monastery, as well as the residences of the abbot, monks, and pilgrims. From the south to the north, the buildings on the central axis are dedicated to the worship of Buddha. This includes the Three Temples. It is important to note that the use of three here is a generic term; the number may vary in different Buddhist monasteries. In Guoqing Si, the Three Temples refers to the main entrance towards the altar, comprised of the Maitreya Hall, the Heavenly King Hall, and the Mahāvīra Hall. The courtyard is called Mingtang (Bright temple), and it is an extension of the main hall space. The role of the courtyard is also reflected in its use in specific Buddhist rituals. For the purposes of comparing the sacred space of Guoqing Si to the Cistercian Church, three halls and two courtyards are considered as a whole volume (Figure 6), which is around 100 m long and around 30 m wide.

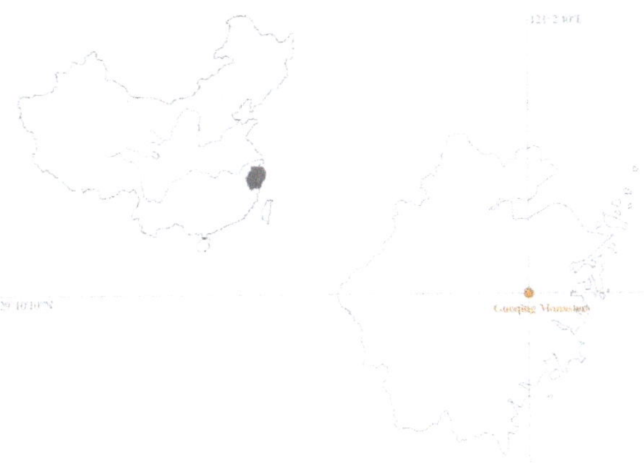

Figure 3. Diagram of the location of Guoqing Si.

Figure 4. Drawing of Guoqing Si in Qing Dynasty.

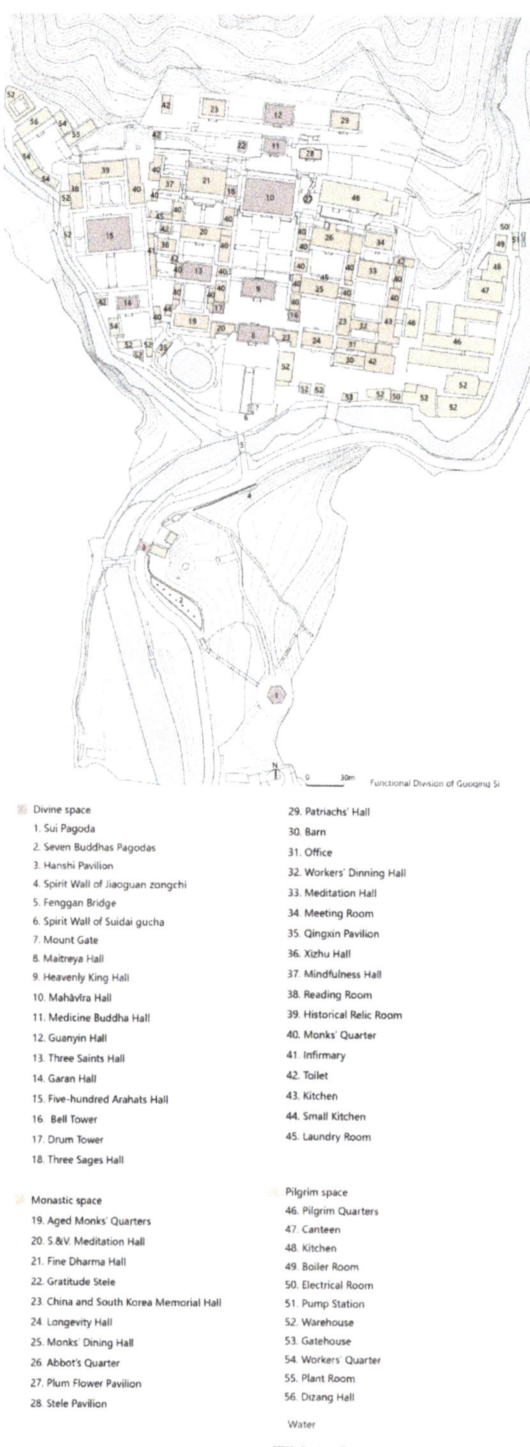

Divine space
1. Sui Pagoda
2. Seven Buddhas Pagodas
3. Hanshi Pavilion
4. Spirit Wall of Jiaoguan zongchi
5. Fenggan Bridge
6. Spirit Wall of Suidai gucha
7. Mount Gate
8. Maitreya Hall
9. Heavenly King Hall
10. Mahāvīra Hall
11. Medicine Buddha Hall
12. Guanyin Hall
13. Three Saints Hall
14. Garan Hall
15. Five-hundred Arahats Hall
16. Bell Tower
17. Drum Tower
18. Three Sages Hall

Monastic space
19. Aged Monks' Quarters
20. S.&V. Meditation Hall
21. Fine Dharma Hall
22. Gratitude Stele
23. China and South Korea Memorial Hall
24. Longevity Hall
25. Monks' Dining Hall
26. Abbot's Quarter
27. Plum Flower Pavilion
28. Stele Pavilion

29. Patriachs' Hall
30. Barn
31. Office
32. Workers' Dinning Hall
33. Meditation Hall
34. Meeting Room
35. Qingxin Pavilion
36. Xizhu Hall
37. Mindfulness Hall
38. Reading Room
39. Historical Relic Room
40. Monks' Quarter
41. Infirmary
42. Toilet
43. Kitchen
44. Small Kitchen
45. Laundry Room

Pilgrim space
46. Pilgrim Quarters
47. Canteen
48. Kitchen
49. Boiler Room
50. Electrical Room
51. Pump Station
52. Warehouse
53. Gatehouse
54. Workers' Quarter
55. Plant Room
56. Dizang Hall

Water

Contour line

Figure 5. Functional division of Guoqing Si.

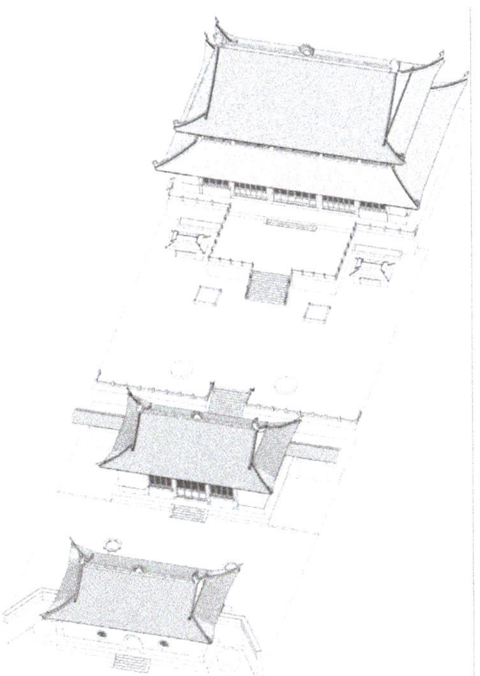

Figure 6. Axonometric aerial view of the Three Temples.

2.2. The Church in Poblet Monastery

Poblet Monastery was built in 1151, and may be seen as the prototype for the Spanish Cistercian Monastery. It developed from French origin, where one can find its motherhouse, Fontfroide Monastery. Poblet Monastery is situated in the city of Vimbodí, Tarragona Province, at the foot of Mount Prade (Figure 7). It is surrounded by mountains on the east, south, and north, though the west side is open (Figure 8). Also, on the east side runs the Francolí River, which provides an abundance of high-quality water flowing from the nearby mountains. The monastery was constructed on the same land upon which the Roman house, la Granja Mitjana, previously stood (Altisent 1974).

The scale of Poblet Monastery is similar to Guoqing Si, in that buildings cover an area of 18,887 square meters. It follows the typical layout of the Cistercian monastery, whereby auxiliary buildings are constructed around the Main Cloister, which includes spaces for use in a pilgrimage such as a hotel, pilgrim house, museum, warehouse, winery, and tourist center constructed between the defensive wall and the Wall of Peter IV. The Main Cloister is mainly composed of the divine space and monastic space. The Church and the New Sacristy are situated in the south side while buildings for monks are mainly set in the north side including the Chapter House, the Monks' Dormitory, the Library, the Warming Room, the Refectory, the Kitchen, the Converts' Locutory, and the Converts' Refectory (Figure 9). Similar to the Three Temples, the Church occupies a volume (Figure 10) of 85 m by 21 m with a central nave and three aisles.

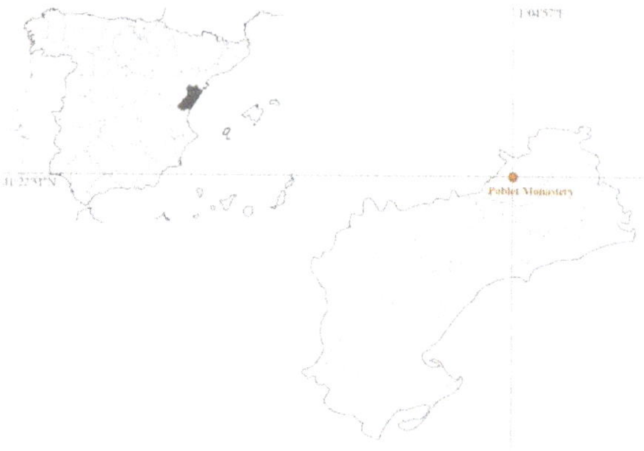

Location: Prades Mountains, Poblet, Tarragona, Spain
Latitude and longitude: 41°22′51″N, 1°04′57″E
Straight-line distance to sea: 33km
Climate: Typical mediterranean climate
Setup: From 12th Century(1151-now)
Religious sect: Roman Catholic Church
Cultural relic level: UNESCO World Heritage Site

Figure 7. Diagram of the location of Poblet Monastery.

Figure 8. General map of Poblet Monastery in 1798.

Figure 9. Functional division of Poblet Monastery.

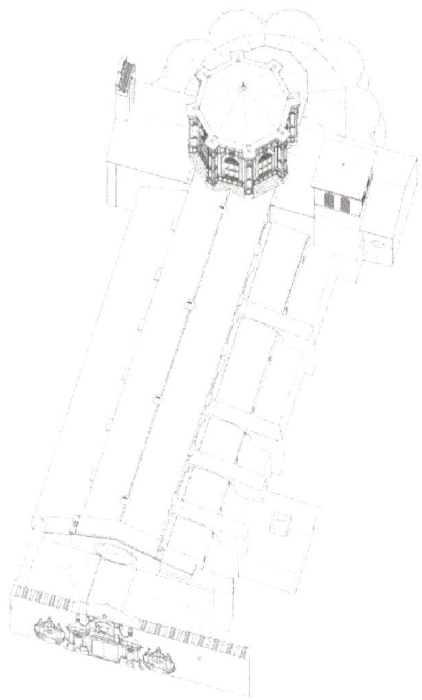

Figure 10. Axonometric aerial view of the Church.

In two such different monasteries, both temple and church serve similar functions. It is worth considering the following two aspects: the roles of worship spaces in the entire monastic layouts, and the configuration of the internal layouts including how they connect deities, monks, and pilgrims.

3. Comparative Analysis

3.1. The Roles of Worship Spaces in the Monastic Layouts

Even though the Three Temples of Guoqing Si and the Church of Poblet Monastery were built in different countries during different periods, they share similarities in monastic layouts, space prototypes, and priority of construction within the overall structure.

3.1.1. Similar Roles in the Generation of Monastic Layouts

Each monastery takes their key space—the Three Temples at Guoqing Si and the Church at Poblet Monastery—as a reference point to develop the remainder of the spatial layout. Guoqing Si takes the Three Temples as the center point and extends the monastic layout in the east-west direction (Figure 11). Poblet Monastery extends towards the north side with the Church as the southern border (Figure 12). In addition to this, the functions of both the Three Temples and the Church are both the most important and most sacred. In the first instance, they are considered the houses of deities. Thus, they are used to hold Buddhist rituals such as the Liberation Rite of Water and Land, blessing ceremonies, holy day ceremonies such as the birthday of Śākyamuni Buddha, and the performance of religious ceremonies to help the souls of the deceased to find peace, as well as Catholic rituals like thanksgiving mass, baptism, funeral mass, marriage, forgiveness, etc. These two spaces also take the role of training schools for monks, by acting as a location where they can conduct prayers like Buddhist chanting and Christian offices, as well as teach doctrine to pilgrims. Thirdly, the spaces perform as a public spiritual place for pilgrims to visit, study, and receive great blessings and merit from the aforementioned ceremonies. In summation, both the Three Temples and the Church are comprised of three common parts: the altar and monks' choir, space for pilgrims, and the entrance.

Figure 11. The Three Temples as the central axis.

Figure 12. Church as the southern border.

3.1.2. Similar Space Prototype

Similar layouts can be seen in the prototypes from which these designs originated. Indeed, both buildings originate from imperial architectural traditions. In the ancient world, imperial space was considered to be the most powerful space, akin to the idea of sacredness. Naturally, they became locations to imitate by both temple and church. Although there is no direct evidence to prove which specific pieces of imperial architecture these structures imitated, they have nonetheless all borrowed the concept of strict hierarchy from imperial buildings to arrange their internal spaces.

The Hall of Supreme Harmony (Taihe Dian) (Figure 13), constructed in the 15th century, is widely considered the most dignified building of the Forbidden City. Its layout set the throne at the geometric centerpoint of the structure, a location where the emperor sat and accepted the pilgrimage of the ministers waiting in the courtyard in front of him. Similarly, Mahāvīra Hall has two sets of Buddhas and Bodhisattvas statues arranged in the altar, separated by a wall reaching to the wooden beams of the roof. When there is a grand ceremony, pilgrims generally light incense and worship Buddha in the courtyard. Another example, Royal Hall (Aula Regia) (Figure 14) in Europe, shares great similarities with the Hall of Supreme Harmony: the King sat in the semicircular apse covered by a dome while the rest stood in front of him. Inside the Church, the altar is set in the east end, symbolizing the direction from whence

God may come. Previously, pilgrims would stand in the vestibule in the west outside of the Church, however today they can sit in the central nave next to the entrance.

Figure 13. Hall of Supreme Harmony, 万国来朝图, from the collection of The Palace Museum.

Figure 14. Aula Regia, Reconstruction Sketch of Flavian Palace, Source J C Golvin.

The key similarity between the two buildings is that deitiesoccupy the highest and most important positions to ensure that they have the best view of the entire situation, thus allowing full control of any situation.

3.1.3. Similar Priority of Construction

Both the Three Temples as well as the Church get priority over other monastic buildings when it comes to construction, renovation, and restoration. The process of construction with regards to the internal layout of the buildings is also similar, with the process in both starting from the altar and extending out towards the entrance. When considering the entire monastery, the order of construction is also based on the division of functions amongst different zones; this is particularly important when economic potential may be restricted in the initial stages of establishing the monastery.

With a history of continuous expansion, destruction, restoration, and development of Guoqing Si (Guanding n.d.), it is clear that seven key construction works played large roles in the expansion and eventual spatial arrangement of the monastery (Figure 15). Initially, in the year 598 of the Sui Dynasty, the Main Hall was constructed at the center of the monastery. However, it was constructed during the humid season and building materials such as bamboo and wood were not used correctly (Guanding n.d.), meaning that in 605, four years after it was established, the buildings began to tilt and suffer from water leakage. It therefore required restoration works. The monastery later suffered during the great Anti-Buddhist Persecution (840–846), and during the late Tang Dynasty in the year 851, Emperor Xuanzong of Tang (685–762) ordered that Guoqing Si should be reconstructed. The Main Hall was to be rebuilt first, then the monks' dormitories(佛殿初营，僧房未置。) (Shen, Huan. 861. 国清寺止观堂记 (The Note of Zhiguan Tang in Guoqing Monastery)). In 1005, Emperor Zhenzong of the Song Dynasty (1010–1063) bestowed a ton of gold to further restore and refurbish the monastery, and the Buddha statues were gilt (次参大佛殿，丈六金色释迦像，左右坐丈六弥陀、弥勒像，烧香礼拜。) (Shi n.d.). It was further renovated after Master Tan E (1285–1373) was reinstated by Emperor Taizu of Ming (1328–1398) to be the Abad of Guoqing Si in 1369 (Ding 1995). In 1733 in the early-mid Qing Dynasty, Emperor Shizong of Qing (1678–1735) issued an Imperial edict to reconstruct it once again, which reshaped the image of the monastery, and resulted in the basic layout of the current Guoqing Si (Ding 1995).

Figure 15. Historical evolution of Guoqing Si.

In 1973, China and Japan resumed diplomatic relations. Japanese Prime Minister Tanaka Kakuei proposed to visit Guoqing Si, meaning the monastery once again had further restoration works under the support of Premier Zhou Enlai. Buddha statues in the Three Temples had their gold surface refurbished and some of them were even replaced by statues coming from Pekin.

A similar history can be seen at the Poblet Monastery (Figure 16). The Church was largely built in two phases (Bassegoda Nonell 1983): the initial phase between 1151 and 1165 driven by the Count of Barcelona, Ramón Berenguer IV (1114–1162), and then again from 1165 through to the 12th century with the support of the King of Aragón and the Count of Barcelona, Alfonso II. This was analyzed in detail by José Luis Bozal González according to the marks left by stonecutters on the surface of the Church (Bozal González 2016). The Church can be divided into 4 parts: head, body, façade, and tower. The tower, situated on the top of the crossing of the central nave, was originally used as a bell tower. It was constructed in 1330 and was used in this way until the south tower of the Church was constructed in 1660, taking over as a bell tower. The form of the central nave was based on that of its motherhouse, the Church of Fontfroide Abbey, but it also took inspiration from the Abbey of Cluny, the grandest, most prestigious, and most well-endowed monastic institution in Europe, which had great influence on Europe from the second half of the 10th century through the early 12th. Alongside this, the Church of Poblet was also influenced by the Abbey of Saint-Denis, the prototype of gothic architecture.

Figure 16. Historical evolution of Poblet Monastery.

It was completed in 1144 under the direction of the Abbot Sugar, then the Abbot Copons decided to reform the south-side aisle, raising the height of the roof and adding seven chapels. After a fire in 1575 and an earthquake in 1792, four buttresses were added to the south of the Church to support the wall of the central nave. However, the imbalance (Portal Liaño and Moreno-Navarro 2013) of the structure caused by this reconstruction lead to cracks on the inner surface of the building, which has required a large amount of restoration in the past decade. All in all, the Church was the first aspect built in the Cloister (Bango Torviso et al. 1998). It also took the longest construction time, and has undergone many transformations and restorations since its initial development.

3.2. Programmed Functional Layouts

3.2.1. Similar Space Dominators

Both the Three Temples (Figure 17) and the Church (Figure 18) are essentially houses of deities, being devoted exclusively to worship and sacred rituals for monks, and serving as silent and sacred spaces for pilgrims. They are both dominated by visible and invisible powers, including supernatural, monastic, and social powers. Differences in their respective hierarchies are marked through the boundaries between their partitioned spaces.

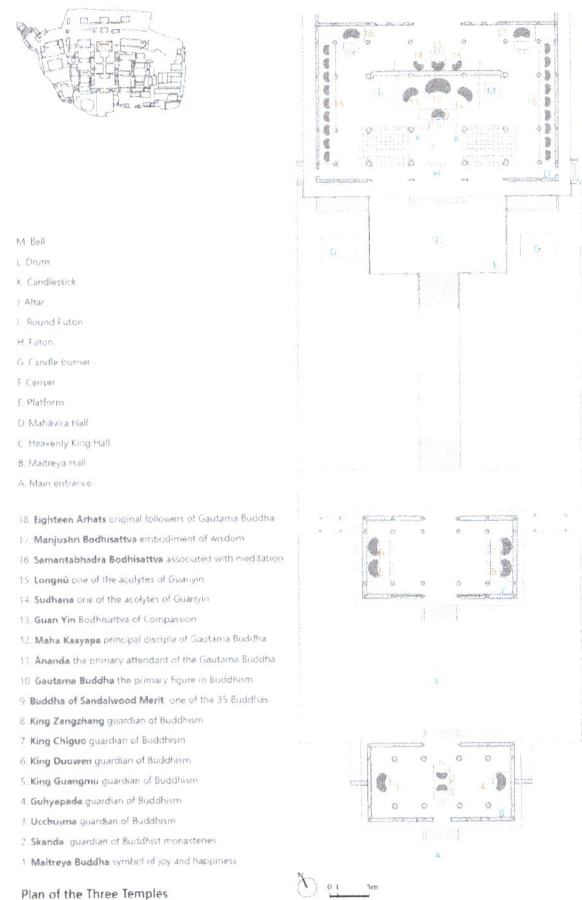

Figure 17. Plan of the Three Temples of Guoqing Si.

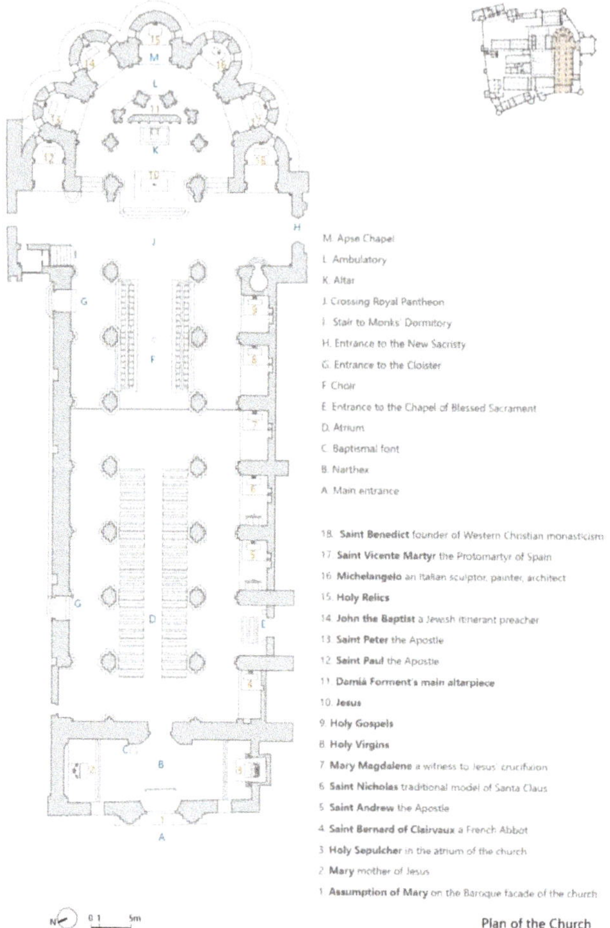

Figure 18. Plan of the Church of Poblet Monastery.

Deities, who have supernatural power as beings of omniscience (all-knowing), omnipotence (all-powerful), and omnipresence (all-present), occupy the most important space in the Three Temples as well as in the Church. In the Three Temples, supernatural power refers to Buddhas, Bodhisattvas, Disciples, and Guardians of Buddhism. Their representatives are set in the corresponding place within the Three Temples according to their place within the hierarchy. Buddhas have to be adored at the end of the sacred axis, accompanied by Bodhisattvas. They are usually set at the center of kernel space in the enclosure, like Mahāvīra Hall. Eighteen Arhats stand on both sides of Buddhas while Guardians like Heng and Ha, and Four Heavenly Kings occupy the outer edge of the sacred space like Maitreya Hall and Heavenly King Hall. All of these symbolic statues in turn constitute a vivid and complete Buddhist world.

In the Church, the eastern end is the most sacred space and is where the cross of Jesus hangs above the altar. This in turn makes the cross the focus of all attention. It points to the east, the direction of Jesus' arrival. The cross is illuminated by the rising sun when monks conduct office in the morning. Seven apses dedicated to saints are added at the head of the Church, which are behind the altarpiece. Seven chapels with holy saints are set in the south side of the Church, again illuminated by windows. In the Narthex, Mary and the Holy Sepulchre are arranged on each of the two sides. The assumption of Mary is located on the Baroque façade of the Church, symbolizing the welcoming of pilgrims to the world of God.

In both the Three Temples and the Church, monks occupy the place closest to the deities. As one of the three powers mentioned above, Monastic power can be roughly understood as comprising the roles of the abbot, monks, and pilgrims. A clear hierarchy exists among these three. Both cenobitic monasteries emphasize the absolute obedience of monks to the Abbots, which is also reflected in the arrangement of the space. The abbots generally occupy the second-best position by being placed below the representation of their respective deities. In Guoqing Si, during morning and evening chanting, the abbot performs the ritual of kneeling on the futon in the center of the Mahāvīra Hall, just in front of the altar, while in the Church the abbot sits behind the altar. Monks in turn stay on both sides of the Abbots, in a position which in Guoqing Si called Liangxu (两序) and in the Church is called Choir.

Pilgrims generally occupy the outermost position closest to the entrance. In the Three Temples, pilgrims normally stand in the courtyard (Mingtang) while in the Church, they sit in the atrium. When speaking about the role of pilgrims, it is worth noting that both monasteries are constructed by monks with the help of donation from pilgrims. With this in mind, some pilgrims hope that monks can sing a daily Mass for their soul (Altisent 1974), since they are the earthly representatives of their respective deities. In the mind of a pilgrim, making an offering to the monks and listening to their teachings is equivalent to making an offering to the deities themselves.

3.2.2. Similar Itineraries

In both monasteries, the itinerary of the journey to the monastery is mainly composed of two routes: mountain, and monastic. In turn, the two routes play different roles. The mountain route acts as a transition between the society and the monastery, and the monastic route becomes a structure which elucidates the relationship between deities and pilgrims. The purpose of the mountain route, with its beautiful natural landscape, is to relax pilgrims and help rid them of distractions. Then, upon entering the monasteries, the monastic route is full of order to reflect the majesty of the deities and the sacredness of the monastery. Usually, symmetrical space can better highlight the importance of the central axis.

A Chinese poem from the Tang Dynasty describes the experience of walking towards the Buddhist monastery in the early morning: *"Early in the morning, I walked into this ancient monastery, the rising sunshine in the mountains. The forested curved path leading to a deeper place, lush and colorful flowers bloom around the Buddhist monastery. The mountain is so bright that the birds are more joyous, the lake is clear and refreshing. At this moment, all things are silent, leaving only the bell chime sound."*[3] Pilgrimage is thus achieved by approaching the monastery step by step.

At the foot of Tiantai Mountain, the first thing pilgrims catch sight of is the large paddy fields. Behind that, Sui Pagoda is the most obvious sign indicating the final destination of Guoqing Si (Figure 19). Pilgrims encounter the Seven Buddhas Pagodas and Hanshi Pavilion on the winding mountain route which leads them towards the monastery. They reach Fenggan Bridge in front of the Mountain Gate and once they pass through the Mountain Gate, a journey of direct Buddha worship begins (Figure 20).

Figure 19. Worship itinerary of Guoqing Si.

Figure 20. Topographic plan of Guoqing Si.

At Poblet Monastery (Figure 21), pilgrimage is a process that takes place from dusk to dawn. Pilgrims from nearby villages walk along the monastic wall and go through the Prades Gate, the Golden Gate, then finally reach the square in front of the Church. In the early morning, dawn vaguely illuminates the outline of the Church, itself located in the wilderness. The main gate of the Church is opened, leading the pilgrims to the altar where the east end of the Church is lit by the sunshine through the window of the thick wall. (Figure 22).

Figure 21. Worship itinerary of Poblet Monastery.

The itinerary set for monks and pilgrims to approach the altars helps to further develop respect and admiration towards their respective deities. It is no exaggeration to say that the longer the route, and the more filled with special and reverent experiences, the more this feeling can be enhanced. Further to this, the spatial layout of the route and monasteries themselves also has a great influence on the feelings of visiting pilgrims.

The terrain of the Three Temples, Maitreya Hall, Heavenly King Hall, and Mahāvīra Hall gradually rises. Such a layout not only corresponds to the mountainous surroundings but also highlights the importance of the Mahāvīra Hall, where the altar is located (Figure 23). In the Church, the position of the altar is higher than that of the Choir, and the position of the Choir is higher still than that of the pilgrims. The height difference, despite only being a single step, allows for the setting up of a separation between the sacred and the secular whilst still being convenient for people to reach (Figure 24).

Figure 22. Topographic plan of Poblet Monastery.

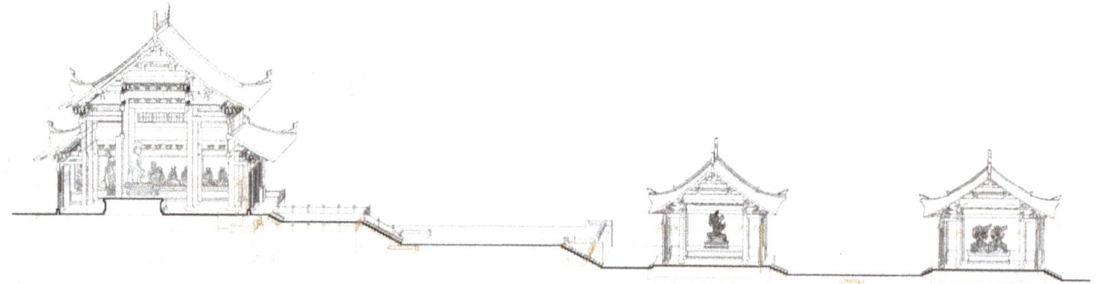

Figure 23. Axonometric section of the Three Temples.

In both cases, deities are arranged at the end of the space, accessed by a one-way route. Spatial layout thus defines the order of importance among deities, monks, and pilgrims and delineates a specific journey to be taken. In their architectural details, both the Three Temples and the Church create a sacred itinerary for pilgrims and monks through height difference, enclosure, and image.

In Buddhist monasteries, the changes in the height of platforms reveal the hierarchy of importance and define the boundary between the sacred and the secular. The buildings of Guoqing Si are situated on eight different platforms of terrain (Figure 25). Take the central axis as an example: there are five temples and a main gate situated across the eight platforms of the terrain. Among them, the Mahāvīra Hall, with the largest dimensions, occupies two platforms. The height difference between the two platforms is around three metres, or approximately the same height as a single floor.

Buildings on two sides of the central axis sit on one to two floors, according to their location on different terrain platforms. The Mahāvīra Hall is placed in the center, under its eaves sitting the eaves of each side building. This, therefore, highlights the importance of the Mahāvīra Hall, whilst steps between terrain platforms are further conducive to leading pilgrims on the continuous process of worshiping the Buddhas.

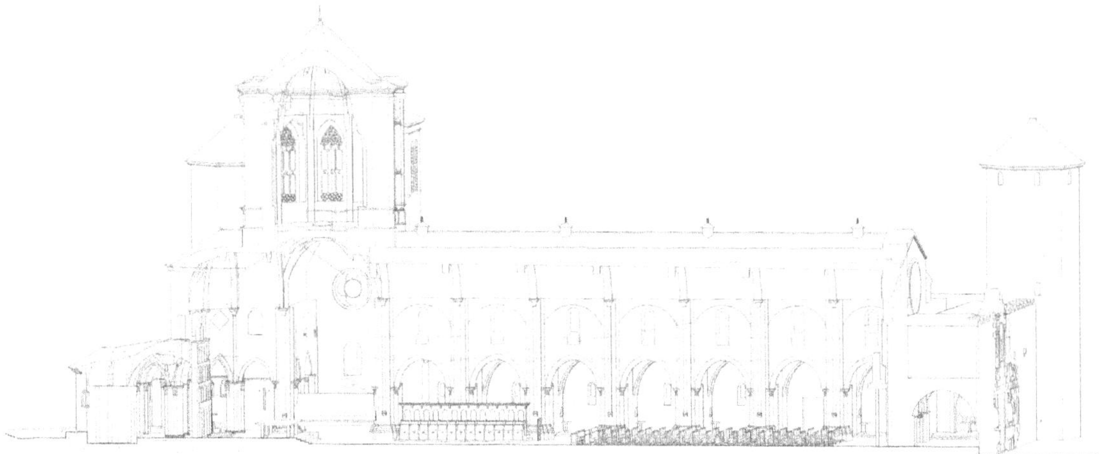

Figure 24. Axonometric section of the Church.

Figure 25. Elevation platforms of Guoqing Si.

In Cistercian monasteries, even though buildings are normally constructed in a flat valley, the Church is similarly set in a higher place compared to the rest of the buildings. The Main Cloister is mainly situated on six terrain platforms (Figure 26). The height difference between each two terrain platforms is around one to two metres. Although the Abbot's Palace is situated on the highest terrain, it is located outside the Main Cloister. Therefore, the Church and the New Sacristy occupy higher terrain than the Cloister, the Refectory, and the Novice House, showing their respective importance.

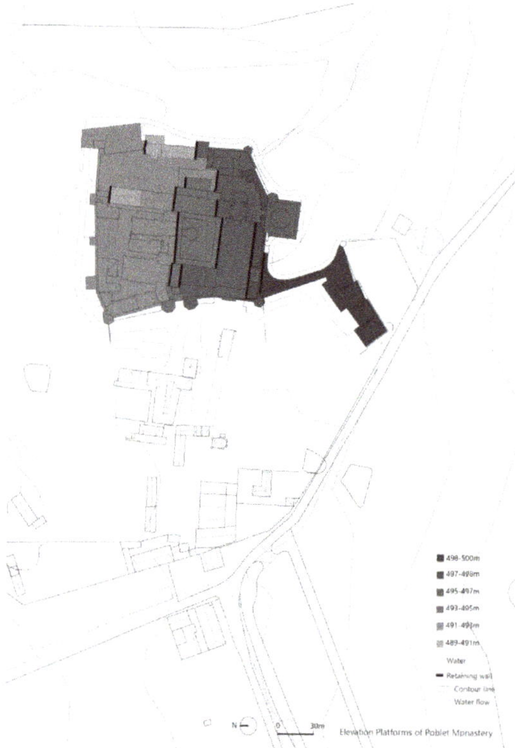

Figure 26. Elevation platforms of Poblet Monastery.

Further to this, on the highest elevated platforms inside both the building of the Three Temples and the Church, one can find holy altars. Their high location both reveals and stresses their sacredness. Indeed, every sacred space defines a sacred object, which in turn marks the environment. The enclosed hall limits the space for monks to worship the Buddha while the niche is a symbolic residence of the Buddha. Pilgrims generally take the route of worship in the central axis, and the corridors on both sides of the Three Temples allow monks to bypass the Heavenly King Hall and avoid overlapping with the route of pilgrims on their journey to worship Buddha—they can enter directly from the side doors of the Mahāvīra Hall. In the Church, the Choir is located next to the altar and is mostly enclosed by huge stone pillars and wooden seats. This means that the monks' daily office will not be interrupted by the pilgrims walking down the two side aisles.

Images are also important in affecting the perception of pilgrims on their itinerary of worship. In the Three Temples, the Maitreya Buddha with a big belly and smiling face sits inside the niche of the Maitreya Hall, which welcomes the coming pilgrims.

The image of the Maitreya Buddha reduces pilgrims' alertness. The solemn expressions of the four heavenly kings in the Heavenly King Hall reshape the majesty and order of the world of Buddhas. When pilgrims enter the Mahāvīra Hall and bow to the Buddha Shakyamuni, they complete their confession through a complete ritual. Similarly, in the Church, the statue of Mary with open hands on the west facade of the Church symbolizes the entrance to the sanctuary where pilgrims are welcome to go inside and receive a blessing from God. At the east end of the Church, above the altar, Christ is nailed to the cross, representing the transmission of life and death and showing the separation between the material world and the heavens. At this, the mood of the pilgrims changes from the initial sense of ease gained from both the mountain route and the entering into the monastery, towards a more serious and reflective nature upon facing the icon of Jesus.

4. Discussion of the Differences behind the Similarities

4.1. Integration and Independence

The way that worship spaces connect to the rest of the monastic layouts are mainly reflected in the following three aspects: open and closed, void and solid, disappearing and emphasizing.

The Three Temples have an image of openness, while the Church appears more closed. The core courtyard is in front of the main hall, which is considered the climax of the central axis sequence (Zhao 2013). It not only constitutes a key aspect of the worship itinerary, but also connects the eastern and western parts of Guoqing Si (Ding 1995).

In contrast, the Church maintains closed in the north and south façades, which are regarded as independently functioning volumes (Domènech i Montaner 1925). The core courtyard in Guoqing Si, as a void space, radiates a certain emptiness to the surrounding space, in contrast to the way in which the Church declares dominance over the entire complex through acting as a solid space.

The courtyard is an introverted external space (Hou 2011), arranged in the traditional layout of Buddhist monasteries (Liang 1961). Indeed, courtyards of different sizes have become the units of the monastery's "organizational procedures" (Li 1982). To some extent, the courtyard is thus highly integrated into the spatial grammar of monastic layout (Ge 1979). The Church has been independent from other aspects of the complex from the beginning (Altisent 1974), and over time the building volume has been added and continuously strengthened (Finestres y de Monsalvo and Fontseré 1947).

In the end, the boundaries of the Three Temples are blurred and melted into the overall spatial layout while the Church is constantly being strengthened, individualized, and emphasized in the monastery complex.

4.2. Different Understanding of Deities

4.2.1. Different Ways of Worship

Even though both cases share a similar functional layout and priority of construction, in considering both monastic traditions, different ways of worship are depicted. In the Three Temples, the courtyard serves as an exterior temple where ceremonies such as incense burning and rituals are held. This means that both exterior and interior spaces function as sacred spaces for monks and pilgrims to worship Buddhas. In contrast, the Church mostly holds ceremonies indoors.

In Guoqing Si, the spacious courtyard not only serves as a location for drying rice for the traditionally agricultural society, but it is also fundamental as a space where monks can perform all of their rituals. Take the three altars of ordination (novice monk, bhikkhu, and bodhisattva) as examples. These processes are usually held at the same time. Monks need to perform confession for at least three days and three nights. During this period, the ordained master must perform related rituals in both the main hall and courtyard, moving back and forth between the spaces as required by the role. It is important to note that both half and full prostrations are used by Buddhist monks. A half prostration involves kneeling and touching the earth with one's hands and forehead. In full prostration, monks have to slide forward until completely lying down, pivoted the forearms on the elbows, and raise the hands with palms up, an action which symbolizes raising the Buddha's feet above one's head. The full prostration shows reverence to the Triple Gem (Buddha, Dharma, and monks). In full prostration, two meters' distance must be maintained between each monk. Thus, half prostrations are generally held inside the temple, while full prostrations are held in the courtyard. Inside the Mahāvīra Hall, half prostration and circumambulation around the altar are usually undertaken by monks during morning and evening chanting. Consistent with the changing rituals, and apart from the fixed position of the Buddha statue, the futons used by monks can be moved according to the needs of different ceremonies.

In contrast, Cistercian monks and pilgrims normally perform rituals solely inside the Church. The exquisite furniture of the Choir is therefore fixed in place front of the altar. The design of the seat is such that it supports the way the monks pray. There is space for them to turn around and bow to the altar. It also satisfies them to sit while looking at each other, whilst at the same time, they can lean on the chairs when they stand up. Such postures are usually performed in daily offices, whilst during misars or certain festivals, the priest also circumambulates the altar. Bread and wine as symbols of the body and blood of Christ are delivered by monks to serve the pilgrims. Generally speaking, pilgrims will sit in the atrium, listen to the prayers of the monks, and stand up at appropriate times or kneel on wooden benches to pray.

In both worship spaces, specific religious behavior affects the definition of how space should be used. In turn, concrete forms of space and furniture regulate and standardize the sacred rituals. Such different ways of worship are the embodiment of the relationship among different space dominators and affect the formation of worship routes, which have to do with different understandings of deities in both monasticisms.

4.2.2. World of Buddhas vs. Immeasurable God

Even though the Three Temples and the Church share a lot of similarities in functions and composition of structure with regards to space dominators, their ways of visualizing the existence of deities are quite different. In expressing sacredness, Buddhist monasteries pay more attention to the depiction of Buddha statues. In contrast, the medieval Cistercian Church was a space devoted to visualization and imagination (Cassidy-Welch 2001).

The depictions of Buddha statues have become important features to demostrate the importance of the deity within different spaces. They are made as symbols to form a world of Buddhas, the size of which normally corresponds with the scale of the architecture. Buddhist monasteries build bodies for Buddhas, and they even have a reference book, *Fo Shuo Zao Xiang Liang Du Jing* (佛说造像量度经) (Gongbuchabu 1874) which defines a suitable size ratio of Buddha statues. Therefore, the space evolves with the scale of the Buddha statues within the monastery. It is worth noting that monks are not allowed to be idolaters. Buddhist monasteries build a world of Buddhas through murals and statues with the goal of breaking the image of Buddhas, which also helps us understand that there are countless incarnations of Buddha; no single image can be worshiped. The worship posture is to remind monks and pilgrims to keep humble at all times. Furthermore, excluding the main Buddha statues which are usually very tall, most Buddha statues are either slightly bigger or similar to human scale, like Luohan statues sitting on the platform in the Mahāvīra Hall. Pilgrims feel less or even no distance from Buddha, since at this scale they seem approachable (Figure 27).

Figure 27. World of Buddhas of Guoqing Si.

For a medieval monk or nun, building an abbey was as much an act of spirituality as the meditation and contemplation carried out within it (Kinder 2000). In the Cistercian church, God Himself is immeasurable. Instead, it is the actual architecture that is considered as the symbol of God (Figure 28). For monks in the Middle Ages, architecture was one of the best tools to express their aesthetic idea, religious practice, and expectations for life. Statues are seen as idols and are not encouraged in the Church, in accordance with God's commandment: *"You shall not make for yourself a graven image or any likeness of anything that is in heaven above, or that is in the earth beneath, or that is in the water under the earth: you shall not bow down to them or serve them"* (Exodus n.d.). With this limitation, therefore, how is the scale of God presented in the Church?

The sacredness of space is represented through the use of light, which symbolizes how God has brought light to the world. Usually, windows in the Church are very small, sealed with a light-transmitting thin marble plate to prevent thieves from entering. On one hand, this is a limitation of the technology and methods in stone construction, but on the other hand, it is also a safety consideration. The interior space of the Church is portrayed as a container of light. The office of Martines takes part in the early morning when the sun rises in the east, while the office of Vispera is performed before sunset as the last sunshine of a day goes through the roseton on the west facade of the Church towards the Choir. Further to all of this, within the Church, the cross of Jesus is abstracted, simplified, and hung in the air to show its importance, uniqueness, and inaccessibility.

The relationships between deities and man are portrayed through space, and finally presented as different itinerary perceptions with different distances to deities.

Figure 28. Immeasurable God of Poblet Monastery.

4.2.3. Different Distances to Deities

Even though they share similar logic in shaping the atmosphere of the worship itinerary, in the two monasteries it is clear that the distance between deities and pilgrims is also different.

In the Buddhist worldview, the term Buddha meaning "enlightened person" means the state of being a Buddha can be reached through practice. The central axis with stairs to climb upward is the symbol of the step-by-step journey towards becoming Buddha. On the contrary, the statue of Jesus is small and is hung in the air, much higher than the flat ground upon which the pilgrims remain. This means the pilgrims' distance to God is insurmountable, confirming the unique and unattainable nature of God and godhood.

In both monasteries, spatial forms perform as effective strategies to assure the legitimacy of sacredness. Pilgrims believe that their proximity to the deities will imbue the individual with power. The closer one can approach the deities, they believe, the stronger and more auspicious the connection between them (Kilde 2014).

In Buddhist monasteries, sacredness is achieved by taking advantage of the height of the natural environment. For example, the central axis of the Three Temples is constructed on the mountain. *"One day I shall climb to the summit, seeing how small surrounding mountain tops appear as they lie below me."* (会当凌绝顶，一览众山小) (Du n.d.) The Cistercians create a contrast between the height of space and the human body, thereby making human surrender and admire. It imagines that the height of the building can be used to narrow the distance between humans and the heavens.

5. Conclusions

This research has considered the specific spatial relationship of deities, monks, and pilgrims, and found the external and internal common ground between two traditional worship spaces. Further on, it triggers more open discussions about the reasons behind the differences and similarities between the two, and considers an historical evolution. Further research also hopes to provide a dialogue bridge and a method for cross-cultural communication in considering the spatial and spiritual aspects.

"The conclusions drawn from an in-depth study of such a small social unit are not necessarily applicable to other units. However, such a conclusion can be used as a hypothesis or comparative material when conducting surveys elsewhere. This is the best way to get true scientific conclusions." (Fei 2001) Inspired by the Peasant Life in China (江村经济), the possibility of popularization can be obtained through in-depth research on individual cases. Both places of worship represent human imaginings of deities. The relationships among pilgrims, monks, and deities are established through the functional layout, space dominators, and itinerary of worship. The research on their grammar structure aims to bring guiding principles for contemporary spiritual space; namely, how to define the scale of deities through the depiction of distance and finally affect pilgrims' methods of adoration.

Externally, the Three Temples and the Church enjoy top priority in terms of construction, occupy key positions in the overall monasteries, and become the main reference for the development of monastic layouts. Internally, they share similar programmed functional layouts; defining the boundary between the sacred and the secular, highlighting the importance of deities within a space, and guiding pilgrims to complete their rituals of worship.

Based on the lives of similar monks, the article has found the external and internal similarities of both worship spaces. However, due to different understandings of deities themselves, both worship spaces and indeed both monastic traditions show different ways of worship, images of deities, and approaches to the distance between them and humanity. To further explore the reasons behind the similarities and differences that the comparative study has found, an open discussion may be considered: how can we reflect the different understandings of deities in both Buddhist and Cistercian monasteries? How have the worship spaces transformed through different periods, and what are their corresponding similarities and differences? In the development process, do they influence each other? These may require the emergence of more historical materials and archaeological results. The importance of research lies in understanding the relationship between deities and pilgrims from the view of the East and the West, and their specific materialization and expression in the architectural space.

The comparative study of the similarities and differences is the cross-cultural contact between the eastern and western spiritual spaces. Such interaction between Buddhism and Christianity helps to recognize the characteristics and differences of their respective cultures, and at the same time provides a bridge for communication and connections by way of their similarities.

Funding: This research was funded by CHINA SCHOLARSHIP COUNCIL, grant number 20140626 0218.

Institutional Review Board Statement: Not applicable.

Informed Consent Statement: Not applicable.

Data Availability Statement: No new data were created or analyzed in this study. Data sharing is not applicable to this article.

Acknowledgments: This paper is conducted under the financial support from the China Scholarship Council, 2014–2018, File No.201406260218. It derives from the dissertation under the direction of José Ignacio Linazasoro and Fangji Wang. I am very grateful to them for their detailed guidance on the structure of the article. Besides, thanks to Aibin Yan and Shuishan Yu who have provided constructive comments on the article. During the process of investigation, Abbot Yunguan, Master Yanru of Guoqing Si and ex-Prior Lluc Torcal, Prior Rafael of Poblet Monastery, and Jordi Portal have generously provided great help and related research materials for the research. Considering the English Editing, Amelia-Rose Rubin has generously offered great help to me, which is appreciated.

Conflicts of Interest: The author declares no conflict of interest.

Notes

[1]   According to the *Rule of Saint Benedict*, there are four kinds of monks: Cenobites, Anchorites (or Hermits), Sarabaites and Landlopers.

[2]   Although Guoqing Si belongs to Tiantai sect, it was converted to Chan Buddhist in Song Dynasty, and later changed back to Tiantai sect.

[3]   常建【唐】题破山寺后禅院"清晨入古寺，初日照高林。曲径通幽处，禅房花木深。山光悦鸟性，潭影空人心。万籁此俱寂，但余钟磬音。"．

References

Altisent, Agustín. 1974. *Història de Poblet*. Tarragona: Tarragona Abadía de Poblet.

Bango Torviso, Isidro G., María Lucía Barrero, Julio David Pizarro, María del Carmen Muñoz Párraga, Concepción Abad Castro, María Teresa López de Guereñoo Sanz, Susana Calvo Capilla, Eduardo Carrero Santamaría, Elena Casas Castells, and Antonio García Flores. 1998. *Monjes y Monasterios. El Cister en el Medievo de Castilla y León*. Valladolid: Junta de Castilla y León.

Bassegoda Nonell, Juan. 1983. *Història de la restauració de Poblet: Destrucció i reconstrucció de Poblet*, 1st ed. Tarragona: Tarragona Abadia de Poblet.

Benedict of Nursia. 516. *The Rule of Saint Benedict*. Tarragona.

Boerschmann, Ernst. 1912. *Chinese Architecture and Its Relation to Chinese Culture*. 1911 Annual report of the Board of Regents of the Smithsonian Institution. Washington, DC: Smithsonian Institution.

Boisvert, Mathieu. 1992. A Comparison of the Early Forms of Buddhist and Christian Monastic Traditions. *Buddhist-Christian Studies* 12: 123–41. [CrossRef]

Bozal González, José Luis. 2016. Iglesia del Monasterio de Santa Maria de Poblet, Una hipótesis sobre las fases de su construcción basada en sus signos de cantero. Available online: http://www.signosdecantero.com/descargas/Santa_Mar%EDa_Poblet_JL_Bozal.pdf (accessed on 4 November 2021).

Braunfels, Wolfgang. 1972. *Monasteries of Western Europe, the Architecture of the Orders*. Princeton: Princeton University Press.

Cassidy-Welch, Megan. 2001. *Monastic Spaces and Their Meanings: Thirteenth-Century English Cistercian Monasteries*. Turnhout: Brepols.

Dehui. 1335. Chi Xiu Baizhang Qing Gui. Available online: http://buddhism.lib.ntu.edu.tw/BDLM/sutra/chi_pdf/sutra19/T48n2025.pdf (accessed on 4 November 2021).

Ding, Tiankui. 1995. *Guo Qing Si Zhi*, 1st ed. Shanghai: Hua Dong Shi Fan Da Xue Chu Ban She.

Domènech i Montaner, Lluís. 1925. *Historia y arquitectura del Monestir de Poblet*. Barcelona: Montaner y Simón.

Du, Fu. n.d. Available online: https://www.shicimingju.com/418.html (accessed on 4 November 2021).

Exodus. n.d. Available online: https://biblia.com/bible/esv/exodus/20/4-5 (accessed on 4 November 2021).

Fei, X. 2001. *Jiangcun Jing Ji: Zhongguo Nong Min De Sheng Huo* (江村经济: 中国农民的生活), 1st ed. Shang Wu Yin Shu Guan Wen Ku. Beijing: Shang Wu Yin Shu Guan.

Finestres y de Monsalvo, Jaime, and Joaquin Guitert i Fontseré. 1947. *Historia del Real Monasterio de Poblet, illustrada con disertaciones curiosas sobre la antiguedad de su fundación*. Barcelona: Editorial Orbis.

Frampton, Kenneth. 2015. *A Genealogy of Modern Architecture: Comparative Critical Analysis of Built Form*. Edited by Ashley Simone. Zürich: Lars Müller.

Ge, Ruliang. 1979. Guoqing Temple in Mount Tiantai. *Journal of Tongji University* 4: 10–22.

Gongbuchabu. 1874. 佛说造像量度经 (Fo Shuo Zao Xiang Liang Du Jing). Available online: http://buddhism.lib.ntu.edu.tw/BDLM/sutra/chi_pdf/sutra10/T21n1419.pdf (accessed on 4 November 2021).

Guanding. n.d. Guoqing Bailu. Available online: http://buddhism.lib.ntu.edu.tw/BDLM/sutra/chi_pdf/sutra19/T46n1934.pdf (accessed on 4 November 2021).

Henry, Patrick Gillespie, and Donald K. Swearer. 1989. *For the Sake of the World: The Spirit of Buddhist and Christian Monasticism*. Minneapolis: Fortress Press, Collegeville: Liturgical Press.

Ho, Puay-peng. 1995. The ideal monastery: Daoxuan's description of the Central Indian Jetavana vihara. *East Asian History (Canberra)* 10: 1–18.

Horn, Walter, Ernest Born, and Saint Adalard. 1979. *The Plan of St. Gall*. California Studies in the History of Art. Berkeley: University of California Press, vol. 19.

Hou, Youbin. 2011. 中国建筑之道 (Zhongguo Jianzhu Zhi Dao). Beijing: China Architecture & Building Press.

Itō, Chūta. 1931. *Shina kenchikushi*. Tōyōshi kōza. Tōkyō: Yūzankaku, vol. 11.

Kilde, Jeanne Halgren. 2014. *Sacred Power, Sacred Space: An Introduction to Christian Architecture and Worship*. New York and Oxford: Oxford University Press.

Kinder, Terryl Nancy. 2000. *Architecture of Silence: Cistercian Abbeys of France*. New York: H.N. Abrams.

King, Peter. 1999. *Western Monasticism: A History of the Monastic Movement in the Latin Church*. Cistercian studies series, no. 185. Kalamazoo: Cistercian Publications.

Li, Yunhe. 1982. *Hua Xia Yi Jiang: Zhongguo Gu Dian Jian Zhu She Ji Yuan Li Fen Xi*. Hong Kong: Guang Jiao Jing Chu Ban She; Hua Feng Shu Ju Fa Xing.

Liang, Sicheng. 1932. 蓟县独乐寺观音阁山门考 (Research on the Gate and Pavilion of Dule Temple in Ji County). 中国营造学社汇刊 (*Journal of the Society for Research in Chinese Architecture*) 3: 1–92.

Liang, Sicheng. 1961. 中国的佛教建筑 (Buddhist architecture in China). 清华大学学报 (*Journal of Tsinghua University*) 12: 53–54.

Liang, Sicheng, and Wilma Fairbank. 1984. *A Pictorial History of Chinese Architecture*. Cambridge: MIT Press.

Liu, Dunzhen. 2007. *Liu Dunzhen Quan Ji*, 1st ed. Beijing: Zhongguo Jian Zhu Gong Ye Chu Ban She.

Moon, Simon Young-suck. 1998. *Korean and American Monastic Practices: A Comparative Case Study*. Lewiston: Edwin Mellen Press.

Portal Liaño, Jorge, and José Luis González Moreno-Navarro. 2013. *La iglesia del Real Monasterio de Santa María de Poblet: pasado y presente de un desequilibrio*. Edited by Santiago, Huertay Fabián and López Ulloa. Madrid: Instituto Juan de Herrera.

Prip-Møller, Johannes. 1937. *Chinese Buddhist Monasteries, Their Plan and Its Function as a Setting for Buddhist Monastic Life*. Copenhagen: G.E.C. Gad, London: Oxford University Press.

Sekino, Tadashi, and Daijō Tokiwa. 1926. *Shina Bukkyō shiseki*. Tōkyō: Bukkyō Shiseki Kenkyūkai.

Shen, Huan. 861. 国清寺止观堂记 (The Note of Zhiguan Tang in Guoqing Monastery).

Shi, Chengxun. n.d. 参天台五台山记 (Can Tiantai Wutai Shan Ji). Available online: http://tripitaka.cbeta.org/B32n0174_001 (accessed on 4 November 2021).

Su, Bai. 2009. *Zhongguo Gu Jian Zhu Kao Gu*, 1st ed. Su Bai Wei Kan Jiang Gao Xi Lie. Beijing: Wen Wu Chu Ban She.

Sun, Dazhang. 2017. *Zhongguo Fo Jiao Jian Zhu*, 1st ed. Beijing: Zhongguo Jian Zhu Gong Ye Chu Ban She.

Walsh, Francis Edward. 1932. *Church Architecture: Building for A Living Faith*. Milwaukee: The Bruce Publishing Company.

Wang, Guixiang. 2011. 中国古建筑测绘十年 (*Ten Years of Surveying and Mapping Ancient Chinese Architecture*). Beijing: Tsinghua University Press.

Wang, Guixiang. 2016. 中国汉传佛教建筑史: 佛寺的建造、分布与寺院格局、建筑类型及其变迁 (*Zhongguo Han Chuan Fo Jiao Jian Zhu Shi: Fo Si De Jian Zao, Fen Bu Yu Si Yuan Ge Ju, Jian Zhu Lei Xing Ji Qi Bian Qian (The History of Chinese Buddhist Architecture))*, 1st ed. Beijing: Tsinghua University Press.

Wang, Weiqao. 2021. Tiempo y espacio: Investigación sobre el espacio monástico a través de la vida en los monasterios budistas y cistercienses = Time and space: Research on Monastic Space through Monks' Life in Buddhist and Cistercian Monasteries. Thesis (Doctoral), E.T.S. Arquitectura (UPM). Available online: https://doi.org/10.20868/UPM.thesis.68991 (accessed on 4 November 2021).

Wang, Weiren, and Zhu Xu. 2011. 中国早期寺院配置的形态演变初探：塔·金堂·法堂·阁的建筑形制 (Topological Study on the evolution of early Chinese Buddhist Monasteries: Layout of Pagoda, Buddha Hall, Lecture Hall and Pavilion Tower). 南方建筑(*South Architecture*) 1: 38–49.

Wang, Yuan, and Bingjie Lu. 2000. 中国古代佛教建筑的场所特征 (Characteristic of Chinese temple architecture place). 华中建筑(*Hua Zhong Jian Zhu*) 18: 131–33.

Zhang, Lianyuan. 1717. *Tiantai Shan Quan Zhi: 18 Juan*. Available online: https://hollis.harvard.edu/primo-ex-plore/fulldisplay?docid=01HVD_ALMA212047190400003941&context=L&vid=HVD2&lang=en_US&search_scope=everything&adaptor=Local%20Search%20Engine&tab=everything&query=any,contains,%E5%A4%A9%E5%8F%B0%E5%B1%B1%E5%85%A8%E5%BF%97&offset=0 (accessed on 4 November 2021).

Zhang, Yuhuan. 2012. *Tu Jie Zhongguo Fo Jiao Jian Zhu*, 1st ed. Beijing: Dang Dai Zhongguo Chu Ban She.

Zhao, Nadong. 2013. 敦煌莫高窟初唐石窟所反映的佛寺核心院落主题 (The theme of the core courtyard of the Buddhist temple reflected in the early Tang Grottoes at Mogao Grottoes at Dunhuang). 中国建筑史论汇刊 (*Journal of Chinese Architecture History*) 1: 212–23.

MDPI

St. Alban-Anlage 66

4052 Basel

Switzerland

Tel. +41 61 683 77 34

Fax +41 61 302 89 18

www.mdpi.com

Religions Editorial Office

E-mail: religions@mdpi.com

www.mdpi.com/journal/religions